HANDBOOK FOR COGENERATION AND COMBINED CYCLE POWER PLANTS

Dr. Meherwan P. Boyce, P.E.

NEW YORK   ASME PRESS

© 2002 by The American Society of Mechanical Engineers
Three Park Avenue, New York, NY 10016

Reprinted with revisions 2004. All rights reserved. Printed in the United States of America. Except as permitted under the United States Copyright Act of 1976, no part of this publication may be reproduced or distributed in any form or by any means, or stored in a database or retrieval system, without the prior written permission of the publisher.

INFORMATION CONTAINED IN THIS WORK HAS BEEN OBTAINED BY THE AMERICAN SOCIETY OF MECHANICAL ENGINEERS FROM SOURCES BELIEVED TO BE RELIABLE. HOWEVER, NEITHER ASME NOR ITS AUTHORS OR EDITORS GUARANTEE THE ACCURACY OR COMPLETENESS OF ANY INFORMATION PUBLISHED IN THIS WORK. NEITHER ASME NOR ITS AUTHORS AND EDITORS SHALL BE RESPONSIBLE FOR ANY ERRORS, OMISSIONS, OR DAMAGES ARISING OUT OF THE USE OF THIS INFORMATION. THE WORK IS PUBLISHED WITH THE UNDERSTANDING THAT ASME AND ITS AUTHORS AND EDITORS ARE SUPPLYING INFORMATION BUT ARE NOT ATTEMPTING TO RENDER ENGINEERING OR OTHER PROFESSIONAL SERVICES. IF SUCH ENGINEERING OR PROFESSIONAL SERVICES ARE REQUIRED, THE ASSISTANCE OF AN APPROPRIATE PROFESSIONAL SHOULD BE SOUGHT.

ASME *shall not be responsible for statements or opinions advanced in papers or . . . printed in its publications* (B7.1.3). Statement from the Bylaws.

For authorization to photocopy material for internal or personal use under those circumstances not falling within the fair use provisions of the Copyright Act, contact the Copyright Clearance Center (CCC), 222 Rosewood Drive, Danvers, MA 01923, tel: 978-750-8400, www.copyright.com.

Library of Congress Cataloging-in-Publication Data

Boyce, Meherwan P.
    Handbook for cogeneration and combined cycle power plants/ Meherwan P. Boyce.
    p.   cm.
Includes bibliographical references and index.
ISBN 0-7918-0169-1
1. Cogeneration of electric power and heat.   2. Combined cycle power plants.
   I. Title.
TK1041 .B68 2002
621.1'99–dc21   2001053958

This book is dedicated to my
Grandfather
Khan Bahdur Jehangir R. Colabawala MBE
(1885–1975)
His dedication to engineering and
his achievements in bringing power to
India and Pakistan
has been the inspiration that has
guided me through my entire career

TABLE OF CONTENTS

DEDICATION   iii

PREFACE   xiii

1. AN OVERVIEW OF POWER GENERATION   1

    Distributed Generation 8
       Diesel and Gasoline Engines 15
       Natural Gas Reciprocating Engines 15
       Gas Turbines 15
       Micro-Turbines 15
       Fuel Cell Technology 17
       Solar Energy-Photovoltaic Cells 20
       BioMass Systems 21
       Wind Energy 21
       River Hydro-Turbines 21
    Cogeneration 23
       Cogeneration Qualifications 25
       Gas Turbine Cycle in Cogeneration Mode 27
    Combined Cycle Plants 30
    Properties of Gas Turbine Exhaust 36
    Steam Generation Calculations 37
    Gas Turbine Heat Recovery 38
    Supplementary Firing of Heat Recovery Systems 40
       Environmental Effects 40

2. CYCLES   43

    Combined Cycle Plant Operation 43
    The Brayton Cycle 43
       Inlet Cooling Effect 45
       Regeneration Effect 52

Increasing the Work Output of the Simple Cycle Gas
　　　Turbine 56
　　　Intercooling and Reheat Effects 57
　　　Injection of Compressed Air, Steam, or Water
　　　for Increasing Power 59
　　　Combination of Evaporative Cooling and Steam
　　　Injection 63
　Advanced Gas Turbine Cycles 63
　　　Compressed Air Energy Storage Cycle 63
　Actual Cycle Analysis 66
　　　The Simple Cycle 66
　　　The Split-Shaft Simple Cycle 68
　　　The Regenerative Cycle 69
　　　The Intercooled Simple Cycle 71
　　　The Reheat Cycle 71
　　　The Intercooled Regenerative Reheat Cycle 73
　　　The Steam Injection Cycle 73
　　　The Evaporative Regenerative Cycle 82
　　　The Brayton-Rankine Cycle 82
　Summation of Cycle Analysis 86
　A General Overview of Combined Cycle Plants 86

3. PERFORMANCE AND MECHANICAL EQUIPMENT STANDARDS　　99

　Major Variables of a Combined Cycle Power Plant 101
　　　Plant Location and Site Configuration 101
　　　Plant Type 104
　Plant Size and Efficiency 104
　　　Type of Fuel 110
　　　Types of HRSG 110
　　　Types of Condensers 111
　　　Enclosures 112
　　　Plant Operation Mode: Base or Peaking 112
　　　Start-up Techniques 113
　Performance Standards 113
　　　ASME, Performance Test Code on Overall Plant Performance 113
　　　ASME, Performance Test Code on Test Uncertainty: Instruments
　　　and Apparatus 114
　　　ASME, Performance Test Code on Gas Turbines 114
　　　ASME, Performance Test Code on Gas Turbine Heat Recovery
　　　Steam Generators 115
　　　ASME, Performance Test Code on Steam Turbines 116
　　　ASME, Performance Test Code on Steam Condensing
　　　Apparatus 116
　　　ASME, Performance Test Code on Atmospheric Water Cooling
　　　Equipment 117

ISO, Natural Gas — Calculation of Calorific Value, Density and Relative Density 117
Table of Physical Constants of Paraffin Hydrocarbons 117
Mechanical Parameters 118
    API Std 616, Gas Turbines for the Petroleum, Chemical and Gas Industry Services 118
    API Std 618, Reciprocating Compressors for Petroleum, Chemical and Gas Industry Services 120
    API Std 619, Rotary-Type Positive Displacement Compressors for Petroleum, Chemical, and Gas Industry Services 120
    API Std 613 Special Purpose Gear Units for Petroleum, Chemical and Gas Industry Services 120
    API Std 677, General-Purpose Gear Units for Petroleum, Chemical and Gas Industry Services 121
    API Std 614, Lubrication, Shaft-Sealing, and Control-Oil Systems and Auxiliaries for Petroleum, Chemical and Gas Industry Services 121
    ANSI/API Std 610 Centrifugal Pumps for Petroleum, Heavy Duty Chemical and Gas Industry Services 121
    API Publication 534, Heat Recovery Steam Generators 122
    API RP 556, Fired Heaters & Steam Generators 122
    ISO 10436:1993 Petroleum and Natural Gas Industries — General Purpose Steam Turbine for Refinery Service 122
    API Std 671, Special Purpose Couplings for Petroleum Chemical and Gas Industry Services 122
    ANSI/API Std 670 Vibration, Axial-Position, and Bearing-Temperature Monitoring Systems 122
    API Std 672, Packaged, Integrally Geared Centrifugal Air Compressors for Petroleum, Chemical, and Gas Industry Services 123
    API Std 681, Liquid Ring Vacuum Pumps and Compressors 123
Gas Turbine 123
Gears 128
Lubrication Systems 130
Vibration Measurements 131
Specifications 133

4. AN OVERVIEW OF GAS TURBINES   137

Industrial Heavy-Duty Gas Turbines 147
Aircraft-Derivative Gas Turbines 150
Medium-Range Gas Turbines 151
Small Gas Turbines 152
Major Gas Turbine Components 155
    Compressors 155
    Regenerators 171

Combustors 172
Combustor Design Considerations 176
Typical Combustor Arrangements 177
Air Pollution Problems 178
Dry Low NO_x Combustor 183
Catalytic Combustion 190
Features of Catalytic Combustion 190
Catalytic Combustor Design 193
Turbine Expander Section 194
Radial-Inflow Turbine 196
Mixed-Flow Turbine 197
Axial-Flow Turbines 197
Impulse Turbine 201
The Reaction Turbine 204
Turbine Blade Cooling Concepts 207
Turbine Blade Cooling Design 209
Cooled-Turbine Aerodynamics 213
Instrumentation and Controls 219

5. AN OVERVIEW OF STEAM TURBINES 221

The Rankine Cycle 221
The Steam Regenerative-Reheat Cycle 223
Heat Rate and Steam Rate 226
Steam Turbine 226
Classification of Steam Turbines 227
Steam Turbine Characteristics 244
Required Material Characteristics 249
Turbine Efficiency 253
Advantages and Disadvantages of
Steam Turbines 253

6. AN OVERVIEW OF PUMPS 255

Range of Operation 255
Pump Selection 259
Pump Materials 260
Types of Pumps 263
Process Pumps 263
Sump Pumps 263
Axial-Flow Pumps 264
Turbine Pumps 264
Regenerative Pumps 264
Gear Pumps 265
Screw Pumps 265
Centrifugal Pumps 265

Pump Application in Combined Cycle Power Plants 270
    The IP-LP Circulating Pump 270
    HP Feed Water Pumps 271
    The HP Circulating Pump 272
    Condenser Pumps 272
    Cooling Water Pumps 272
    Lubrication Pumps 272
    Fuel Pumps 273

7. HEAT RECOVERY STEAM GENERATORS   275

The Horizontal Type HRSG 276
The Vertical Type HRSG 277
Once Through Steam Generator 279
Design Considerations 281
Multi-Pressure Steam Generators 281
    Pinch Point 282
    Approach Temperature 282
Casing of the HRSG 283
Finned Tubing 283
Off-Design Performance 283
Evaporators 284
Forced Circulation System 284
Back Pressure Considerations (Gas Side) 284
Supplementary Firing of Heat Recovery Systems 284
Design Features 286
Once Through Steam Generators 290
HRSG Operational Characteristics 294
HRSG Effectiveness 296
Water Chemistry 296
Vibration and Noise 296
Filter Housing, Duct Work and Insulation 297
Filter Housing Duct Work 298
Diverters, Silencers and Burners 299
    Diverters 299
    A Typical Case 301
Insulation 303
Duct Burners 304
Silencers 304
HRSG Reliability and Durability 304

8. CONDENSERS AND COOLING TOWERS   307

Condensers 307
    Types of Condensers 309

x • Handbook for Cogeneration and Combined Cycle Power Plants

  Condensate-Polishing Systems 313
  Condenser Fouling 313
 Cooling Towers 314
  Design of Cooling Towers 316
 Chemical Water Treatment 321

9. GENERATORS, MOTORS AND SWITCH GEARS   323

 Motors 323
  Constant Speed Motors 324
  Alternating Current Squirrel-Cage Induction Motors 324
  Synchronous Alternating-Current Motors 325
  Power-Factor Correction 326
 Generator 327
  Design Characteristics 330
 Switchgear 345
 Electrical Single Line Diagram 351

10. FUELS, FUEL PIPING AND FUEL STORAGE   353

 Fuel Specifications 355
 Fuel Properties 358
 Heavy Fuels 370
 Cleaning of Turbine Components 374
  Turbine Wash 374
  Compressor Washing 376
 Fuel Economics 377
 Heat Tracing of Piping Systems 378
  Types of Heat-Tracing Systems 380
  Choosing the Best Tracing System 386
 Storage of Liquids 387
  Atmospheric Tanks 387
  Elevated Tanks 388
  Open Tanks 388
  Fixed Roof Tanks 388
  Floating Roofs Tanks 389
  Pressure Tanks 389
  Calculation of Tank Volume 389
  Container Materials, Insulation and Support 392

11. BEARINGS, SEALS AND LUBRICATION SYSTEMS   395

 Bearings 395
  Rolling Bearings 395

Journal Bearings 402
Thrust Bearings 410
Seals 414
Non-contacting Seals 415
Lubrication Oil System 420
Lubricant Selection 425
Oil Sampling and Testing 425
Oil Contamination 426
Filter Selection 426
Cleaning and Flushing 428
Coupling Lubrication 429
Lubrication Management Program 430

12. CONTROL SYSTEMS AND CONDITION MONITORING 433

Control Systems 433
Condition Monitoring Systems 444
Identification of Losses in a Combined
Cycle Power Plants 446
The Gas Turbine 446
HRSG 449
Steam Turbine 450
Condenser 450
Design of a Condition-Monitoring System 452
Monitoring Software 453
Implentation of a Condition-Monitoring System 455
Plant Power Optimization 457
Life Cycle Costs 459

13. PERFORMANCE TESTING OF COMBINED CYCLE POWER PLANT 463

Gas Turbine 466
Air Inlet Filter Module 467
Compressor Module 467
Combustor Module 469
Expander Module 469
Life Cycle Consideration of Various Critical Hot-Section
Components 469
HRGS Calculations Module 470
Steam Turbine Calculations 472
Condenser Calculations 474
Performance Curves 474
Performance Computations 476
General Governing Equations 477
Gas-Turbine Performance Calculation 482

xii • Handbook for Cogeneration and Combined Cycle Power Plants

      Heat-Recovery Steam Generator 492
    Steam Turbines 495
      Plant Losses 499
Nomenclature 500

14. MAINTENANCE TECHNIQUES   503

Philosophy of Maintenance 503
Maximization of Equipment Efficiency and Effectiveness 505
Organization Structures 508
    Performance-Based Total Productive Maintenance Program 508
    Implementation of a Performance-Based Total Productive
    Maintenance 508
    Maintenance Department Requirements 511
    Spare Parts Inventory 516
    Inspections 517
    Condition and Life Assessment 517
    Redesign for Higher Machinery Reliability 518
    Maintenance Scheduling 526
    Maintenance Communications 527

APPENDIX A   531

APPENDIX B   535

BIBLIOGRAPHY   537

INDEX   553

ABOUT THE AUTHOR   569

PREFACE

Handbook for Cogeneration and Combined Cycle Power Plants discusses the design, fabrication, installation, operation, and maintenance of combined cycle power plants. The book has been written to provide an overall view for the experienced engineer working in a specialized aspect of the subject and for the young engineering graduate or undergraduate student who is being exposed to the field of power plants for the first time. The book will be very useful as a textbook for undergraduate courses as well as for in-house company training programs related to power generation.

Cogeneration and combined cycle power plants are not new but with the improvement of the gas turbine technology; efficiencies in the mid-50s are common, and with a little bit of ingenuity, efficiencies in the low 60s will be possible. These high efficiencies have totally revolutionized the industry, making the old steam plants a thing of the past.

The use of cogeneration and combined cycle power plants in all industries, and in the power generation field, has mushroomed in the past few years. It is to these users and manufacturers of combined cycle power plants that this book is directed. The book will give the manufacturer a glimpse of some of the problems associated with his equipment in the field and help the user to achieve maximum performance efficiency and high availability for his plant.

I have been involved in the research, design, operation, and maintenance of various types of combined cycle power plants since the early 1960s. I have also taught courses at the graduate and undergraduate levels at the University of Oklahoma and Texas A&M University, and now, in general, to the industry for the past 30 years. I have taught over 3000 students from over 400 corporations around the world. The enthusiasm of the students associated with these courses gave me the inspiration to undertake this endeavor. The many courses I have taught over the past 37 years have been an educational experience for me, as well as for the students. The Texas A&M University Turbomachinery Symposium, which I had the privilege to organize and chair for 7 years, is a great contributor to the operational and maintenance sections of this book. The discussions and consultations that resulted from my association with highly professional individuals have been a major contribution to both my personal and professional life as well as to this book.

In this book, I have tried to assimilate the subject matter of various papers, and sometimes diverse views, into a comprehensive, unified treatment of combined cycle power plants. Many illustrations, curves, and tables are employed to broaden the understanding of the descriptive text. Mathematical treatments are deliberately held to a minimum so that the reader can identify and resolve any problems before he is ready to execute a specific design. In addition, the references direct the reader to sources of information that will help him to investigate and solve his specific problems. It is hoped that this book will serve as a reference text after it has accomplished its primary objective of introducing the reader to the broad subject of combined cycle power plants.

I wish to thank the many engineers whose published work and discussions have been a cornerstone to this work.

Lastly, I wish to acknowledge and give special thanks to my wife, Zarine, for her readiness to help and her constant encouragement throughout this project.

I sincerely hope that this book will be as interesting to read as it was for me to write and that it will be a useful reference to the fast-growing field of combined cycle power plants.

MEHERWAN P. BOYCE
HOUSTON TEXAS
2001

Chapter 1

AN OVERVIEW OF POWER GENERATION

Energy costs in the past decade have risen dramatically. With this large increase in energy costs, the acceptability of inefficient engine systems is very limited. Turbine and diesel engine efficiencies range from the 30% to 45% range (with some even lower), which implies that between 55% and 70% of the energy supplied is wasted. The privatization of large central energy corporations run by large government bureaucracies throughout the world has been a major incentive in the search for more efficient techniques of generation of power. The United Kingdom spearheaded the privatization schemes in the 1980s and 1990s; other countries, small and large, have followed. The United States of America is opening up its large power market, and by the year 2010, a very open market will be in place in the U.S.

The energy market place of the first quarter of the new century (2000 to 2025) will be very different from the last quarter of the 1900s. Competition for the energy market will be very fierce and non-traditional, with many new and efficient energy conversion systems in the market place. The traditional utilities of the 1900s will not exist in the 2000s. The traditional utilities, which were generating power, transmitting power and then distributing the power will be broken up into three separate companies in these areas. These companies will be autonomous and will have no relationship with each other than what the market place will exert on them. The transmitting companies will be transmitting power purchased by the distributing companies from various power generation companies. Power will be a commodity, like grain, and will be traded freely, allowing consumers to buy from various power sources.

The total production capacity of electric power in the world exceeds 3000 GW, with the power in the U.S. in 1998 being 891 GW, and the component of combined cycle is about 300 GW, which amounts to about 10% of the existing capacity. Figure 1-1 shows the various power systems which make up the overall capacity. Steam plants account for about 56% of the capacity, nuclear plants account for about 12%, hydro plants about 20%, combined cycle and gas turbine power plants about 10%, diesel plants about 2%, renewable energies amount to about a tenth of 1%. Some predict that the expected power by 2050 would be about 10,000 GW. This would mean that there would have to be an addition of 140 GW/year at an investment of US$100 to $150 billion/year. A more conservative estimate would be that in the next 20 years, through 2020, the

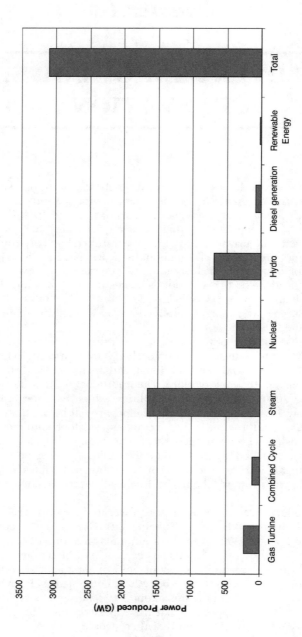

Figure 1-1. World Power Outlook

new additional power in the world would amount to about 400 GW, from which it is expected that the combined cycle component could be as high 100 GW. Throughout the world, there are over 300 GW of gas turbine and combined cycle power plants, which have been brought on line mostly in the 1990s in electrical utility service.

The growth of the power industry over a 10-year period indicates that there would be about 642 GW of additional power, an increase of about 20% on the existing world capacity. Figure 1-2 shows the type of power that is expected to make up the new power. The industry has seen and will see some very large hydroelectric projects, such as America's Grand Coulee (6480 MW). In Egypt, 20,000 MW of power is supplied by the Aswan Dam and the Aswan High Dam on the Nile. The Guri hydroelectric plant, with over 10,000 MW of installed capacity (the world's fourth largest capacity), became Venezuela's largest single source of electricity upon its completion in 1986. The Guri Dam, located on the Río Caroní, saved the country the equivalent of 300,000 barrels of oil a year. The next major hydroelectric project coming on line is China's Three Gorges Dam, an 18,000-MW project scheduled for operation in the year 2009. The Three Gorges Dam, which is located on the Yangtze River, the third longest river in the world, measures approximately 6390 km from its headwaters at the Gelandandong Glacier in Tibet to its mouth near Shanghai on China's eastern seaboard.

In many countries, where there is an abundance of coal, steam plants will grow. China, India and the U.S. are three major countries that have vast coal

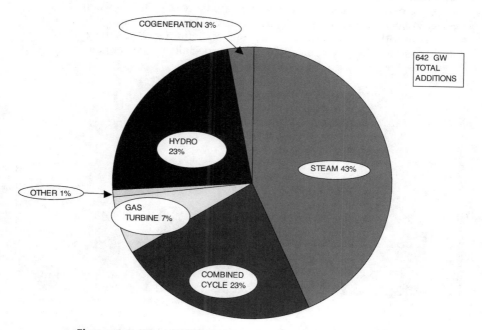

Figure 1-2. World-Wide Generation Additions 1997–2006

reserves. Conventional steam turbine plants are base-load operated and use coal as fuel. The choice between conventional steam plants and combined cycle plants is fuel. If natural gas is available, the combined cycle plant will be the plant of choice. It is estimated that 43% of the power in the years from 1997 to 2006 will be from steam plants mainly operating on coal. The development of new abatement techniques make steam plants environmentally friendly, as compared with their predecessors.

The shape of this market place will be very dependent of the availability and cost of fossil fuel. In the majority of the 1900s, coal was the preferred fuel. Today, natural gas is by far the preferred fuel; that is not to say that coal and oil will not be used, rather than natural gas, for its minimum pollution and very low maintenance cost, will be the preferred fuel as long as it is available and the price is affordable.

Natural gas is the fuel of choice wherever it is available because of its clean burning and its competitive pricing as seen in Figure 1-3. Prices for uranium, the fuel of nuclear power stations, and coal, the fuel of the steam power plants, have been stable over the years and have been the lowest. Environmental, safety concerns, high initial cost, and the long time from planning to production have hurt the nuclear and steam power industries. Whenever oil or natural gas is the fuel of choice, gas turbines and combined cycle plants are the power plants of choice, as they convert the fuel into electricity very efficiently and cost effectively. It is estimated that from 1997 to 2006, 23% of the plants will be combined cycle power plants, and that 7% will be gas turbines. It should be noted that about 40% of gas turbines are not operated on natural gas.

The use of natural gas has increased and in the year 2000, has reached prices as high as $4.50 in certain parts of the U.S. This will bring other fuels onto the market to compete with natural gas as the fuel source. Figure 1-4 shows the growth of the natural gas as the fuel of choice in the U.S., especially for power generation. This

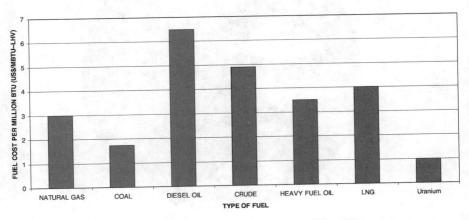

Figure 1-3. Typical Fuel Costs per Million BTUs

An Overview of Power Generation • 5

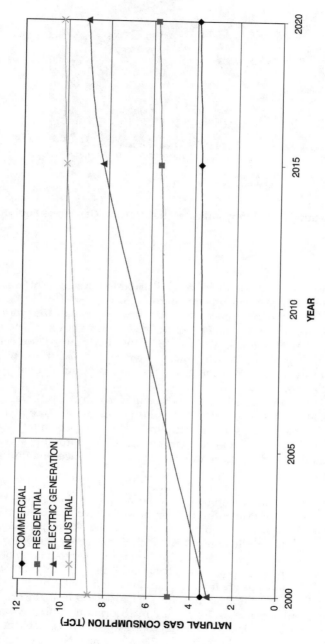

Figure 1-4. Projected Natural Gas Consumption 2000–2020

6 • COGENERATION AND COMBINED CYCLE POWER PLANTS

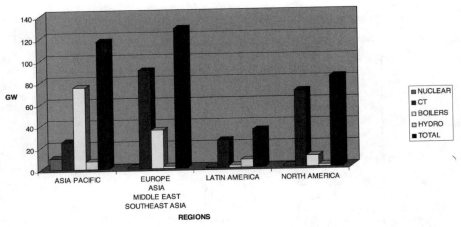

Figure 1-5. Technology Trends Indicate That Natural Gas is the Fuel of Choice

growth is based on completion of a good distribution system. This signifies the growth of combined cycle power plants in the U.S.

Figure 1-5 shows the preference of natural gas throughout the world. This is especially true in Europe, where 71% of the new power is expected to be fueled by natural gas, Latin America, where 73% of the new power is expected to be fueled by natural gas, and North America, where 84% of the new power is expected to be fueled by natural gas. This means a substantial growth of combined cycle power plants.

Figure 1-6 shows the growth of the power industry in the U.S, through 2020. In the year 2000, there is about 14% of the power in the U.S. generated by combined cycle power plants and gas turbines. This is expected to rise to about 43% by the year 2020. This is dependent on the use of natural gas as the main fuel. It will require new pipelines to bring competitive-priced natural gas to all parts of the U.S.

The last two decades, the 1980s and 1990s, have seen the growth of cogeneration systems and combined cycle power plants throughout the world. In the next 20 years, wherever natural gas or oil is available, combined cycle power plants and high-efficiency gas turbine plants will be the plants of choice.

Cogeneration is the production of two or more forms of energy from a single plant. The most common application of the term is for the production of electrical power and steam for use in process applications. This does not mean that other types of cogeneration plants are not being designed and used. Cogeneration plants are used to produce power, and use the direct exhaust gases from the prime movers for preheating air in furnaces, or for the use in absorption cooling systems, or for heating various types of fluids in different process applications. Between 1996 and 2006, cogeneration plants will account for 3% of the new power being generated worldwide, this amounts to 19.6 GW of power. Cogeneration systems are also used

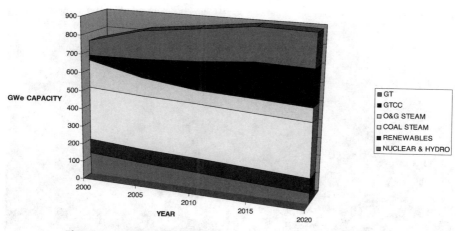

Figure 1-6. Projected U.S. Generating Capacity 2000–2020

in petrochemical plants, where the prime-mover drives are used to drive compressors to compress process gasses, and then the heat is used to either produce steam for process use, or for use direct in processes.

Combined cycle power plant is a term usually associated with electrical power plants, which uses the waste heat from the prime mover for the production of steam and, consequently, the steam is used in a steam turbine for the production of additional power. This is usually a combination of the Brayton Cycle (Gas Turbine) as the topping cycle , and the Rankine Cycle (Steam Turbine) as the bottoming cycle. However, technically, the term can be used for any combination of cycles. Many small plants use the Diesel Cycle as the topping cycle, with the Rankine Cycle as the bottoming cycle. Plants are also using the Brayton Cycle as both the topping and the bottoming cycles.

The fossil power plants of the 1990s and into the early part of the new millennium will be the combined cycle power plants. It is estimated that between 1997 and 2006, there will be an addition of 147.7 GW of power. These plants have replaced the large steam turbine plants, which were the main fossil power plants through the 1980s. Figure 1-7 shows a large combined cycle power plant. The combined cycle power plant is not new in concept, since some have been in operation since the mid-1950s. These plants came into their own with the new high-capacity and high-efficiency gas turbines.

The new market place of energy conversion will have many new and novel concepts in combined cycle power plants. Figure 1-8 shows the heat rates of the plants of the present and future, and Figure 1-9 shows the efficiencies of the same plants. The plants referenced are the simple cycle gas turbine (SCGT), with firing temperatures of 2400°F (1315°C), recuperative gas turbine (RGT), the steam turbine plant (ST), the combined cycle power plant (CCPP), and the advanced combined cycle power plants (ACCP), such as combined cycle power plants using advanced gas turbine cycles, and, finally, the hybrid power plants (HPP).

8 • COGENERATION AND COMBINED CYCLE POWER PLANTS

Figure 1-7. A Typical Combined Cycle Facility (Courtesy of Enron Corp.)

The demographics, both local and worldwide, will determine the capital available for investment. Countries, such as India, starting the new century with a power capacity of about 95,000 MW, will require another 100,000 MW through the year 2050. The world population will continue to increase for the next three to four generations, reaching a peak of about 11 to 12 billion people in the year 2150 as per World Bank figures.

Distributed Generation

The very large fossil plants ranging upwards of 1500 MW will be fewer, and plants between 150 and 300 MW will dot the landscape and will reduce transmission losses, which can reach as high as 20% to 30% due to electrical losses, as well as theft and other parasite losses. In fact, there has been a growth in the late 1990s at the very low end of the power spectrum with the advent of the micro-gas turbine, a turbine that produces power in the 50 to 100 kW range, leading to the concept of distributed generation. Distributed generation (DG) is the integrated or stand-alone use of small, modular electricity generation resources by utilities, industry and retail centers, in applications that benefit the entire electric system.

The growth of centralized power plants in the 1950s and 1960s was due to aging equipment, plus the growth of the U.S. Rural Electrification Agency (REA), and abundant supply of low cost fuels. A growth of distributed generation is expected in the next 20 to 30 years. Historically, the distributed power plants were the "norm"

An Overview of Power Generation • 9

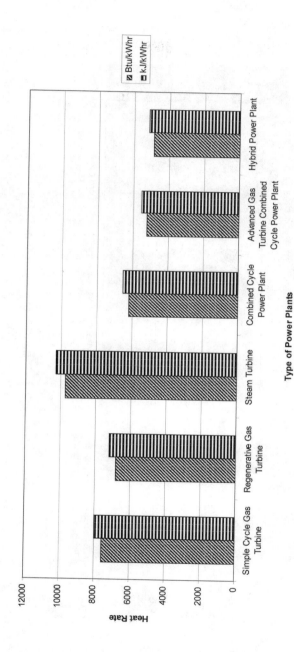

Figure 1-8. Typical Heat Rates of Various Types of Plants

10 • COGENERATION AND COMBINED CYCLE POWER PLANTS

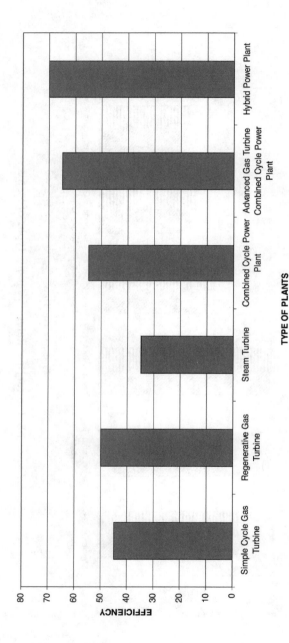

Figure 1-9. Typical Efficiencies of Various Types of Plants

in the 1920s, 1930s, and 1940s. Cities, especially in Europe, have both large and small municipal plants with district heating from the waste energy.

Distributed energy would find the best expectance in the following areas:

- Emergency generation
- Peak shaving
- Special base load application.

There are many barriers to distributed generation that have to be overcome to enable distributed generation to be widely used:

1. Technical Requirements for Interconnection
    1.1. The adoption of uniform standards for interconnecting distributed power to the grid. "Plug and play" standards are key to opening markets for distributed generation.
    1.2. Adoption of IEEE standards for interconnection equipment, implementation, testing, and certification would assure utilities.
    1.3. The technology of various distributed power technologies must be improved
        1.3.1. Reduce installed cost from about $1400 to about $400/kW
        1.3.2. Reduce NO_x output to less than 9 ppm
        1.3.3. Improve plant efficiency to about 39% to 45%

2. Business Practices and Contractual Requirements
    2.1. Standard commercial practices will have to be adopted in dealing with utilities to review interconnection
    2.2. Establish standard business plan for interconnection agreements
    2.3. Help utilities to assess the value impact of distributed power at any point on the grid

3. Regulatory Barriers
    3.1. New regulatory principles will have to be adopted favorable to distributed generation in both competitive and utility markets
    3.2. Conditions for a right to interconnect must be spelled out
    3.3. Expedited dispute resolutions processes
    3.4. New tariffs must be established in the following traditional tariff areas
        3.4.1. Demand charges
        3.4.2. Backup tariffs
        3.4.3. Buy back rates
        3.4.4. Uplift tariffs.

Distributed generation provides benefits for both the user and the utility. The benefits to the user would be:

- Power reliability — reduced power outages, and thus the minimization of the cost associated with power outages in many chemical processes.

Finally, the safety and loss of production associated with power loss from the grid.
- Improved power quality — voltage fluctuations are minimized, which negatively effect many machinery, increasing maintenance costs.
- Reduced total energy costs — these systems can produce hot or cold water for heating or cooling. Steam for other processes in the plant. Efficiencies could reach 80% with combined heat and power (CHP).
- Barrier against price volatility — the price of electric supply in many countries varies significantly from year to year.
- Source of revenue — the ability to sell any excess power to other neighboring plants, as well as in some cases to the utility itself.
- Environmental — reduction of NO_x can be reduced, as most of these smaller plants run at lower firing temperatures.

There are also many benefits for the utility to consider in the area of distributed power generation:

- Reduced transmission and distribution losses — these loses can run as high as 30%, continuous losses are between 12% and 15%.
- Avoid or defer increase in capacity — since this gives an increase in power source, the need for large capital investment is greatly reduced.
- Deferral of transmission and distribution upgrades — use of DG reduces the need, or defers the need, for upgrading existing transmission or distribution lines. Grid extension can cost anywhere from $8000 to $18,000/km depending on the terrain. Distribution capacity can be estimated to cost in major metropolitan areas as much as $400 to $500/kW.
- VAR support — provide reactive power (VAR) that helps utilities maintain system voltage.
- Peaking power — the DG power can reduce the need for peaking plants, which are expensive to operate and maintain.
- Improved power quality — DG plants can eliminate the demand that negatively impacts the power quality of the grid system.
- Reduced reserve margin — by lowering the overall power demand, reserve margins can be reduced.
- Transmission congestion — by generating the power near the point of consumption, the effectiveness of the transmission and distribution system for all customers.
- Increased power reliability — reduction of power outages in certain part of the system that are overloaded.
- Environmental impact — societal concerns of transmission lines is growing, thus DG reduces the necessity of new lines to be constructed.

The most common type of distributed generation in the near future can be classified into four major categories:

- Emergency generation
  - Hospitals
  - Nursing homes

- Peak shaving
  - Power plants
  - Offices
  - Hotels
  - Manufacturing operations

- Special base-load application
  - Colleges and schools
  - Convenience stores
  - Supermarkets

- Remote operations
  - Mining operations

The effect of distributed generation on emissions could also be beneficial on the environment. The US utility average is about 3.4 lb/MW·h. Large coal plants have emissions of about 5.6 lb/MW·h. Table 1-1 shows the emissions of some types of power generation units.

There are many sources of energy available for the production of electric power for a distributed network of energy generation systems. A number of the technologies that will make DG a very viable alternative will depend on the new technologies, such as fuel cells and micro-turbines, proving themselves. Distribution system stability and safety, as well as interconnection standards, have yet to be settled. The following is a list of some of the common power generation technologies available today and will be available in the next 20 years, which can be used, and are being used, for distributed generation throughout the world:

1. Diesel and gasoline engines
2. Natural gas reciprocating engines
3. Gas turbines
4. Micro-turbines
5. Fuel cell technology
6. Solar energy-photovoltaic cells
7. Biomass systems
8. Wind energy
9. River hydro-turbines

Figure 1-10 shows the initial cost of the various technologies as they exist in the year 2000. It is obvious that for many of these technologies to be competitive, the

Table 1-1. NO_x Emissions From Various Plants

Combined Cycle (lb/MW·h)	Large Gas Turbines (lb/MW·h)	Advanced Gas Turbines (lb/MW·h)	Small Turbines (lb/MW·h)	Micro Turbines (lb/MW·h)	Diesel Engines (lb/MW·h)	Gas Engines (lb/MW·h)
0.06	0.6	0.3	1.0	0.4	12.0	0.4

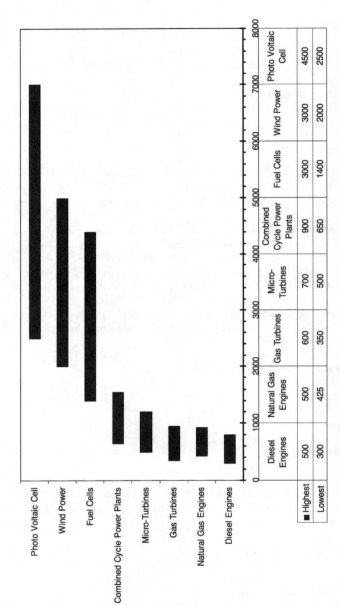

Figure 1-10. Typical First Cost of Various Types of Plants

values of these initial cost have to be greatly reduced before some of these technologies become competitive.

Diesel and Gasoline Engines

The diesel and gasoline engine has been providing power from the early 1900s, especially for the rural areas, through the 1950s. The diesel engines have been relatively robust and have efficiencies which vary from 30% to 37%. Diesel engines can be classified into three groups: low-speed diesels (200 to 400 rpm), medium-speed diesels (800 to 1200 rpm), and high-speed diesels diesel (1500 to 3600 rpm). The low-speed diesel engines can be very large units, ranging in size to 100 MW. The medium-speed diesel engines range from 600 to 7000 kW, while the high-speed diesel is between 200 and 1800 kW. Medium-speed diesels are generally used in the power generation in distributed generation systems. Diesel fuel is easily available and is, thus, still the major source of independent and back-up power used around the world. It is not uncommon to see these engines being used as the power source for hotels, hospitals and shopping centers.

Natural Gas Reciprocating Engines

The natural gas reciprocating engine is getting more and more popular and, in many areas, is replacing the diesel engine. Natural gas is a cleaner burning fuel than diesel oil, and these engines have slightly higher efficiencies, as well as lower maintenance costs than the diesel engines.

Gas Turbines

The simple cycle gas turbine is classified into three groups: industrial type, aeroderivative gas turbines, frame type. The industrial-type gas turbine varies in range from about 500 to 15,000 kW. This type of turbine has been used extensively in many petrochemical plants, and is the source of remote power. The efficiencies of these units is the low 30s. Aeroderivative, as the name indicates, are power generation units, which have origin in the aerospace industry as the prime mover of aircraft. These units have been adapted to the electrical generation industry by removing the by-pass fans and adding a power turbine at their exhaust. These units range in power from 2.5 to about 50 MW. The efficiencies of these units can range from 35% to 42%. The frame units are the large power generation units ranging from 3 to 350 MW in a simple cycle configuration, with efficiencies ranging from 30% to 43%.

Micro-Turbines

Micro-turbines are usually referred to units of less than 350 kW. These units are usually powered by either diesel fuel or natural gas. They

16 • COGENERATION AND COMBINED CYCLE POWER PLANTS

utilize technology already developed. The micro-turbines can be either axial flow or centrifugal-radial inflow units. The initial cost, efficiency, and emissions will be the three most important criteria in the design of these units.

The micro-turbines, to be successful, must be compact in size, have low-manufacturing cost, high efficiencies, quiet operation, quick startups and minimal emissions. These characteristics, if achieved, would make micro-turbines excellent candidates for providing base load and cogeneration power to a range of commercial customers. The micro-turbines are largely going to be a collection of technologies that have already been developed. The challenges are in economically packaging these technologies.

The micro-turbines on the market today range from about 20 to 350 kW. Today's micro-turbines are using radial-flow turbines and compressors, as seen in Figure 1-11. To improve the overall thermal efficiency, regenerators are used in the micro-turbine design, and in combination with absorption coolers or other thermal loads, very high efficiencies can be obtained. Figure 1-12 shows a

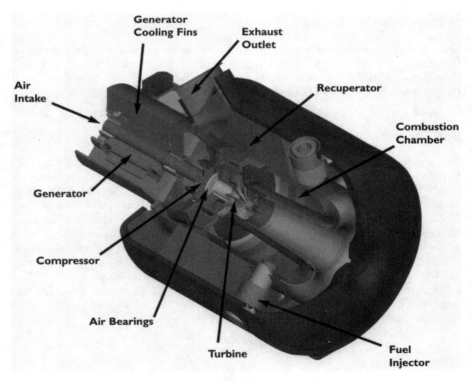

**Figure 1-11. A Compact Micro-Turbine Schematic
(Courtesy of Capstone Corporation)**

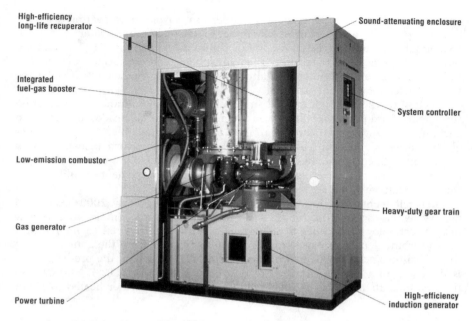

**Figure 1-12. A Cogeneration Micro-Turbine System Package
(Courtesy of Ingersoll Rand Corporation)**

typical cogeneration system package using a micro-turbine. This compact form of distributed power systems has great potential in the years to come.

Fuel Cell Technology

The general concept of a fuel battery, or fuel cell, dates back to the early days of electrochemistry. A fuel cell is an electrochemical device that combines hydrogen fuel and oxygen from the air to produce electricity, heat, and water. Fuel cells operate without combustion, so they are virtually pollution-free. Since the fuel is converted directly to electricity, a fuel cell can operate at much higher efficiencies than most combustion engines, extracting more energy from the same amount of fuel. The fuel cell itself has no moving parts, making it a quiet and reliable source of power.

The fuel cell is composed of an anode (a negative electrode that repels electrons), an electrolyte membrane in the center, and a cathode (a positive electrode that attracts electrons). As hydrogen flows into the fuel cell anode, platinum coating on the anode helps to separate the gas into protons (hydrogen ions) and electrons. The electrolyte membrane in the center allows only the protons to pass through the membrane to the cathode side of the fuel cell. The electrons cannot pass through this membrane and flow through an external circuit in the form of electric current.

18 • COGENERATION AND COMBINED CYCLE POWER PLANTS

As oxygen flows into the fuel cell cathode, another platinum coating helps the oxygen, protons, and electrons combine to produce pure water and heat. Individual fuel cells can be then combined into a fuel cell "stack". The number of fuel cells in the stack determines the total voltage, and the surface area of each cell determines the total current. Multiplying the voltage by the current yields the total electrical power generated.

A cell of this sort is built around an ion-conducting membrane such as Nafion (trademark for a perfluorosulfonic acid membrane). The electrodes are catalyzed carbon, and several construction alignments are feasible. Solid polymer electrolyte cells function well (as attested to by their performance in Gemini spacecraft), but cost estimates are high for the total system compared to the types described above. Engineering or electrode design improvements could change this disadvantage.

Fuel cell technology is rapidly growing, and in the early 2000, successful factory testing of the first fuel cell/gas turbine has been completed. The new hybrid technology combines a pressurized solid oxide fuel cell (SOFC) with a micro-turbine generator as seen in Figure 1-13. The air from the compressor of the micro-turbine is pre-heated in a recuperator, from which the pre-heated air is then sent to a duct burner, where it is heated to about 1112°F (600°C), and the gases are then sent to the fuel cell, where they are further heated to about

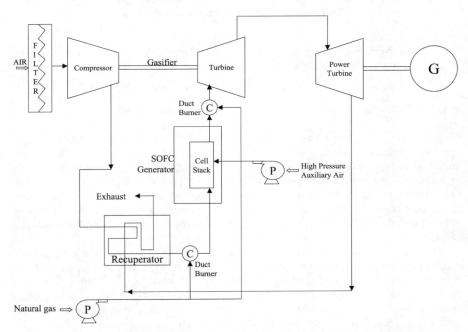

Figure 1-13. Simplified Process Flow Diagram for the SOFC/Micro-Turbine Hybrid System

1562°F (850°C), and if necessary, they are then sent through another duct burner where the gases are further reheated to about 1832°F (1000°C). The gases then enter the gasifier turbine, which drives the micro-turbine compressor. The gases then enter the power turbine, where the gases are further expanded and drive the generator. The gases from the power turbine, which exit at about 1150°F (621°C), enter the recuperator to pre-heat the compressed air.

SOFC systems have been operated at about 100 kW successfully. In some ways, solid oxide fuel cells are similar to molten carbonate devices. Most of the cell materials, however, are special ceramics with some nickel. The electrolyte is an ion-conducting oxide, such as zirconia, treated with yttria. The fuel for these experimental cells is hydrogen combined with carbon monoxide, just as for molten carbonate cells. While internal reactions would be different in terms of path, the cell products would be water vapor and carbon dioxide. Because of the high operating temperature, 1500°F to 1850°F (815°C to 1010°C), the electrode reactions proceed very readily. As in the case of the molten carbonate fuel cell, there are many engineering challenges involved in creating a long-lived containment system for cells that operate at such a high-temperature range. Solid oxide fuel cells have been designed for use in central power-generation stations, where temperature variation could be controlled efficiently, and where fossil fuels would be available. The system would, in most cases, be associated with the so-called bottoming steam (turbine) cycle — i.e., the hot gas product, 1850°F (at 1010°C), of the fuel cell could be used to generate steam to run a turbine and extract more power from heat energy. The major characteristic of the fuel cell in the final completed project will be the very high efficiency of over 60% and extremely low NO_x emission (<0.5 ppm), and no measurable SO_x.

The direct fuel cell (DFC) application being developed by companies such as FuelCell Energy, Inc. use carbonate fuel cell technology, in which the reforming action occur within the fuel cell stacks. This version of the carbonate technology is referred to as the direct fuel cell because of the internal reforming aspect of the design. The process involves treating natural gas to remove impurities, after which it is mixed with steam and sent to the fuel cell stacks. The fuel/steam mixture is reformed in the stacks, providing the hydrogen, which is consumed in the fuel cell anodes (at a hydrogen utilization rate of up to 80%). The anode reaction also produces CO_2, which is required in the fuel cell cathodes. The cathode feed is produced by taking the anode exhaust, catalytically reacting any residual fuel with air, and sending the flue gas with excess air to the cathodes. Cathode exit gases are sent to a packaged heat recovery unit to supply heat for steam generation, fuel pre-heat, and final heating of the steam/fuel anode feed gas.

A hybrid system which utilizes DFC/gas turbine is shown schematically in Figure 1-14. The system concept is as follows. Fuel and water are sent to the system heat recovery unit (HRU) unit, where steam is produced and mixed with heated fuel for use as the fuel cell gas feed. As with the simple-cycle system, residual fuel from the anode exit is consumed in an anode exhaust oxidizer. Air is compressed in an inter-cooled compressor to the desired turbine pressure, 235 psia (16 bar), in the baseline hybrid system,

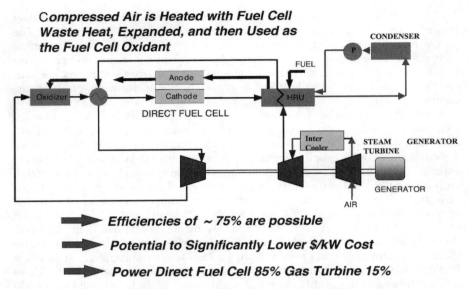

Figure 1-14. Schematic of a Fuel Cell (Courtesy FuelCell Energy, Inc.)

heated with system exhaust in the HRU, heated further with exhaust from the anode exhaust oxidizer, and expanded in a turbine to produce additional electricity. The expanded, low-pressure air leaving the turbine is used as the oxidant in the anode exhaust oxidizer. Flue gas leaving the oxidizer is first cooled by the turbine air, and then sent as the cathode feed gas to the fuel cells. The cathode exhaust gas is sent through the HRU to provide the required pre-heat and water vaporization. A typical 20-MW system would have an efficiency of 71% based on the lower heating value (LHV) of the fuel, the direct fuel cell power is 17 MW, the gas turbine power is 3.44 MW, and the parasitic loss is 0.03 MW.

Installed costs for fuel cells in the year 2000 are still very expensive, averaging over $1500/kW. For the fuel cell to be competitive, costs will have to be reduced to around $400 to $600/kW.

Solar Energy-Photovoltaic Cells

Photovoltaic process is a process in which two dissimilar materials in close contact act as an electric cell when struck by light or other radiant energy. Light striking such crystals as silicon or germanium, in which electrons are usually not free to move from atom to atom within the crystal, provides the energy needed to free some electrons from their bound condition. Free electrons cross the junction between two dissimilar crystals more easily in one direction than in the other, giving one side of the junction a negative charge and, therefore, a negative voltage with respect to the other side, just as one electrode of a battery has a negative voltage with respect to the other. The photovoltaic battery can continue to provide voltage and current as long as light continues to fall on the two materials. This current can be a source of power in an electrical circuit, as in the modern solar battery. Tropical islands like Hawaii may favor rooftop photovoltaics, which compete economically with fossil fuels for power generation in those more isolated regions. Presently, the cost of these systems is prohibitive, but in the next 10 to 20 years, the costs will be more competitive. Hybrid systems using this technology have a very bright future.

BioMass Systems

Biomass plants have been getting more popular, as have the river hydro-projects in different areas of the world. Biomass turbines using methane gas being produced by garbage fills are starting to get more common. These are low-Btu gas turbines using gas with a lower heating value (LHV) of about 300 to 450 Btu/cu ft (11,183 to 16,775 kJ/m^3); they have an efficiency about 30% and can produce between 4 and 10 MW.

Wind Energy

Wind energy is also an area of remote power, and large wind energy farms have been growing throughout the world. In 1999, there were more than 3600 MW of new wind generating capacity, bringing total worldwide installed capacity to a 13,400-MW range. This an increase of 36% over the 1998 figures. Wind turbines generating 2.5 MW/turbine are being presently designed with a rotor diameter of 262 ft (80 m); the prototype turbines are to be installed by early 2000.

River Hydro-Turbines

Small-scale hydroelectric systems capture the energy in flowing water, and convert it to electric power. The potential for these systems depends on the availability of suitable water flow; where the resource is available, it provides cheap and reliable power. Norway is one of the leading countries in the technology of river hydro projects. These systems blend with their surroundings and have minimal negative environmental impact.

Table 1-2 gives an economic comparison of distributed generation technologies from the initial cost of such systems to the operating costs of these systems. Because distributed generation is very site-specific, the cost will vary and the justification of installation of these type of systems will also vary. Sites for distributed generation vary from large metropolitan areas to the slopes of the Himalayan mountain range. The economics of power generation depend on the fuel cost, running efficiencies, maintenance cost, and first cost, in that order. Site

Table 1-2. Economics of Distributed Generation Technologies

Technology Comparison	Diesel Engine	Gas Engine	Simple Cycle Gas Turbine	Micro-Turbine	Fuel Cell	Solar Energy Photovoltaic Cell	Wind	BioMass	River Hydro
Product Rollout	Available	Available	Available	Available	1996-2010	Available	Available	2,020	Available
Size Range (kW)	20-25,000+	50-7,000+	500-100,000+	30-200	50-1,000+	1+	10-2,500	NA	20-1,000+
Efficiency (%)	36-43%	28-42%	21-40%	25-30%	35-54%	NA	45-55%	25-35%	60-70%
Gen Set Cost ($/kW)	125-300	250-600	300-600	350-800	1,500-3,000	NA	NA	NA	NA
Turnkey Cost, No-Heat Recovery ($/kW)	200-500	600-1,000	400-650	475-900	1,500-3,000	5,000-10,000	700-1,300	800-1,500	750-1,200
Heat Recovery, Added Cost ($/kW)	75-100	75-100	150-300	100-250	1,900-3,500	NA	NA	150-300	NA
O&M Cost ($/kW·h)	0.007-0.015	0.005-0.012	0.003-0.008	0.006-0.010	0.005-0.010	0.001-0.004	0.007-0.012	0.006-0.011	0.005-0.010

selection depends on environmental concerns such as emissions and noise, fuel availability, and size and weight.

Cogeneration

An ideal cogeneration situation exists when there is an equality between power and thermal demands. In several cases, the heat demand may vary with seasonal changes (greater heat loads during the winter and less during the summer). The choice of equipment (type, size, etc.) is strongly influenced by the load demand pattern.

The following data is required to evaluate cogeneration feasibility:

- Power demand with variations (seasonal and daily)
- Required heat loads (heat content and mass flow)
- Peak requirements for heat and power
- Special plant requirements — heaters, chillers, plant air requirements, etc.
- Fuel availability
- Environmental impact
- Reliability, availability, and maintainability considerations.

The recovery of waste heat has become economically feasible and is being addressed today on a very large scale. Industrial plants throughout the world are considering cogeneration and combined cycle power plants. Bottoming cycles for gas turbine drives are being investigated with many working fluids, such as steam, freon, ammonia, butane, ethylene, to name just a few. The use of heavy crude and other dirty fuels, such as pulverized coal and waste products, such as sawdust, support the use of externally fired gas turbines to overcome the reduced hot gas path life caused by contaminants in the fuel. The use of waste heat in plants from various processes, whether they be in the petrochemical, iron/steel, or the paper industry, require the design of modified power generation units, which can directly convert waste energy into useful energy.

Cogeneration, or the combined generation of both power and heat, has been utilized for over a hundred years and has been given a number of names (total energy or combined heating and power). As a result of the energy awareness that started around 1973, United States industry has shown an increasing interest in the cogeneration concept. Early cogeneration systems did not tie into the electric utilities. This created problems in steadily maintaining the demand for electricity and heat. Because of this, some of the early cogeneration systems were not totally successful.

The underlying concept involving cogeneration is that current day prime movers have low efficiencies, which implies that a greater part of the fuel energy is being converted to heat rather than shaft horsepower. Cogeneration would then involve the sequential utilization of the waste heat for some process-related needs such as drying, steam production, absorption cooling, and auxiliary heat to furnaces.

In 1983, about 5% of the power in the United States was cogenerated; in the year 2010, over 20% of the power will be cogenerated. In Europe, where energy costs have been historically higher, cogeneration has been well established. For example, in Germany, about 25% to 35% of the power consumed is cogenerated. In the fall of 1978, the United States Congress enacted the Public Utility Regulatory Policies Act, commonly known as PURPA. Part of PURPA required that the Federal Energy Regulatory Commission (FERC) develop rules to encourage cogeneration. This resulted in the February 1980 ruling that qualified cogeneration facilities may parallel utility grids and should be paid rates for electricity that are equal to the cost that the utility avoids by not having to generate or obtain power from another source, i.e., the utilities *avoided cost*. Utilities also had to provide standby power at non-discriminatory rates and were obligated to interconnect. State regulatory commissions were then required to implement FERC's rules. By March 1981, most states (and unregulated utilities) had some rules in place. Several states simply rubber-stamped the ideas and adopted FERC rules, while other states developed very detailed *specific* documents, including methods for measurement of avoided costs, setting rate levels and defining contract terms. The rubber-stamp approach left it to the utilities to determine how avoided costs were to be measured and how rates were to be structured. Initially, the utilities were very much against this law, however, most utilities have now their own cogeneration divisions, and are now some of the largest independent power producers (IPP) in the world.

There are several unresolved legal, environmental, interconnection and regulatory problems that make implementation of cogeneration difficult. The best advice for any potential cogenerator is to get *expert* help in dealing and negotiating with the state agencies and utilities, and to keep informed of changes occurring in the laws. There are several firms that specialize in the legal, regulatory and procedural steps of initiating a cogeneration project, and these consultants should be utilized. There are several cases in which obstacles will seem insurmountable, and professional help is the best way to solve this problem.

The term avoided cost is deceptively simple to define, but it is very difficult to find much agreement on the term. Several legal battles have occurred, and probably millions of dollars have been spent in legal fees on this issue. As utilities utilize economic dispatch programs, they use their most efficient units first, and their least efficient plants last. Therefore, as a cogenerator comes on-line, the utility would shut down the least efficient unit. The avoided cost would thus relate to this unit, which explains why avoided costs are high (typically $0.05 to $0.07/kW·h).

Gas turbine cogeneration is far more efficient than the typical steam utility central plants of the 1970s. About 75% heat utilization can be realized for power and heat, with about 25% leaving in the exhaust gases. In a fossil steam plants, only 35% of the fuel energy is obtained as power with condenser losses and boiler losses accounting for 48% and 15%, respectively.

Over 19.6 GW in cogeneration capacity is estimated by 2010. Gas turbine-based cogeneration and combined cycles will dominate this market. Some interesting projections regarding gas turbine cogeneration predict that about 70% of cogeneration market growth will take place in the developing nations of the world, with about half of the sales occurring in the petrochemical industry. Most of the systems will be for units with steam capabilities exceeding 100 million Btu/hr.

Cogeneration systems are so varied that classification is not an easy task. One way of classifying the systems is as follows:

- Utility cogeneration — these are funded and operated by a municipality, usually involving large units with district heating and cooling. This concept has been and is still being widely used in Europe. In the U.S., new sports complexes, as well as downtown areas of large cities, are looking into similar heating and cooling complexes with power generation utilities.
- Industrial cogeneration — this is operated for a private sector industry, i.e., a petrochemical plant, paper mill, glass factories, textile mills, and many other industrial complexes. The popularity of the "inside the fence" power plants especially in the developing world will be large, as not only does it provide cheap energy, but very importantly, it produces reliable energy.
- Desalination plants — desalting costs are reduced by using cogeneration and hybrid processes. Cogeneration desalination plants are large-scale facilities that produce both electric power and desalted seawater. Distillation methods, in particular, are suitable for cogeneration. The high-pressure steam that runs electric generators can be recycled in the distillation unit's brine heater. This significantly reduces fuel consumption compared with what is required if separate facilities are built. Cogeneration desalination plants using gas turbines are very common in the Middle East and North Africa, where they have been in operation for the past 30 years.

Cogeneration Qualifications

Cogeneration qualifications in the U.S. were designed to promote meaningful energy conservation and to ensure that the cogeneration concept is not abused by the creation of facilities that produce power and a trivial amount of heat. To qualify as a cogenerator, the system designed (utilizing fuel oil or natural gas) must have a "PURPA efficiency" of at least 42.5% on an annual basis. If the thermal energy output is less than 15% of the total energy output, the PURPA efficiency to be met increases to 45%,

$$\text{PURPA Efficiency} = \frac{\text{Electric O/P} + 1/2 \text{ Thermal O/P}}{\text{Energy Input (LHV)}} \qquad (1-1)$$

where
    O/P = output power

For example, assume the cogeneration system as shown in Figure 1-15. In this system, the exhaust gas from the gas turbine enters the duct burner, where the exhaust gas is further heated. The heated gas then enters the HRSG, where steam is created. This steam is sent to an extracting condensing steam turbine, where the steam is extracted for the process purpose, and the remainder steam is sent through the steam turbine to generate additional power if needed. A simpler system would have no steam turbine, and all the steam produced

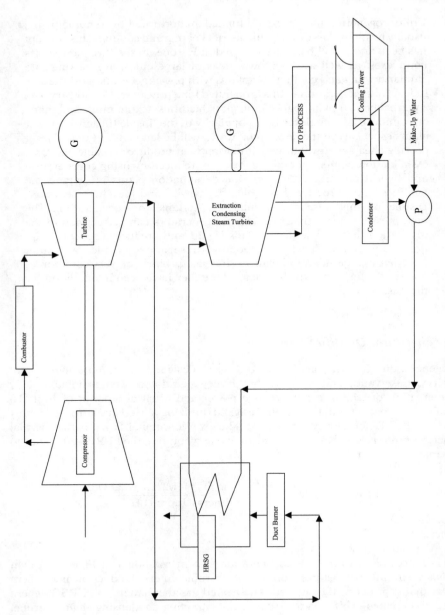

Figure 1-15. A Schematic of a Typical Cogeneration Facility

would be used for the process. Consider a system that involves a gas turbine that produces 3825 kW of electricity at a 95% utilization (i.e., 8322 hr/year). The gas turbine heat rate is 12,186 Btu/kW·h. The system also utilizes duct burners that are used for 6146 hr/year, and are rated at 13.6 (106 Btu/hr). The system generates 31,5540 lb/hr of 125-psig saturated steam. The electric power output is therefore:

$$E = 3825 \text{ kW} \times 8322 \text{ hr} \times 3412.2 \text{ Btu/kW·h} = 108,616 \times 10^6 \text{ Btu} \quad (1-2)$$

The system also generates 31,540 lb/hr steam. Thermal output is given by:

$$T = 31,540 \text{ lb/hr} \times 8322 \text{ hr} \times 1011.7 \text{ Btu/lb} = 265,547 \times 10^6 \text{ Btu} \quad (1-3)$$

The energy input (EI), based on both gas turbine and duct burner fuel flows, is given by:

$$\begin{aligned} EI &= (3890 \text{ kW} \times 8322 \text{ hr} \times 12,186 \text{ Btu/kW·h}) \\ &\quad + (13.6 \times 10^6 \text{ Btu/hr} \times 6146 \text{ hr}) \\ &= 394,492 \times 10^6 \text{ Btu} + 83,585 \times 10^6 \text{ Btu} = 478,078 \times 10^6 \text{ Btu} \quad (1-4) \end{aligned}$$

PURPA efficiency, therefore, is given by:

$$\text{PURPA}(\eta) = \frac{E + 1/2\, T}{EI}$$

$$\text{PURPA}(\eta) = \frac{108,616 + 1/2(265,547)}{478,078} = 50.5\% \quad (1-6)$$

As this exceeds 42.5%, the system is "qualified" in this aspect. It must be noted that there are other qualifiers. The operating standard qualifier requires that a topping cycle produce a minimum of 5% of the total energy output as useful thermal energy. The ownership qualification states that a utility may not own more than 51% of a cogeneration plant. These laws are rapidly changing, and with the new deregulation laws being instigated worldwide, these laws will be obsolete in a few years as the power industry will undergo major changes.

Gas Turbine Cycle in Cogeneration Mode

The utilization of gas turbine exhaust gases, for steam generation or the heating of other heat-transfer mediums, or in the use of cooling or heating buildings or parts of cities, is not a new concept and is currently being exploited to its full potential.

Regenerators are commonly used with gas turbines to capture waste heat and increase the air inlet temperature to the combustor. This reduces the amount of heat necessary to be added to the combustor, thus increasing the overall turbine efficiency. Turbine efficiency for simple cycle units is dependent on the pressure ratio developed in the compressor and the cycle peak temperature of the combustion gases. The relationship between power and efficiency for simple cycle gas turbines, and of regenerative cycles is described in the following chapters. It is important to note that *not all turbines* would benefit from regenerators, especially those operating at high-pressure ratios and low-firing temperatures. In most cases, the additional cost of a regenerator installation can be recovered in the first 12 to 18 months of operation. It is this large return on investment that makes regeneration a very interesting proposal. It is important to note that while regeneration involves waste heat recovery, it does *not* constitute cogeneration. Several cogeneration facilities, however, use regenerated gas turbines. Gas distribution companies have also used waste heat recovery successfully since as much as 20% to 30% of the gas transported is utilized in running the pumping stations.

Gas turbine efficiencies can be increased by 10% to 20% with the right pressure ratio and temperature combination. Increase in gas turbine cycle efficiency could provide fuel savings of about $1.64 million for every 10% increase in efficiency in a 100-MW power plant (assuming a 100-MW unit, operating base-loaded for a year with an availability of 97%, and a fuel cost of $2.50/MBtu or $2.37/MkJ).

The typical cogeneration plant uses the waste gases from the gas turbine to produce steam in a heat recovery steam generator (HRSG), or waste heat boilers (WHB); these terms are often used interchangeably for use in various chemical processes. A typical cogeneration plant generates high-pressure steam, which then is used in an extracting condensing steam turbine. It usually extracts part of the steam at a lower pressure for use in various chemical processes; the rest of the steam goes through the second part of the steam turbine, and then to the condenser. The steam turbines usually drive separate generators; however, the system can be designed in a way that both the gas turbine and the steam turbine drive the same generator. Dampers may be used between the gas turbine exhaust and the heat recovery steam generator. In this type of configuration, the turbines can be used easily in a simple cycle mode, thus allowing great flexibility in operation. The use of dampers must be very carefully evaluated since the effectiveness of the dampers, in most cases, after short operation times to totally shut-off the flow, is very poor. Many plants have done away with their dampers. It is very strongly advised not to have any work done on systems downstream of the dampers, while the upstream systems are operating.

In most of these applications, the gas turbine is a single shaft unit, exhausting gas at a temperature between 900°F (482°C) and 1100°F (593°C), depending on the turbine efficiency and the turbine inlet temperature. The hot gases (approximately 90 lb/sec (40.8 kg/sec) for a 15-MW turbine, to about 1400 lb/sec (636 kg/sec) for a 200-MW turbine) are piped into a boiler, where steam is generated for use as process heat, or for use in an extraction or backpressure steam turbine.

In combined heat power (CHP) plants, the exhaust gases produce hot or chilled water using boilers or absorption chillers, respectively. The hot or chilled water produced is then circulated through various buildings at temperatures between 150°F and 160°F (65.6°C to 71.1°C) for heating purposes, and at temperatures between 45°F and 55°F (7.5°C to 12.7°C) for cooling purposes. The water loops in the building, the temperatures of the water between the entering and leaving, are maintained at about 10°F to 12°F (5.6°C to 6.7°C). These applications are being more commonly used in large shopping centers, universities, hospitals, stadiums, and even entire city blocks. In some cogeneration petrochemical applications, the waste heat is used for various other chemical processes and for heating other fluids.

The use of bottoming cycles other than steam are being investigated since they offer high-energy transfer in the temperature ranges lower than those in which gas or steam turbines operate efficiently. The cost of installing these organic bottoming cycles can sometimes be very high, between $350/kW and $700/kW, due to the high cost of the heat exchangers, condensers, pumps, and piping required in addition to the turbine. The turbine or expander cost would be approximately $30/kW to $50/kW. There are many organic fluids that can be considered. The following are some of the major thermodynamic criteria that an ideal fluid must have:

- It must be stable over all operating conditions.
- The specific heat must be low and the critical temperature must be high, probably over 500°F (260°C).
- The inlet pressure should be below 2000 psia (135 bar) in order to keep costs within reason and above 500 psia (34 bar) to produce a reasonable amount of work per pound of fluid.
- The saturation pressure must not be below atmospheric conditions at 32°F (0°C) in order to avoid a vacuum situation during shutdown. The saturation pressure, however, must not be too high at operating conditions, thus avoiding excessive backpressure on the turbine and in the condenser.

Typical bottoming cycle fluids and their corresponding molecular weights are shown in Table 1-3. The major advantage for the use of organic bottoming

Table 1-3. Bottoming Cycle Fluid Characteristics

Fluids	Formula	Molecular Weight	Working Pressure and Temperature
Steam	H_2O	18	1000–1500 psia (68–102 bar), 800–1200°F (426–649°C)
FL-85	0.85 CF_3CH_2OH + 0.15 H_2O	87.7	500–700 psia (34–48 bar), 450–600°F (232–316°C)
Ammonia	NH_3	17	1000–1500 psia (68–102 bar), 450–800°F (232–426°C)
n-Butane	C_4H_{10}	58.1	500–600 psia (34–41 bar), 450–600°F (232–316°C)
Iso-Butane	C_4H_{10}	58.1	500–600 psia (68–41 bar), 450–600°F (232–316°C)

30 • COGENERATION AND COMBINED CYCLE POWER PLANTS

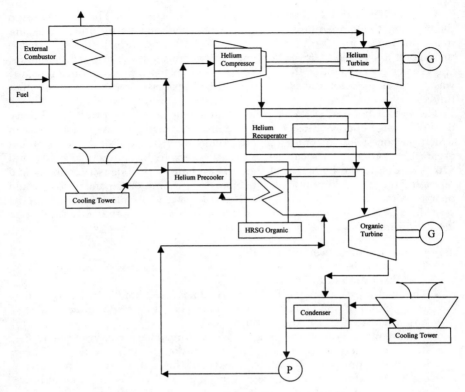

Figure 1-16. Closed Cycle Gas Turbine Application

cycles rather than steam is the low temperature of the gas. Closed cycles using fluids such as helium, can also be coupled to an organic bottoming cycle as shown in Figure 1-16.

Combined Cycle Plants

Two types of cycles are possible, a topping and a bottoming cycle. If power is generated first and the rejected energy is used as the heat energy for another prime mover, that first system is known as a topping cycle. The secondary prime mover powered by the energy of the first system generating electricity is known as the bottoming cycle. Gas turbine-based combined cycle systems fall into this category. The discussion here will focus on gas turbine-based topping cycles and steam turbine-based bottoming cycles. However, any heat transfer fluids may be utilized as the medium for the bottoming cycle.

The combined cycles used in power plants have generally been limited to the use of steam as the bottoming cycle. A typical combined cycle power plant uses the exhaust gases from a gas turbine to produce steam in a boiler (which may have the capability of supplementary firing), for further utilization in a condensing steam turbine.

The combined cycle, in most cases, consists of the combination of the Brayton and Rankine Cycles, is one of the most efficient cycles in operation for practical power generation systems. The Brayton Cycle is the gas turbine cycle, and the Rankine Cycle is the steam turbine cycle. In most combined cycle applications, the gas turbine is the topping cycle, and the steam turbine is the bottoming cycle. Thermal efficiencies of the combined cycles can reach as high as 60%. In the typical combination, the gas turbine produces about 60% of the power, and the steam turbine, about 40%. Individual unit thermal efficiencies of the gas turbine and the steam turbine are between 30% and 40%. The steam turbine utilizes the energy in the exhaust gas of the gas turbine as its input energy. The energy transferred to the heat recovery steam generator (HRSG) by the gas turbine is usually equivalent to about the rated output of the gas turbine at design conditions. At off-design conditions, the inlet guide vanes (IGV) are used to regulate the air so as to maintain a high temperature to the HRSG. Figure 1-17 shows the distribution of the energy entering a combined cycle power plant. About 40% of the energy is converted to power by the gas turbine, and about 20% of the energy is converted to power by the steam turbine.

In the traditional combined cycle plant, air enters the gas turbine, where it is initially compressed, and then enters the combustor, where it undergoes a very high rapid increase in temperature at constant pressure. The high-temperature and high-pressure air then enters the expander section, where it is expanded to nearly atmospheric conditions. This expansion creates a large amount of energy, which is used to drive the compressor used in compressing the air, plus the generator, where power is produced. The compressor in the gas turbine uses about 50% and 60% of the power generated by the expander.

The air, upon leaving the gas turbine, is essentially at atmospheric pressure conditions, and at a temperature between 950°F and 1200°F (510°C and 650°C). This air enters the heat recovery steam generator (HRSG), where the energy is transferred to the water to produce steam. There are many different HRSG units. Most HRSG units are divided into the same amount of sections as the steam turbine. In most cases, each section of the HRSG has a pre-heater, an economizer and feedwater, and then a superheater. The steam entering the steam turbine is superheated.

In most large plants, the steam turbine consists of three sections: a high-pressure turbine stage (HP) with pressures between 1500 and 1700 psia (100.7 and 114.2 bar), an intermediate-pressure turbine stage (IP) with pressures between 350 and 550 psia (23.51 and 36.9 bar), and a low-pressure turbine stage (LP) with pressures between 90 and 60 psia (6 and 4 bar). The steam exiting from the high-pressure stage is usually reheated to about the same temperature as the steam entering the HP stage before it enters the intermediate stage. The steam exiting the IP stage enters directly into the LP stage after it is mixed with steam coming from the LP superheater.

The steam from the LP turbine enters the condenser. The condenser is maintained at a vacuum of between 2 and 0.5 psia (0.13 and 0.033 bar). The increase in backpressure in the condenser will reduce the power produced. Care is

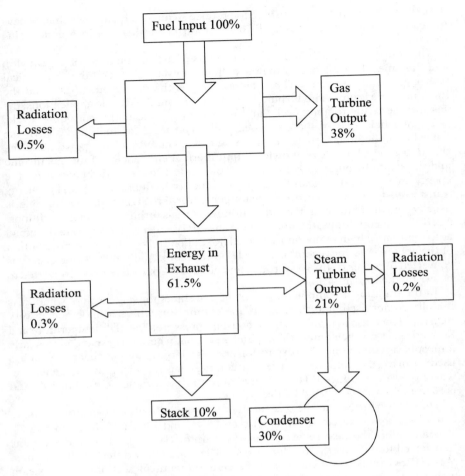

Figure 1-17. Energy Flow Diagram

taken to ensure that the steam leaving the LP stage blades has not a high content of liquid in the steam to avoid erosion of the LP blading.

The new gas turbines also utilize low NO_x combustors to reduce the NO_x emissions, which otherwise would be high due to the high firing temperature of about 2300°F (1260°C). These low NO_x combustors require careful calibration to ensure an even firing temperature in each combustor. New type of instrumentation, such as dynamic pressure transducers, have been found to be effective in ensuring steady combustion in each of the combustors.

Combined cycle plants have several advantages. These include: (1) high thermal efficiencies (50% to 65%), (2) rapid startup (2-hour cold start), and (3) low first installed costs ($600 to $900/kW). Maintenance costs for combined cycles range from $0.003 to $0.007/kW·h (similar to cogeneration plants).

An Overview of Power Generation

As several gas turbine cogeneration plants utilize steam turbines (both back-pressure and condensing), cogeneration and combined cycles are similar in several aspects — design and operation lessons learned in one can be applied to the other.

The new combined cycle power plants being put into operation in the late 1990s have reached 2800 MW. Each module, which consists of the gas turbine and steam turbine on a single shaft, produces 350 MW. This is a single shaft turbine; advantages and disadvantages of these types of turbines will be studied later on. Another combined cycle gas turbine operates at a 30:1 pressure ratio and at turbine inlet temperature of 2400°F (1315°C). The concept of reheat applied here involves the use of two combustors. The second combustor is used to reheat the air between the intermediate-pressure and low-pressure turbines.

Several studies indicate that coal gasification combined cycles will be of importance in the near future, especially in many developing nations. Several demonstration plants have been developed, including: (1) coal gasification plants deriving fuel for use in a gas turbine combined cycle, (2) cycles utilizing atmospheric and pressurized fluidized beds (AFB and PFB), and (3) gas turbine reheat cycles that will operate at efficiencies around 55%.

Combined cycle power plants and cogeneration projects are capital intensive with installed first costs ranging from $800 to $1000/kW. The choice of plants depends on a host of factors including:

- Location of the plant
- Annual utilization
- Power and heat demand ratios
- Utility electric rates
- Net fuel rate
- Type of fuel
- Fuel cost
- Rate of return criteria
- Plant first cost ($/kW)
- Operations and maintenance costs.

Table 1-4 is an analysis of the competitive standing of the various types of power plants, their capital cost, heat rate, operation and maintenance costs, availability and reliability, and time for planning. Examining the capital cost and installation time of these new power plants, it is obvious that the gas turbine is the best choice for peaking power. Steam turbine plants are about 50% higher in initial costs, $800 to $1000/kW, than combined cycle plants, which are about $600 to $900/kW. Nuclear power plants are the most expensive. The high initial costs and the long time in construction make such a plant unrealistic for a deregulated utility.

In the area of performance, the steam turbine power plants have an efficiency of about 35%, as compared to combined cycle power plants, which have an efficiency of about 55%. Newer gas turbine technology will make combined cycle efficiencies range between 60% and 65%. As a rule of thumb, a 1% increase in efficiency could mean that 3.3% more capital can be invested.

The time taken to install a steam plant from conception to production is about 36 to 42 months as compared to 22 to 24 months for combined cycle

Table 1-4. Economic and Operation Characteristics of Plant

Type of Plant	Capital Cost ($/kW)	Heat Rate (Btu/kW·h; kJ/kW·h)	Net Efficiency	Variable Operation and Maintenance ($/MW·h)	Fixed Operation and Maintenance ($/MW·h)	Availability (%)	Reliability (%)	Time from Planning to Completion (Months)
Simple Cycle Gas Turbine (2500°F/1371°C), Natural Gas-Fired	300–350	7,582; 8,000	45	5.8	0.23	88–95	97–99	10–12
Simple Cycle Gas Turbine, Oil-Fired	400–500	8,322; 8,229	41	6.2	0.25	85–90	95–97	12–16
Simple Cycle Gas Turbine, Crude-Fired	500–600	10,662; 11,250	32	13.5	0.25	75–80	90–95	12–16
Regenerative Gas Turbine, Natural Gas-Fired	375–575	6,824; 7,200	50	6.0	0.25	86–93	96–98	12–16
Combined Cycle Gas Turbine	600–900	6,203; 6,545	55	4.0	0.35	86–0.93	95–98	22–24
Advanced Gas Turbine, Combined Cycle	800–1,000	5,249; 5,538	65	4.5	0.4	84–90	94–96	28–30
Combined Cycle Power Plant	1,200–1,400	6,950; 7,332	49	7.0	1.45	75–85	90–95	30–36
Combined Cycle Coal Gasification	1,200–1,400	7,300; 7,701	47	7.0	1.45	75–85	90–95	30–36
Combined Cycle Fluidized Bed	1,800–200	10,000; 10,550	34	8	2.28	80–89	92–98	48–60
Nuclear Power	800–1,000	9,749; 10,285	35	3	1.43	82–89	94–97	36–42
Steam Plant, Coal-Fired	400–500	7,582; 8,000	45	6.2	4.7	90–95	96–98	12–16
Diesel Generator, Diesel-Fired	600–700	8,124; 8,570	42	7.2	4.7	85–90	92–95	16–18
Diesel Generator, Power Plant Oil-Fired								
Gas Engine Generator Power Plant	650–750	7,300; 7,701	47	5.2	4.7	92–96	96–98	12–16

power plants. The time taken for construction affects the economics of a unit; the longer the capital is employed without return, interest, insurance, and taxes accumulate.

It is obvious from this that as long as natural gas or diesel fuel is available, the choice of combined cycle power plants is obvious.

The initial cost of a combined cycle plant as shown in Figure 1-18 is made up of various components such as the gas turbine (30%), steam turbine (10%), the HRSG (10%), mechanical systems (15%), electrical systems (12%), controls (3%), and the civil structures and infrastructures (20%). Figure 1-19 shows the cost distribution over the life cycle of combined cycle power plant. It is interesting to note that the initial cost runs about 8% of the total life-cycle cost, and the operational and maintenance cost is about 17%, and the fuel cost is abut 75%.

Worldwide experience in combined cycles power plants indicate high availability and efficiency.

The Availability of a power plant is the percent of time the plant is available to generate power in any given period at its acceptance load. The Acceptance Load or the Net Established Capacity would be the net electric power generating capacity of the Power Plant at design or reference conditions established as result of the Performance Tests conducted for acceptance of the plant. The actual power produced by the plant would be corrected to the design or reference conditions and is the actual net available capacity of the Power Plant. Thus it is necessary to calculate the effective forced outage hours which are based on the maximum load the plant can produce in a given time interval when the plant is unable to produce the power required of it. The effective forced outage hours is based on the following relationship:

$$EFH = HO \times \frac{(MW_d - MW_a)}{MW_d} \quad (1-6)$$

where
MW_d = Desired Output corrected to the design or reference conditions. This must be equal to or less than the plant load measured and corrected to the design or reference conditions at the acceptance test.
MW_a = Actual maximum acceptance test produced and corrected to the design or reference conditions.
HO = Hours of operation at reduced load.

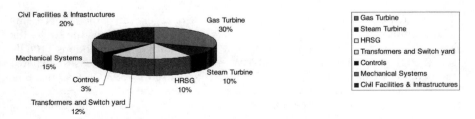

Figure 1-18. Cost Components of Different Plant Areas in a Combined Cycle Power Plant

36 • COGENERATION AND COMBINED CYCLE POWER PLANTS

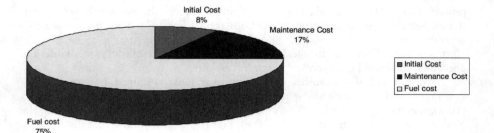

Figure 1-19. Plant Life Cycle Cost for a Combined Cycle Power Plant

The Availability of a plant can now be calculated by the following relationship, which takes into account the stoppage due to both forced and planed outages, as well as the forced effective outage hours:

$$A = \frac{(PT - PM - FO - EFH)}{PT} \qquad (1-7)$$

where
  PT = Time Period (8760 hrs/year)
  PM = Planned Maintenance Hours
  FO = Forced Outage Hours
  EFH = Equivalent forced outage hours

The reliability of the plant is the percentage of time between planed overhauls and is defined as:

$$R = \frac{(PT - FO - EFH)}{PT} \qquad (1-8)$$

Availability and reliability have a very major impact on the plant economy. Reliability is essential in that when the power is needed, it must be there.

Properties of Gas Turbine Exhaust

The exhaust from gas turbines in the simple cycle or regenerative mode represents a large amount of waste heat. Mass flow rates can vary depending on the size of the gas turbine from 4 lb/sec (1.8 kg/sec), for a 180-kW unit, to a tremendous 1100 lb/sec (500 kg/sec), for a large 143-MW gas turbine. In most cases, exhaust temperatures will be between 900°F (482°C) and 1100°F (593°C). Gas turbines utilizing a regenerator may have exhaust gas temperatures of around 950°F (510°C). As gas turbines operate with large amounts of excess air, about 18% oxygen is available in the exhaust, and this allows supplemental firing.

Temperatures after supplemental firing can be as high as 1600°F (871°C), and it is often possible to double the steam production by supplemental firing. Variation in the exhaust gas specific heat (c_p) for combustion is linear with temperature, varying from 0.259 Btu/lb/R (1.084 kJ/kg/K) at 800°F (427°C) to about 0.265 Btu/lb/R (1.1094 kJ/kg/K) at 900°F (482°C).

When computing exhaust heat usable for a cogeneration project, capability variations in the gas turbine performance must be taken into account. These include ambient temperature variations, altitude effects, and humidity. The effect of backpressure on the gas turbine capability is also important. Backpressure from heat recovery equipment should not exceed 10 in. of water (gage). The effect of hot ambient temperatures on gas turbine performance can be significant from a cogeneration perspective, as both electrical output and steam generation may drop off. Typically, there is about a 0.5% drop in power output for every 1°F rise in temperature. This drop has led to a number of schemes, such as evaporative cooling and chilling of the inlet air, to overcome the problem.

Steam Generation Calculations

Most gas turbine manufacturers have a set of curves for their different gas turbine models, showing steam-generating capabilities under different conditions. If curves are not available, a relatively simple calculation with regard to the waste heat boiler can be done to get a ballpark estimate. In order to compute the amount of steam generated, the following items are required:

- Gas turbine mass flow rate
- Exhaust gas temperature
- Steam conditions pressure and temperature required
- Feedwater temperature.

Some general rules of thumb may be stated, as they are implicit in the computation. First, the steam outlet temperature will be 50°F (28°C) or more below the exhaust gas temperature. Also, the exhaust gas leaving the boiler evaporator must be 40°F (22°C) greater than the saturation temperature. The pinch point is taken to be not less than 40°F (22°C) and the water entering the evaporator must be 15°F (8°C) below the saturation temperature. Based on a simple heat balance the steam flow-rate is:

$$\dot{m}_{steam} = \frac{\dot{m}_{air} \times c_p \times (T_2 - T_1) \times K}{(h_2 - h_1)} \quad (1-9)$$

where

$\dot{m}_{steam}$ = steam flow rate, lb/hr, kg/hr
$\dot{m}_{air}$ = gas turbine mass flow rate, lb/hr, kg/hr
T_2 = exhaust gas temperature
T_1 = stack exhaust temperature [saturation + 40°F (22°C); it also will vary depending on the type of fuel, fuel with low sulfur or no sulfur can have lower exhaust temperatures]

c_p = specific heat at constant pressure (average between T_1 and T_2)
h_2 = steam final enthalpy (leaving HRSG)
h_1 = enthalpy of water entering evaporator
K = radiation loss factor (approximately 0.985)

More details on heat recovery unit design are presented in a later section. It must be noted that any process fluid may be utilized in the heat recovery unit.

Gas Turbine Heat Recovery

The waste heat recovery system is a critically important subsystem of a cogeneration system. In the past, it was viewed as a separate "add-on" item. This view is being changed with the realization that good performance, both thermodynamically and in terms of reliability, grows out of designing the heat recovery system as an integral part of the overall system.

Some important points and observations relating to gas turbine waste heat recovery are:

- Multipressure steam generators — these are becoming increasingly popular. With a single pressure boiler, there is a limit to the heat recovery because the exhaust gas temperature cannot be reduced below the steam saturation temperature. This problem is avoided by the use of multipressure levels.
- Pinch point — this is defined as the difference between the exhaust gas temperature leaving the evaporator section and the saturation temperature of the steam. Ideally, the lower the pinch point, the more heat recovered, but this calls for more surface area and, consequently, increases the backpressure and cost. Also, excessively low pinch points can mean inadequate steam production if the exhaust gas is low in energy (low mass flow or low exhaust gas temperature). General guidelines call for a pinch point of 15°F to 40°F (8°C to 22°C). The final choice is obviously based on economic considerations.
- Approach temperature — this is defined as the difference between the saturation temperatures of the steam and the inlet water. Lowering the approach temperature can result in increased steam production, but at increased cost. Conservatively, high approach temperatures ensure that no steam generation takes place in the economizer. Typically, approach temperatures are in the 10°F to 20°F (5.5°C to 11°C) range. Figure 1-20 is the temperature energy diagram for a system and also indicates the approach and pinch points in the system.
- Off-design performance — this is an important consideration for waste heat recovery boilers. Gas turbine performance is affected by load, ambient conditions and gas turbine health (fouling, etc.). This can affect the exhaust gas temperature and the air flow rate. Adequate considerations must be given to bow steam flows (low pressure and high pressure), and superheat temperatures vary with changes in the gas turbine operation.
- Evaporators — these usually utilize a fin-tube design. Spirally finned tubes of 1.25 to 2 in. outer diameter (OD) with three to six fins per inch are

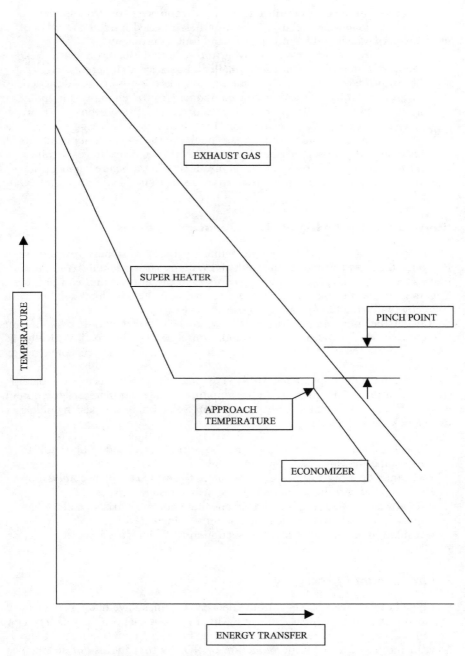

Figure 1-20. Energy Transfer Diagram in an HRSG of a Combined Cycle Power Plant

common. In the case of unfired designs, carbon steel construction can be used and boilers can run dry. As heavier fuels are used, a smaller number of fins per inch should be utilized to avoid fouling problems.
- Forced circulation system — using forced circulation in a waste heat recovery system allows the use of smaller tube sizes with inherent increased heat transfer coefficients. Flow stability considerations must be addressed. The recirculating pump is a critical component from a reliability standpoint and standby (redundant) pumps must be considered. In any event, great care must go into preparing specifications for this pump.
- Backpressure considerations (gas side) — these are important, as excessively high backpressures create performance drops in gas turbines. Very low-pressure drops would require a very large heat exchanger and more expense. Typical pressure drops are 8 to 10 in. of water.

Supplementary Firing of Heat Recovery Systems

There are several reasons for supplementary firing a wasteheat recovery unit. Probably, the most common is to enable the system to track demand (i.e., produce more steam when the load swings upwards than the unfired unit can produce). This may enable the gas turbine to be sized to meet the base load demand with supplemental firing taking care of higher load swings.

Raising the inlet temperature at the waste heat boiler allows a significant reduction in the heat transfer area and, consequently, the cost. Typically, as the gas turbine exhaust has ample oxygen, duct burners can be conveniently used.

An advantage of supplemental firing is the increase in heat recovery capability (recovery ratio). A 50% increase in heat input to the system increases the output 94%, with the recovery ratio increasing by 59%. Some important design guidelines to ensure success include:

- Special alloys may be needed in the superheater and evaporator to withstand the elevated temperatures.
- The inlet duct must be of sufficient length to ensure complete combustion and avoid direct flame contact on the heat transfer surfaces.
- If natural circulation is utilized, an adequate number of risers and feeders must be provided as the heat flux at entry is increased.
- Insulation thickness on the duct section must be increased.

Environmental Effects

Combined cycle power plants with the new gas turbines have had a very positive impact on the environment as compared with other types of power plants. The new low NO_x combustors have reduced NO_x levels below 10 ppm. Figure 1-21 shows how, in the past 30 years, the reduction of NO_x by first the use of steam (wet combustors) injection in the combustors, and then in the 1990s, the dry low NO_x combustors, have greatly reduced the NO_x output. New units under development

An Overview of Power Generation • 41

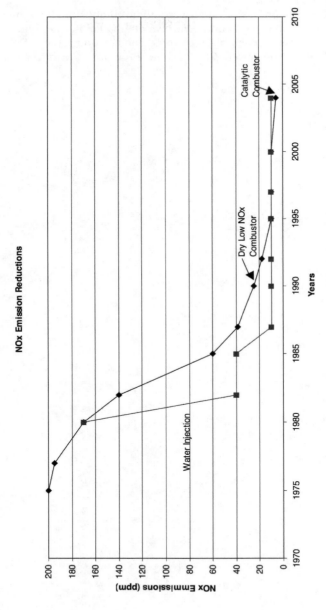

Figure 1-21. Control of Gas Turbine NO$_x$ Emissions Over the Years

have goals, that would reduce NO_x levels below 9 ppm. Catalytic converters have also been used in conjunction with both these types of combustors to even further reduce the NO_x emissions.

New research in combustors such as catalytic combustion have great promise, and values of as low as 2 ppm can be attainable in the future. Catalytic combustors are already being used in some engines under the U.S. Department of Energy's (DOE), Advanced Gas Turbine Program, and have obtained very encouraging results.

Chapter 2

CYCLES

Combined Cycle Plant Operation

The performance of the combined cycle plant is a function of the Brayton and Rankine Cycles. The Gas Turbine operates under the Brayton Cycle, and the Steam Turbine operates under the Rankine Cycle. The heat rejected by the Brayton Cycle in an isobaric process is the energy, which is used in the Rankine Cycle to produce the steam. Both cycles accept and reject heat in an isobaric process.

The Brayton Cycle

The Brayton cycle in its ideal form consists of two isobaric processes and two isentropic processes. The two isobaric processes consist of the combustor system of the gas turbine and the gas side of the HRSG. The two isentropic processes represent the compression (Compressor) and the expansion (Turbine Expander) processes in the gas turbine. Figure 2-1 shows the ideal Brayton Cycle.

A simplified application of the first law of thermodynamics to the air-standard Brayton Cycle in Figure 2-1 (assuming no changes in kinetic and potential energy) has the following relationships:

Work of compressor

$$W_c = \dot{m}_a(h_2 - h_1) \qquad (2-1)$$

Work of turbine

$$W_t = (\dot{m}_a + \dot{m}_f)(h_3 - h_4) \qquad (2-2)$$

Total output work

$$W_{cyc} = W_t - W_c \qquad (2-3)$$

Heat added to system

$$Q_{2,3} = \dot{m}_f \times LHV_{fuel} = (\dot{m}_a + \dot{m}_f)(h_3) - \dot{m}_a h_2 \qquad (2-4)$$

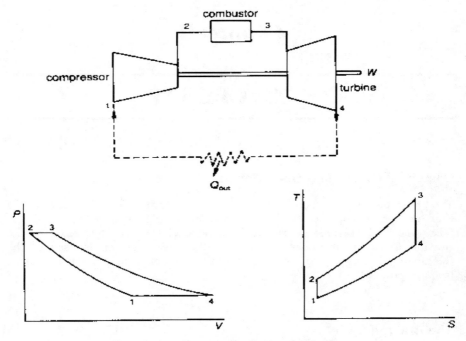

Figure 2-1. The Air-Standard Brayton Cycle

Thus, the overall cycle efficiency is

$$\eta_{\text{cyc}} = W_{\text{cyc}}/Q_{2,3} \qquad (2-5)$$

Increasing the pressure ratio and the turbine firing temperature increases the Brayton Cycle efficiency (2-6). This relationship of overall cycle efficiency is based on certain simplification assumptions such as: (1) $\dot{m}_a \gg \dot{m}_f$; (2) the gas is caloricaly and thermally perfect, which means that the specific heat at constant pressure (c_p) and the specific heat at constant volume (c_v) are constant, thus, the specific heat ratio γ remains constant throughout the cycle; (3) the pressure ratio in both the compressor and the turbine are the same; and (4) all components operate at 100% efficiency. With these assumptions, the effect on the ideal cycle efficiency as a function of pressure ratio for the ideal Brayton Cycle operating between the ambient temperature and the firing temperature is given by the following relationship:

$$\eta_{\text{ideal}} = \left(1 - \frac{1}{Pr^{\frac{\gamma-1}{\gamma}}}\right) \qquad (2-6)$$

where Pr is the pressure ratio; and γ is the ratio of the specific heats. The above equation tends to go to very high numbers as the pressure ratio is increased. In the case of the actual cycle, the effect of the turbine compressor (η_c) and expander (η_t) efficiencies must also be taken into account to obtain the overall cycle efficiency between the firing temperature (T_f) and the ambient temperature (T_{amb}) of the turbine. This relationship is given in the following equation:

$$\eta_{cycle} = \left(\frac{\eta_t T_f - \dfrac{T_{amb} Pr^{\left(\frac{\gamma-1}{\gamma}\right)}}{\eta_c}}{T_f - T_{amb} - T_{amb} \left(\dfrac{Pr^{\left(\frac{\gamma-1}{\gamma}\right)} - 1}{\eta_c} \right)} \right) \left(1 - \frac{1}{Pr^{\left(\frac{\gamma-1}{\gamma}\right)}} \right) \qquad (2-7)$$

Figure 2-2 shows the effect on the overall cycle efficiency of the increasing pressure ratio and the firing temperature. The increase in pressure ratio increases the overall efficiency at a given firing temperature; however, increasing the pressure ratio beyond a certain value at any given firing temperature can actually result in lowering the overall cycle efficiency. It should also be noted that the very high pressure ratios tend to reduce the operating range of the turbine compressor. This causes the turbine compressor to be much more intolerant to dirt buildup in the inlet air filter and on the compressor blades and creates large drops in cycle efficiency and performance. In some cases, it can lead to compressor surge, which in turn can lead to a flameout, or even serious damage and failure of the compressor blades and the radial and thrust bearings of the gas turbine.

The optimum pressure ratio for maximum output for a turbine taking into account the efficiencies of the compressor and the turbine expander section can be expressed by the following relationship:

$$Pr_{opt} = \left[\left(\frac{T_{amb}}{T_f} \right) \left(\frac{1}{\eta_c \eta_t} \right) \right]^{\frac{\gamma}{2-2\gamma}} \qquad (2-8)$$

Figure 2-3 shows that the maximum work per kilogram of air occurs at a much lower pressure ratio than the point of maximum efficiency at the same firing temperature.

Thus, a cursory inspection of the efficiency indicates that the overall efficiency of a cycle can be improved by increasing the pressure ratio, decreasing the inlet temperature, or increasing the turbine inlet temperature. However, these relationships do not indicate the effect of quantity on the efficiency of the cycle nor the inaccuracies inherent in the relationships at high pressure ratios, high turbine inlet temperatures, and the inefficiencies in the components.

Inlet Cooling Effect

There are several techniques that are used for cooling the inlet:

- **Evaporative methods** — Either conventional evaporative coolers or direct water fogging

46 • **COGENERATION AND COMBINED CYCLE POWER PLANTS**

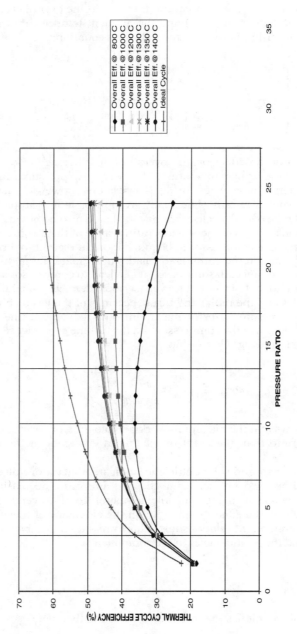

Figure 2-2. Overall Cycle Efficiency of the Pressure Ratio

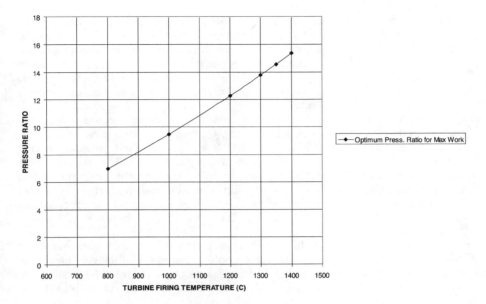

Figure 2-3. Pressure Ratio for Maximum Work Per KG of Air

- **Refrigerated inlet cooling systems** — Utilizing absorption or mechanical refrigeration
- **Combination of evaporative and refrigerated inlet systems** — The use of evaporative cooler to assist the chiller system to attain lower temperatures of the inlet air
- **Thermal energy storage systems** — These are intermittent use systems where the cold is produced off-peak and then used to chill the inlet air during the hot hours of the day.

Evaporative Cooling of the Turbine. Traditional evaporative coolers that use media for evaporation of the water have been widely used in the gas turbine industry over the years, especially in hot climates with low humidity areas. The low capital cost, installation and operating costs make it attractive for many turbine-operating scenarios. Evaporation coolers consist of water being sprayed over the media blocks, which are made of fibrous corrugated material. The airflow through these media blocks evaporates the water; as water evaporates, it consumes about 1059 Btu (1117 kJ) (latent heat of vaporization) at 60°F (15°C). This results in the reduction of air temperature entering the compressor from that of the ambient air temperature. This technique is very effective in low-humidity regions.

The work required to drive the turbine compressor is reduced by lowering the compressor inlet temperature, thus increasing the output work of the turbine. Figure 2-4 is a schematic of the evaporative gas turbine and its effect on the Brayton Cycle. The volumetric flow of most turbines is constant and therefore by increasing the mass flow, power increases in an inverse proportion to the temperature of the inlet air. The psychometric chart shown reveals that cooling is limited especially in highly humid conditions. It is a very low-cost option and can be installed very easily. This technique does not, however, increase the efficiency of the turbine. The turbine inlet temperature is lowered by about 18°F (10°C), if the outside temperature is around 90°F (32°C). The cost of an evaporative cooling system runs around US$50/kW.

Direct inlet fogging is a type of evaporative cooling method where demineralized water is converted into a fog by means of high-pressure nozzles operating at 1000 to 3000 psi. (67 to 200 Bar) This fog then provides cooling when it evaporates in the air inlet duct of the gas turbine. The air can attain 100% relative humidity at the compressor inlet, and thereby gives the lowest temperature possible without refrigeration (the web bulb temperature). Direct high-pressure inlet fogging can also be used to create a compressor intercooling effect by allowing excess fog into the compressor, thus boosting the power output further.

Refrigerated Inlets for the Gas Turbines. The refrigerated inlets are more effective than the previous evaporative cooling systems as they can lower the temperatures by about 45-55°F (25-30°C). Two techniques for refrigerating the inlet of a gas turbine are vapor compression (mechanical refrigeration) and absorption refrigeration.

Mechanical Refrigeration. In a mechanical refrigeration system, the refrigerant vapor is compressed by means of a centrifugal, screw, or reciprocating compressor. Figure 2-5 is a schematic of a mechanical refrigeration intake for a gas turbine. The psychometric chart included shows that refrigeration provides considerable cooling and is very well suited for hot humid climates.

Centrifugal compressors are typically used for large systems in excess of 1000 ton (12.4 × 10^6 Btu/13.082 × 10^6 kJ) and would be driven by an electric motor. Mechanical refrigeration has significantly high auxiliary power consumption for the compressor driver and pumps required for the cooling water circuit. After compression, the vapor passes through a condenser where it gets condensed. The condensed vapor is then expanded in an expansion valve and provides a cooling effect. The evaporator chills cooling water that is circulated to the gas turbine inlet chilling coils in the air stream. Chlorofluorocarbon (CFC)-based chillers are now available and can provide a large tonnage for a relatively smaller plot space and can provide cooler temperature than the lithium-bromide (Li-Br) absorption based cooling systems. The drawbacks of mechanical chillers are high capital and operation and maintenance (OM) costs, high power consumption, and poor part load performance.

Direct expansion is also possible wherein the refrigerant is used to chill the incoming air directly without the chilled water circuit. Ammonia, which is an excellent refrigerant, is used in this sort of application. Special alarm systems would have to be utilized to detect the loss of the refrigerant into the combustion air and to shut down and evacuate the refrigeration system.

Absorption Cooling Systems. Absorption systems typically employ lithium-bromide (Li-Br) and water, with Li-Br being the absorber and water acting as the refrigerant. Such systems can cool the inlet air to 50°F (10°C). Figure 2-6 is a schematic of an absorption refrigerated inlet system for the gas turbine. The

Cycles • 49

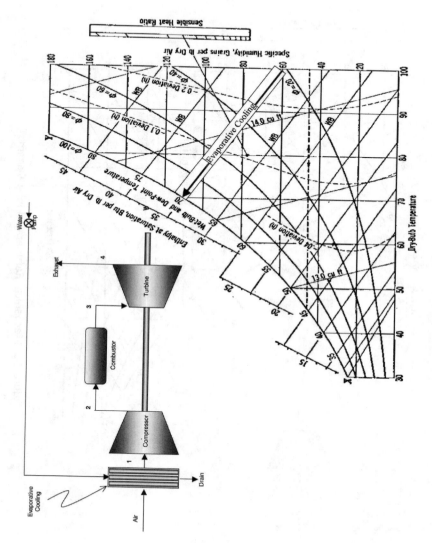

Figure 2-4. Schematic of Evaporative Cooling

50 • COGENERATION AND COMBINED CYCLE POWER PLANTS

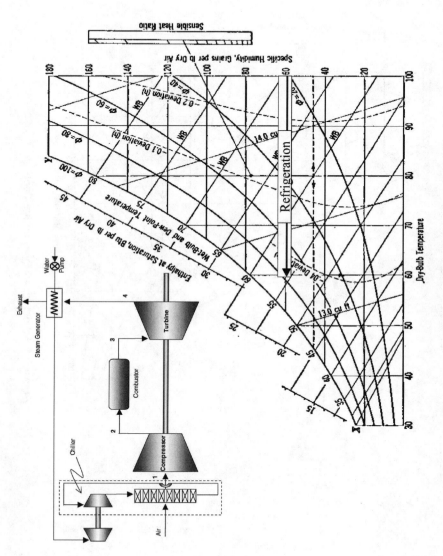

Figure 2-5. Mechanical Refrigerated Inlet System

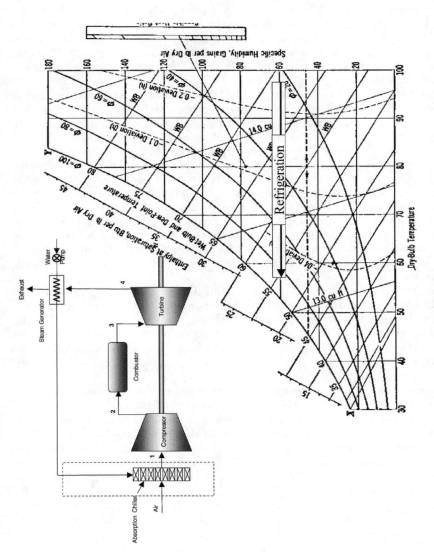

Figure 2-6. Absorption Refrigerated Inlet Cooling System

cooling shown on the psychometric chart is identical to the one for the mechanical system. The heat for the absorption chiller can be provided by gas, steam or gas turbine exhaust. Absorption systems can be designed to be either single or double effect. A single effect system will have a coefficient of performance (COP) of 0.7 to 0.9, and in a double effect unit, a COP of 1.15. Part load performance of absorption systems is relatively good, and efficiency does not drop off at part load like it does with mechanical refrigeration systems. The costs of these systems are much higher than the evaporative cooling system; however, refrigerated inlet cooling systems in hot humid climates are more effective due to the very high humidity.

Combination of Evaporative and Refrigerated Inlet Systems. Depending on the specifics of the project, location, climatic conditions, engine type, and economic factors, a hybrid system utilizing a combination of the above technologies may be the best. The possibility of using fogging systems ahead of the mechanical inlet refrigeration system should be considered, as seen in Figure 2-7. This may not always be intuitive since evaporative cooling is an adiabatic process that occurs at constant enthalpy. When water is evaporated into an air stream, any reduction in sensible heat is accompanied by an increase in the latent heat of the air stream (the heat in the air stream being used to effect a phase change in the water from liquid to the vapor phase). If fog is applied in front of a chilling coil, the temperature will be decreased when the fog evaporates, but since the chiller coil will have to work harder to remove the evaporated water from the air steam, the result would yield no thermodynamic advantage.

To maximize the effect, the chiller must be designed in such a manner that in combination with evaporative cooling, the maximum reduction in temperature is achieved. This can be done by designing a slightly undersized chiller which is not capable of bringing the air temperature down to the ambient dew point temperature, but in conjunction with evaporative cooling, the same effect can be achieved — Thus, taking the advantage of evaporative cooling to reduce the load of refrigeration.

Thermal Energy Storage Systems. These systems are usually designed to operate the refrigeration system at off-peak hours and then use the refrigerated media at peak hours. The refrigerated media in most cases is ice and the gas turbine air is then passed through the media, which lowers the inlet temperature as seen in Figure 2-8. The size of the refrigeration system is greatly reduced as it can operate for 8-10 hours at off-peak conditions to make the ice, which is then stored, and when air passed through it at peak operating hours, that rate may only be for about 4-6 hours.

The cost for such a system runs about $90-$110/kW, such systems cost around $90-110/kW, and have been successfully employed for gas turbines producing 100-200 MW.

Regeneration Effect

In a simple gas turbine cycle, the turbine exit temperature is nearly always appreciably higher than the temperature of the air leaving the compressor.

Cycles • 53

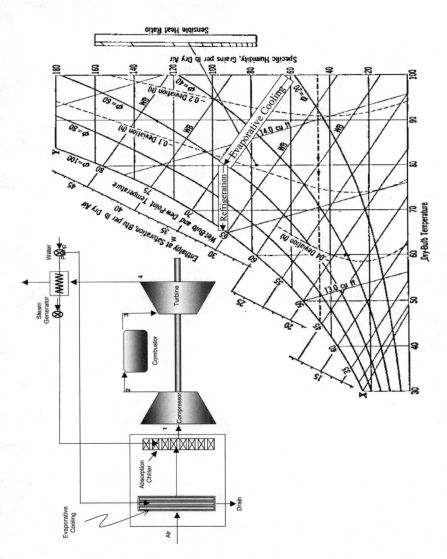

Figure 2-7. Evaporative and Refrigerated Inlet Systems

54 • COGENERATION AND COMBINED CYCLE POWER PLANTS

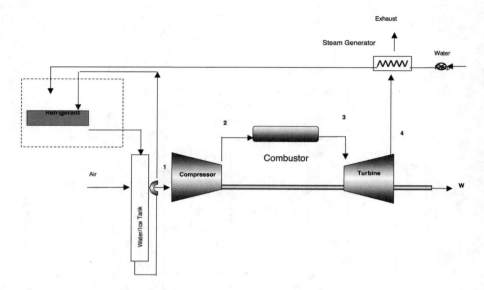

Figure 2-8. Thermal Storage Inlet System

Obviously, the fuel requirement can be reduced by the use of a regenerator in which the hot turbine exhaust gas preheats the air between the compressor and the combustion chamber. Figure 2-9 shows a schematic of the regenerative cycle and its performance in the *T-S* diagram. In an ideal case, the flow through the regenerator is at constant pressure. The regenerator effectiveness is given by the following relationship:

$$\eta_{\text{reg}} = \frac{T_3 - T_2}{T_5 - T_2} \qquad (2-9)$$

Thus, the overall efficiency for this system's cycle can be written as

$$\eta_{\text{RCYC}} = \frac{(T_4 - T_5) - (T_2 - T_1)}{(T_4 - T_3)} \qquad (2-10)$$

Increasing the effectiveness of a regenerator calls for more heat transfer surface area which increases the cost, the pressure drop, and the space requirements of the unit.

Figure 2-10 shows the improvement in cycle efficiency because of heat recovery with respect to a simple open-cycle gas turbine of 4.33:1 ratio pressure and 1200°F inlet temperature. Cycle efficiency drops with an increasing pressure drop in the regenerator.

There are two types of heat exchangers, regenerative and recuperative. The term "regenerative heat exchanger" is used for a system in which the heat transfer between two streams is affected by the exposure of a third medium alternately to the two flows. The heat flows successively into and out of the third medium,

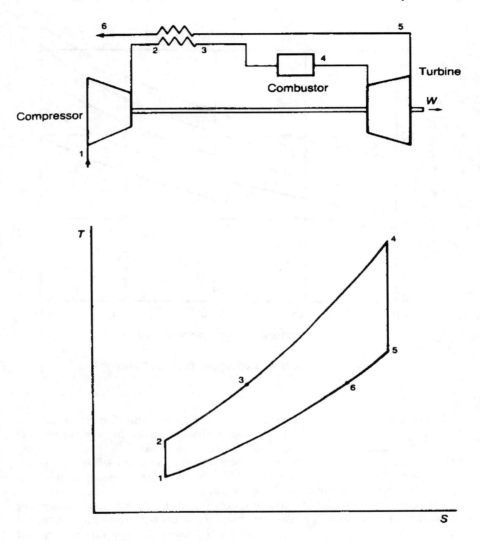

Figure 2-9. The Regenerative Gas Turbine Cycle

which undergoes a cyclic temperature. These types of heat exchangers are widely used where compactness is essential. The automotive regenerators consisted of a large circular drum with honeycombed ceramic passages. The drum was rotated at a very low rpm (10 to 15 rpm). The drum surface was divided into two halves by an air seal. The hot air would pass through one half of the circular drum heating the honeycombed passages the air would encounter, then the cooler air would pass through these same passages as the drum was rotated and would be heated.

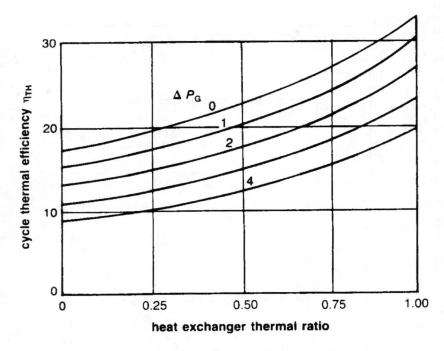

Figure 2-10. Heat Exchanger Thermal Ratio

In a recuperative heat exchanger, each element of heat-transferring surface has a constant temperature and, by arranging the gas paths in contra-flow, the temperature distribution in the matrix in the direction of flow is that giving optimum performance for the given heat-transfer conditions. This optimum temperature distribution can be achieved ideally in a contra-flow regenerator and approached very closely in a cross-flow regenerator.

The matrix permitting the maximum flow per unit area will yield the smaller regenerator for a given thermal and pressure drop performance. A material with a high heat capacity per unit volume is preferred, since this property of the material will increase the switching time and tend to reduce carry-over losses. Another desirable property of the arrangement is low thermal conductivity in the direction of the gas flow. All leakages within the regenerator must be avoided. A leakage of 3% reduces the regenerator effectiveness from 80% to 71%.

Increasing the Work Output of the Simple Cycle Gas Turbine

The way to enhance the power output of a gas turbine can be achieved by intercooling and reheat.

Intercooling and Reheat Effects

The net work of a gas turbine cycle is given by

$$W_{cyc} = W_t - W_c \qquad (2-11)$$

and can be increased either by decreasing the compressor work or by increasing the turbine work. These are the purposes of intercooling and reheating, respectively.

Multi-staging of compressors is sometimes used to allow for cooling between the stages to reduce the total work input. Figure 2-11 shows a polytropic compression process 1-a on the P-V plane. If there is no change in the kinetic energy, the work done is represented by the area 1-a-j-k-1. A constant temperature line is shown as 1-x. If the polytropic compression from State 1 to State 2 is divided into two parts, 1-c and d-e with constant pressure cooling to $T_d = T_1$ between them, the work done is represented by area 1-c-d-e-I-k-1. The area c-a-e-d-c represents the work saved by means of the two-stage compression with intercooling to the initial temperature. The optimum pressure for intercooling for specified values P_1 and P_2 is

$$P_{OPT} = \sqrt{P_1 P_2} \qquad (2-12)$$

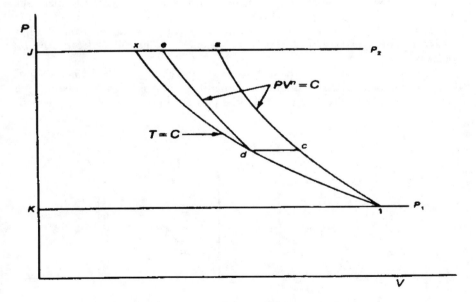

Figure 2-11. Multistage Compression with Intercooling

58 • COGENERATION AND COMBINED CYCLE POWER PLANTS

Therefore, if a simple gas turbine cycle is modified with the compression accomplished in two or more adiabatic processes with intercooling between them, the net work of the cycle is increased with no change in the turbine work.

The thermal efficiency of an ideal simple cycle is decreased by the addition of an intercooler. Figure 2-12 shows the schematic of such a cycle. The ideal simple gas turbine cycle is 1-2-3-4-1, and the cycle with the intercooling added is 1-*a*-*b*-*c*-2-3-4-1. Both cycles in their ideal form are reversible and can be simulated by a number of Carnot cycles. Thus, if the simple gas turbine cycle 1-2-3-4-1 is divided into a number of cycles like *m*-*n*-*o*-*p*-*m*, these little cycles approach the Carnot cycle as their number increases. The efficiency of such a Carnot cycle is given by the relationship

$$\eta_{\text{CARNOT}} = 1 - \frac{T_m}{T_p} \qquad (2-13)$$

Notice that if the specific heats are constant, then

$$\frac{T_3}{T_4} = \frac{T_m}{T_p} = \left(\frac{P_2}{P_1}\right)^{\frac{\gamma-1}{\gamma}} \qquad (2-14)$$

All the Carnot cycles making up the simple gas turbine cycle have the same efficiency. Likewise, all of the Carnot cycles into which the cycle *a*-*b*-*c*-2-*a* might similarly be divided have a common value of efficiency lower than the Carnot cycles which comprise cycle 1-2-3-4-1. Thus, the addition of an intercooler, which adds *a*-*b*-*c*-2-*a* to the simple cycle, lowers the efficiency of the cycle.

The addition of an intercooler to a regenerative gas turbine cycle increases the cycle's thermal efficiency and output work because a larger portion of the heat required for the process *c*-3 in Figure 2-12 can be obtained from the hot turbine exhaust gas passing through the regenerator instead of from burning additional fuel.

The reheat cycle increases the turbine work, and consequently the net work of the cycle, can be increased without changing the compressor work or the turbine inlet temperature by dividing the turbine expansion into two or more parts with

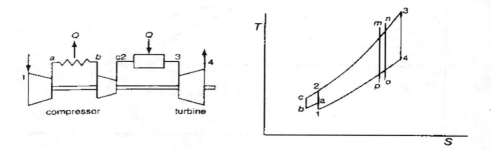

Figure 2-12. Air-Standard with Intercooling Cycle

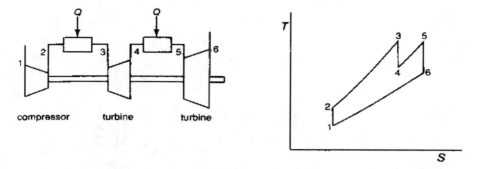

Figure 2-13. Reheat Cycle and *T-S* Diagram

constant pressure heating before each expansion. This cycle modification is known as reheating as seen in Figure 2-13. By reasoning similar to that used in connection with Intercooling, it can be seen that the thermal efficiency of a simple cycle is lowered by the addition of reheating, while the work output is increased. However, a combination of regenerator and reheater can increase the thermal efficiency.

Injection of Compressed Air, Steam, or Water for Increasing Power

- **Injection of Humidified and Heated Compressed Air** — Compressed Air from a separate compressor is heated and humidified to about 60% relative humidity by the use of an HRSG and then injected into the compressor discharge.
- **Steam Injection** — Injection of the steam, obtained from the use of a low-pressure single stage heat recovery steam generator (HRSG), at the compressor discharge and/or injection in the combustor.
- **Water Injection** — Mid compressor flashing is used to cool the compressed air and add mass flow to the system.

Mid compressor Flashing of Water. In this system, the water is injected into the mid-stages of the compressor to cool the air and approach an isothermal compression process as shown in Figure 2-14. The water injected is usually mechanically atomized so that very fine droplets are entered into the air. The water is evaporated as it comes in contact with the high-pressure and -temperature air stream. As water evaporates, it consumes about 1058 Btu (1117 kJ) (latent heat of vaporization) at higher pressure and temperature, resulting in lowering the temperature of the air stream entering the next stage. This lowers the work required to drive the compressor.

The intercooling of the compressed air has been very successfully applied to high-pressure engines. This system can be combined with any of the previously described systems.

60 • COGENERATION AND COMBINED CYCLE POWER PLANTS

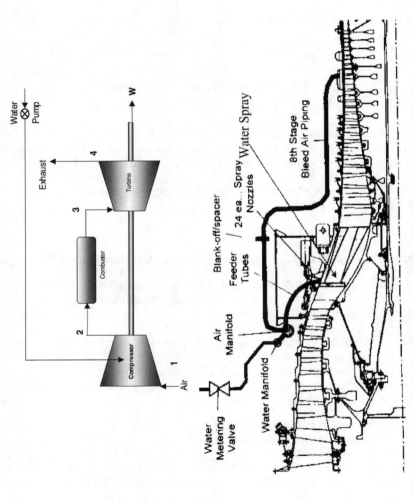

Figure 2-14. Mid Compressor Cooling Showing a Schematic as well as an Actual Application in a GE LM 6000 Engine (Courtesy of GE Corporation)

Cycles • 61

Injection of Humidified and Heated Compressed Air. Compressed air from a separate compressor is heated and humidified to about 60% relative humidity by the use of an HRSG and then injected into the compressor discharge. Figure 2-15 is a simplified schematic of a compressed air injection plant, which consists of the following major components:

1. A commercial combustion turbine with the provision to inject, at any point upstream of the combustor, the externally supplied humidified and pre-heated supplementary compressed air. Engineering and mechanical aspects of the air injection for the compressed air injection plant concepts are similar to the steam injection for the power augmentation, which has accumulated significant operating experience.
2. A supplementary compressor (consisting of commercial off-the-shelf compressor or standard compressor modules) that provides the supplementary airflow up-stream of combustors.
3. A saturation column for the supplementary air humidification and pre-heating.
4. Heat recovery water heater and the saturated air pre-heater.
5. Balance-of-plant equipment and systems including interconnecting piping, valves, controls, etc.

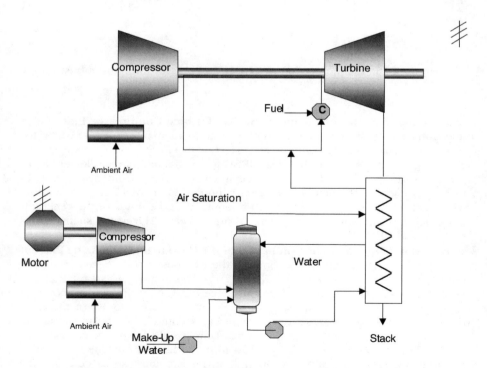

Figure 2-15. Heated and Humidified Compressed Air Injection

62 • COGENERATION AND COMBINED CYCLE POWER PLANTS

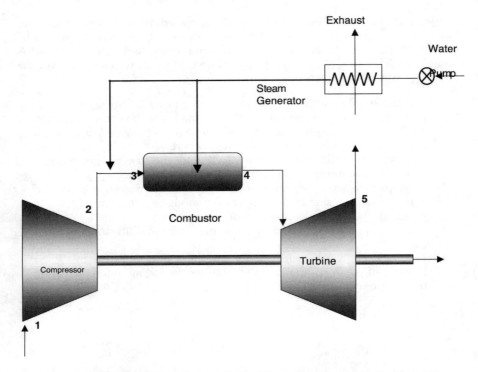

Figure 2-16. Steam Injection at Exit of Compressor and in Combustor

Injection of Water or Steam at the Gas Turbine Compressor Exit. Steam injection or water injection has been often used to augment the power generated from the turbine as seen in Figure 2-16. Steam can be generated from the exhaust gases of the gas turbine. The HRSG for such a unit is very elementary as the pressures are low. This technique augments power and also increases turbine efficiency. The amount of steam is limited to about 12% of the airflow, which can result in a power increase of about 25%. The limits of the generator may restrict the amount of power, which can be added. The cost of such systems runs around $100/kW.

Injection of Steam in the Combustor of the Gas Turbines Utilizing Present Dual Fuel Nozzles. Steam injection in the combustor has been commonly used for NO_x control as seen in Figure 2-17. The amount of steam that can be added is limited due to combustion concerns. This is limited to about 2% to 3% of the airflow. This would provide an additional 3% to 5% of the rated power. The dual fuel nozzles on these turbines could easily be retrofitted to achieve the goal of steam injection. The steam would be produced using an HRSG. The HRSG could be designed to utilize exhaust from one gas turbine for providing steam for all units. Multiple turbines could also be tied into one HRSG.

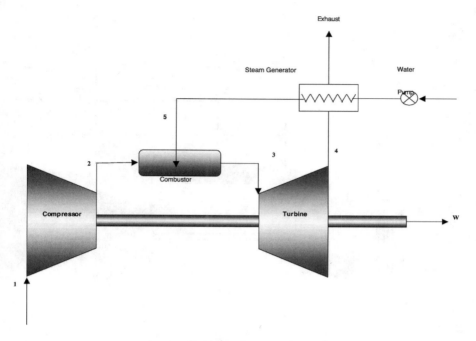

Figure 2-17. Steam Injection in Combustor

Parameters studied and constraints:

- Steam Temperature = $T_1 + 50°C$
- Steam Pressure = $P_1 + 6$ Bar
- Relative Humidity, $RH = 60\%$
- Firing Temperature, $TIT = 1100°C$ (GE), $1050°C$ (Siemens)

Combination of Evaporative Cooling and Steam Injection

The combination of the above techniques must also be investigated as none of these techniques is exclusive of the other techniques and can be easily used in conjunction with each other. Figure 2-18 is a schematic of a combination of the inlet evaporative cooling with injection of steam in both the compressor exit and the combustor.

Advanced Gas Turbine Cycles

Compressed Air Energy Storage Cycle

The compressed air energy storage (CAES) cycle is used as a peaking system that uses off-peak power to compress air into a large solution-mined

64 • COGENERATION AND COMBINED CYCLE POWER PLANTS

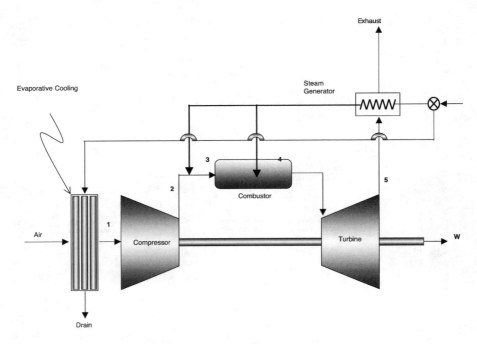

Figure 2-18. Evaporative Cooling and Steam Injection

underground cavern and withdraws the air to generate power during periods of high system power demand. Figure 2-19 is a schematic of such a typical plant being operated by Alabama Electric Cooperative Inc, with the plant heat and mass balance diagram, with generation-mode parameters at rated load and compression-mode parameters at average cavern conditions.

The compressor train is driven by the motor/generator, which has a pair of clutches that enable it to act as a motor when the compressed air is being generated for storage in the cavern, and declutches it from the expander train and connects it to the compressor train. The compressor train consists of a three-section compressor, each section having an intercooler to cool the compressed air before it enters the other section, thus reducing the overall compressor power requirements.

The power train consists of an HP and LP expander arranged in series, thus driving the motor/generator, which in this mode is declutched from the compressor train and is connected by clutch to the HP and LP expander train. The HP expander receives air from the cavern that is regeneratively heated in a recuperator utilizing exhaust gas from the LP expander, and then further combusted in combustors before entering the HP expander. The expanded air from the HP expander exhaust is reheated in combustors before entering the LP expander. Can-type combustors of similar design are employed in both the HP and LP expanders. The HP expander, which

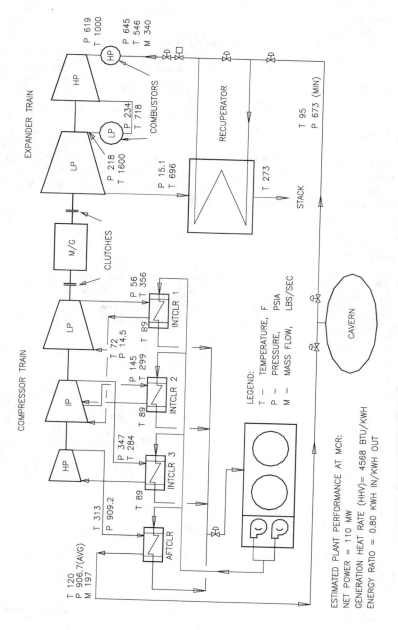

Figure 2-19. Schematic of a Compressed Air Energy Storage Plant (Courtesy ASME Paper No. 2000-GT-0595)

produces about 25% of the power, utilizes two combustors, while the LP expander, producing 75% of the power, has eight. The plant is designed to operate with either natural gas or No. 2 distillate oil fuels and operates over a range of 10 to 110 MW.

The generator is operated as a motor during the compression mode. The system is designed to operate on a weekly cycle, which includes power generation 5 days/week, with cavern recharging during weekday nights and weekends. The plant was commissioned in May 1991.

Actual Cycle Analysis

The previous section dealt with the concepts of the various cycles. Work output and efficiency of all actual cycles are considerably less than those of the corresponding ideal cycles because of the effect of compressor, combustor, and turbine efficiencies and pressure losses in the system.

The Simple Cycle

The simple cycle is the most common type of cycle being used in gas turbines in the field today. The actual open simple cycle as shown in Figure 2-20, indicates the inefficiency of the compressor and turbine and the loss in pressure through the burner. Assuming that compressor efficiency is η_c and turbine

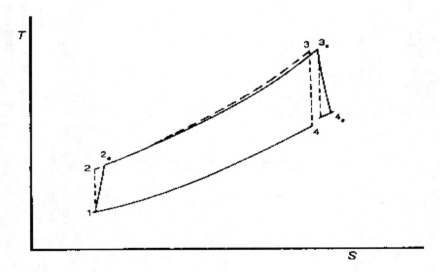

Figure 2-20. *T-S* Diagram of the Actual Open Simple Cycle

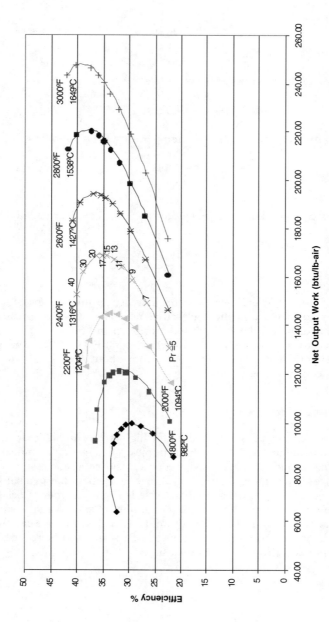

Figure 2-21. The Performance Map of a Simple Cycle Gas Turbine

efficiency is η_1, then the actual compressor work and the actual turbine work are given by:

$$W_{ca} = \dot{m}_a(h_2 - h_1)/\eta_c \qquad (2-15)$$

$$W_{ta} = (\dot{m}_a + \dot{m}_f)(h_{3a} - h_4)\eta_t \qquad (2-16)$$

Thus, the actual total output work is

$$W_{act} = W_{ta} - W_{ca} \qquad (2-17)$$

The actual fuel required to raise the temperature from 2a to 3a is

$$\dot{m}_f = \frac{h_{3a} - h_{2a}}{(LHV)\eta_b} \qquad (2-18)$$

Thus, the overall adiabatic thermal cycle efficiency can be calculated from the following equation:

$$\eta_c = \frac{W_{act}}{\dot{m}_f(LHV)} \qquad (2-19)$$

Analysis of this cycle indicates that an increase in inlet temperature to the turbine causes an increase in the cycle efficiency. The optimum pressure ratio varies with the turbine inlet temperature from an optimum of about 17:1 at 1800°F (982°C) to a pressure ratio of 40:1 at 3000°F (1649°C). The pressure ratio for maximum work, however, varies from about 15:1 to about 30:1 for the same respective temperatures.

The Split-Shaft Simple Cycle

The split-shaft simple cycle is mainly used for high torque and large load variant. Figure 2-22 is a schematic of the two-shaft simple cycle. The first turbine drives the compressor; the second turbine is used as a power source. If one assumes that the number of stages in a split-shaft simple cycle are more than that in a simple shaft cycle, then the efficiency of the split-shaft cycle is slightly higher at design loads because of the reheat factor. However, if the number of stages are the same, then there is no change in overall efficiency. From the H-S diagram, one can find some relationships between turbines. Since the job of the high-pressure turbine is to drive the compressor, the equations to use are:

$$h_{4a} = h_3 - W_{ca} \qquad (2-20)$$

$$h_{4a} = h_3 - (W_{ca}/\eta_t) \qquad (2-21)$$

Thus, the output work can be represented by the relationship:

$$W_a = (\dot{m}_a + \dot{m}_f)(h_{4a} - h_5)\eta_t \qquad (2-22)$$

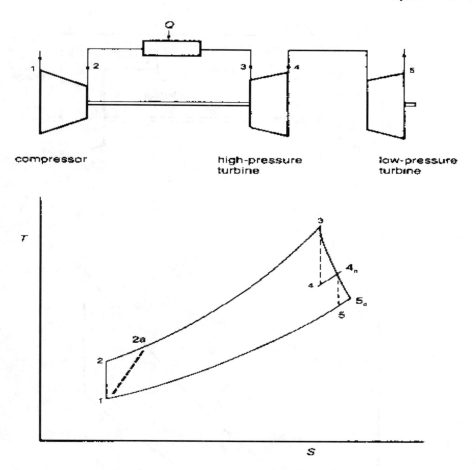

Figure 2-22. The Split-Shaft Gas Turbine Cycle

In the split-shaft cycle, the first shaft supports the compressor and the turbine that drives it, while the second shaft supports the free turbine that drives the load. The two shafts can operate at entirely different speeds. The advantage of the split-shaft gas turbine is its high torque. A free-power turbine gives a very high torque at low rpm. Very high torque at low rpm is convenient for automotive use, but with constant full-power operation, it is of little or no value. Its use should be limited to variable mechanical-drive applications.

The Regenerative Cycle

The regenerative cycle is becoming prominent in these days of tight fuel reserves and high fuel costs. The amount of fuel needed can be reduced by the use of a regenerator in which the hot turbine exhaust gas is used to pre-heat the air

70 • **COGENERATION AND COMBINED CYCLE POWER PLANTS**

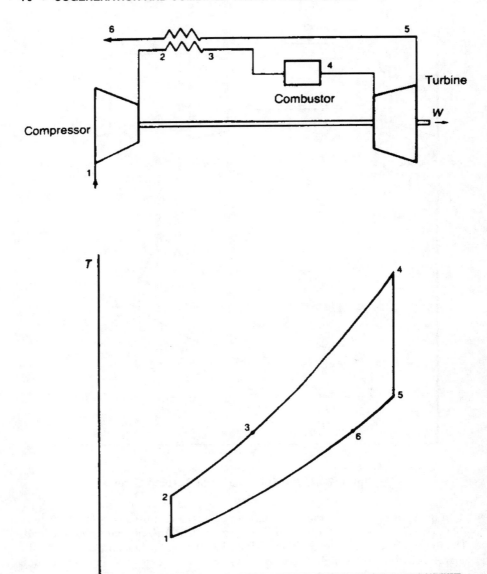

Figure 2-23. The Regenerative Gas Turbine Cycle

between the compressor and the combustion chamber. From Figure 2-23 and the definition of a regenerator, the temperature at the exit of the regenerator is

$$T_3 = T_{2a} + \eta_{\text{reg}}(T_5 - T_{2a}) \qquad (2-23)$$

where T_{2a} is the actual temperature at the compressor exit. The regenerator increases the temperature of the air entering the burner, thus reducing the fuel-to-air ratio and increasing the thermal efficiency.

For a regenerator assumed to have an effectiveness of 80%, the efficiency of the regenerative cycle is about 40% higher than its counterpart in the simple cycle, as seen in Figure 2-24. The work output per pound of air is about the same or slightly less than that experienced with the simple cycle. The point of maximum efficiency in the regenerative cycle occurs at a lower pressure ratio than that of the simple cycle, but the optimum pressure ratio for the maximum work is the same in the two cycles. Thus, when companies are designing gas turbines, the choice of pressure ratio should be such that maximum benefit from both cycles can be obtained, since most offer a regeneration option. It is not correct to say that a regenerator at off-optimum would not be effective, but a proper analysis should be made before a large expense is incurred.

The Intercooled Simple Cycle

A simple cycle with intercooler can reduce total compressor work and improve net output work. Figure 2-25 shows the simple cycle with intercooling between compressors. The assumptions made in evaluating this cycle are: (1) compressor interstage temperature equals inlet temperature, (2) compressor efficiencies are the same, (3) pressure ratios in both compressors are the same and equal to $\sqrt{P_2 P_1}$.

The intercooled simple cycle reduces the power consumed by the compressor. A reduction in consumed power is accomplished by cooling the inlet temperature in the second or other following stages of the compressor to the same as the ambient air and maintaining the same overall pressure ratio. The compressor work then can be represented by the following relationship:

$$W_c = (h_a - h_1) + (h_c - h_1) \qquad (2-24)$$

This cycle produces an increase of 30% in work output, but the overall efficiency is slightly decreased as seen in Figure 2-26. An intercooling regenerative cycle can increase the power output and the thermal efficiency. This combination provides an increase in efficiency of about 12% and an increase in power output of about 30%. Maximum efficiency, however, occurs at lower pressure ratios, as compared with the simple or reheat cycles.

The Reheat Cycle

The regenerative cycles improve the efficiency of the split-shaft cycle, but do not provide any added work per pound of air flow. To achieve this latter goal, the concept of the reheat cycle must be utilized. The reheat cycle, as shown in Figure 2-27, consists of a two-stage turbine with a combustion chamber before each stage. The assumptions made in this chapter are that the high-pressure turbine's only job is to drive the compressor and that the gas leaving this turbine

72 • COGENERATION AND COMBINED CYCLE POWER PLANTS

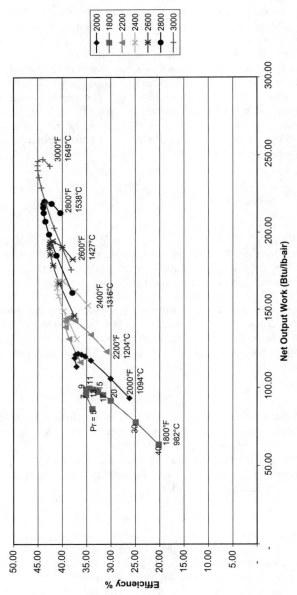

Figure 2-24. The Performance Map of a Regenerative Gas Turbine Cycle

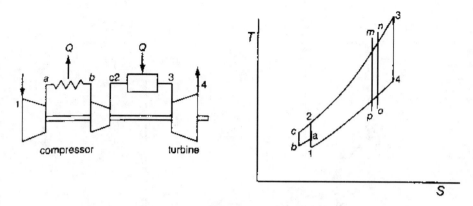

Figure 2-25. The Intercooled Gas Turbine Cycle

is then reheated to the same temperature as in the first combustor before entering the low-pressure or -power turbine. This reheat cycle has an efficiency lower than that encountered in a simple cycle, but produces about 35% more shaft output power, as shown in Figure 2-28.

The Intercooled Regenerative Reheat Cycle

The Carnot cycle is the optimum cycle and all cycles incline toward this optimum. Maximum thermal efficiency is achieved by approaching the isothermal compression and expansion of the Carnot cycle, or by intercooling in compression and reheating in the expansion process. Figure 2-29 shows the intercooled regenerative reheat cycle, which approaches this optimum cycle in a practical fashion.

This cycle achieves the maximum efficiency and work output of any of the cycles described up to this point. With the insertion of an intercooler in the compressor, the pressure ratio for maximum efficiency moves to a much higher ratio, as indicated in Figure 2-30.

The Steam Injection Cycle

Steam injection has been used in reciprocating engines and gas turbines for a number of years. This cycle may be an answer to the present concern with pollution and higher efficiency. Corrosion problems are the major factor to hurdle in such a system. The concept is simple and straightforward: water is injected into the compressor discharge air and increases the mass flow rate through the turbine, as shown in the schematic in Figure 2-31. The steam being injected downstream from the compressor does not increase the work required to drive the compressor.

74 • COGENERATION AND COMBINED CYCLE POWER PLANTS

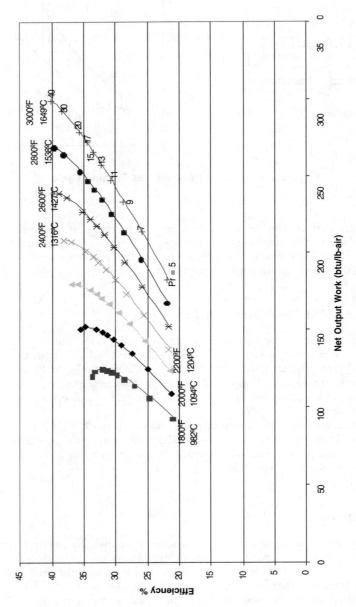

Figure 2-26. The Performance Map of an Intercooled Gas Turbine Cycle

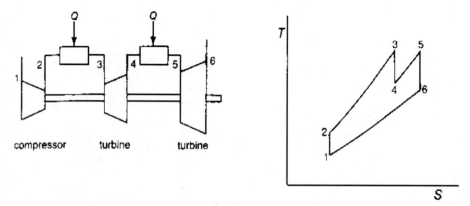

Figure 2-27. Reheat Cycle and T-S Diagram

The steam used in this process is generated by the turbine exhaust gas. Typically, water at 14.7 psia (1 Bar) and 80°F (26.7°C) enters the pump and regenerator, where it is brought up to 60 psia (4 Bar) above the compressor discharge and the same temperature as the compressor discharged air. The steam is injected after the compressor but far upstream of the burner to create a proper mixture which helps to reduce the primary zone temperature in the combustor and the NO_x output. The enthalpy of State 3 (h_3) is the mixture enthalpy of air and steam. The following relationship describes the flow at that point:

$$h_3 = (\dot{m}_a h_{2a} + \dot{m}_s h_{3a})/(\dot{m}_a + \dot{m}_s) \qquad (2-25)$$

The enthalpy entering the turbine is given by the following:

$$h_4 = ((\dot{m}_a + \dot{m}_f)h_{4a} + \dot{m}_s h_{4s})/(\dot{m}_a + \dot{m}_f + \dot{m}_s) \qquad (2-26)$$

with the amount of fuel needed to be added to this cycle as

$$\dot{m}_f = \frac{h_4 - h_3}{\eta_b (LHV)} \qquad (2-27)$$

The enthalpy leaving the turbine is

$$h_5 = ((\dot{m}_a + \dot{m}_f)h_{5a} + \dot{m}_s h_{5s})/(\dot{m}_a + \dot{m}_f + \dot{m}_s) \qquad (2-28)$$

Thus, the total work by the turbine is given by

$$W_t = (\dot{m}_a + \dot{m}_s + \dot{m}_f)(h_4 - h_5)\eta_t \qquad (2-29)$$

And the overall cycle efficiency is

$$\eta_{cyc} = \frac{W_t - W_c}{\dot{m}_f (LHV)} \qquad (2-30)$$

76 • COGENERATION AND COMBINED CYCLE POWER PLANTS

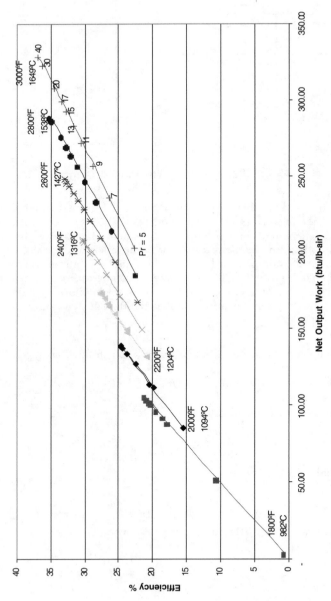

Figure 2-28. The Performance of a Reheat Gas Turbine Cycle

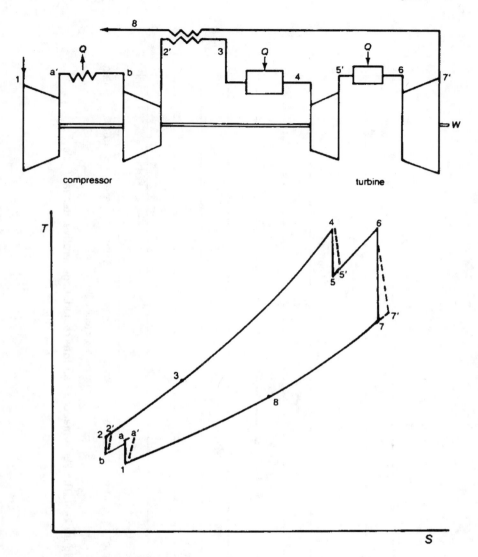

Figure 2-29. The Intercooled Regenerative Reheat Split-Shaft Gas Turbine Cycle

The cycle leads to an increase in output work and an increase in overall thermal efficiency.

Figure 2-32 show the effect of 5% by weight of steam injection at a turbine inlet temperature of 2400°F (1316°C) on the system. With about 5% injection at 2400°F (1316°C) and a pressure ratio of 17:1, an 8.3% increase in work output is noted with an increase of about 19% in cycle efficiency over that experienced in the simple cycle. The assumption here is that steam is injected

78 • COGENERATION AND COMBINED CYCLE POWER PLANTS

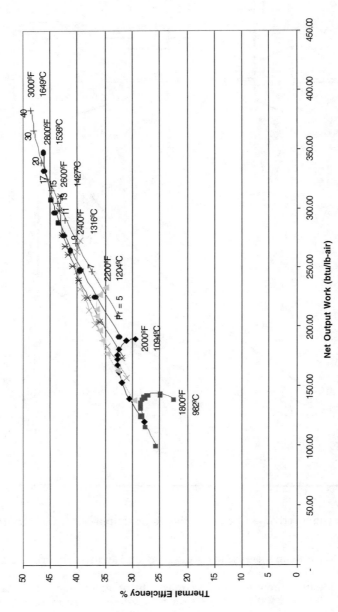

Figure 2-30. The Performance of an Intercooled, Regenerative, Reheat Cycle

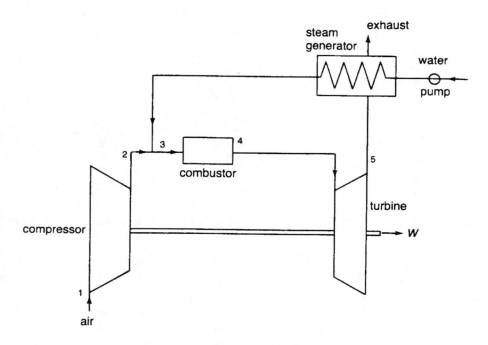

Figure 2-31. The Steam Injection Cycle

at a pressure of about 60 psi (4 Bar) above the air from the compressor discharge and that all the steam is created by heat from the turbine exhaust. Calculations indicate that there is more than enough waste heat to achieve these goals.

Figure 2-33 shows the effect of 5% steam injection at different temperatures and pressures. Steam injection for power augmentation has been used for many years and is a very good option for plant renewable.

Figure 2-33 is the performance map of a steam injected gas turbine. This cycle's great advantage is in the low production level of nitrogen oxides. That low level is accomplished by the steam being injected in the compressor discharge diffuser wall, well upstream from the combustor, creating a uniform mixture of steam and air throughout the region. The uniform mixture reduces the oxygen content of the fuel-to-air mixture and increases its heat capacity, which in turn reduces the temperature of the combustion zone and the NO_x formed. Field tests show that the amount of steam equivalent to the fuel flow by weight will reduce the amount of NO_x emissions to acceptable levels. The major problem encountered is corrosion. The corrosion problem is being investigated, and progress is being made. The attractiveness of this system lies in the premise that major changes are not needed to add this feature into an existing system. The location of the water injector is crucial for the proper operation of this system and cycle.

80 • COGENERATION AND COMBINED CYCLE POWER PLANTS

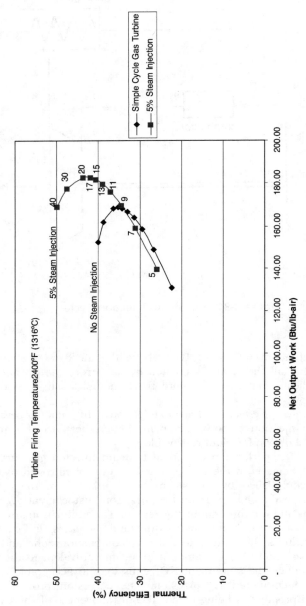

Figure 2-32. Comparison Between 5% Steam Injection and Simple Cycle Gas Turbine

Cycles • 81

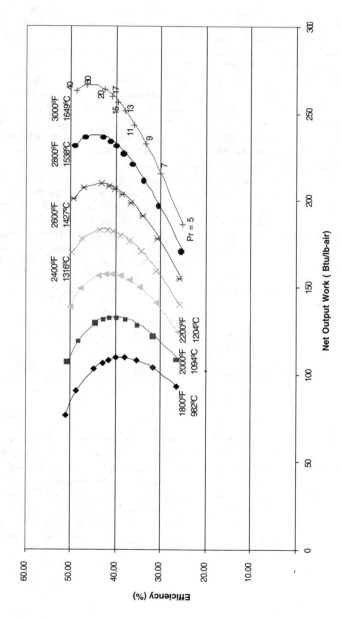

Figure 2-33. The Performance Map of a Steam Injected Gas Turbine

The Evaporative Regenerative Cycle

This cycle, as shown in Figure 2-34, is a regenerative cycle with water injection. Theoretically, it has the advantages of both the steam injection and regenerative systems reduction of NO_x emissions and higher efficiency. The work output of this system is about the same as that achieved in the steam injection cycle, but the thermal efficiency of the system is much higher.

A high-pressure evaporator is placed between the compressor and the regenerator to add water vapor into the air steam and, in the process, lower the temperature of this mixed stream. The mixture then enters the regenerator at a lower temperature, increasing the temperature differential across the regenerator. Increasing the temperature differential reduces the temperature of the exhaust gases considerably so that these exhaust gases, which are otherwise lost, are an indirect source of heat used to evaporate the water. Both the air and the evaporated water pass through the regenerator, combustion chamber, and turbine. The water enters at 80°F (26.7°C) and 14.7 psia (1 Bar) through a pump into the evaporator, where it is discharged as steam at the same temperature as the compressor discharged air and at a pressure of 60 psia (4 Bar) above the compressor discharge. It is then injected into the air stream in a fine mist where it is fully mixed. The governing equations are the same as in the previous cycle for the turbine section, but the heat added is altered because of the regenerator. The following equations govern this change in heat addition. From the first law of thermodynamics, the mixture temperature (T_4) is given by the relationship:

$$T_4 = \frac{\dot{m}_a c_{pa} T_2 + \dot{m}_s c_{pw}(T_s - T_3) - \dot{m}_s h_{fg}}{\dot{m}_a c_{pa} + \dot{m}_s c_{ps}} \quad (2-31)$$

The enthalpy of the gas leaving the regenerator is given by the relationship:

$$h_5 = h_4 + \eta_{reg}(h_7 - h_4) \quad (2-32)$$

Similar to the regenerative cycle, the evaporative regenerative cycle has higher efficiencies at lower pressure ratios. Figures 2-32 and 2-33 show the performance of the system at various rates of steam injection and turbine inlet temperatures. Similar to the steam injection cycle, the steam is injected at 60 psi (4 Bar) higher than the air leaving the compressor. Corrosion in the regenerator is a problem in this system. When not completely clean, regenerators tend to develop hot spots that can lead to fires. This problem can be overcome with proper regenerator designs. The NO_x emission level is low and meets EPA standards.

The Brayton-Rankine Cycle

The combination of the gas turbine with the steam turbine is an attractive proposal, especially for electric utilities and process industries where steam is

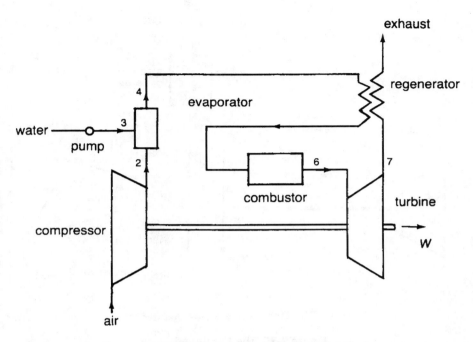

Figure 2-34. The Evaporative Regenerative Cycle

being used. In this cycle, as shown in Figure 2-35, the hot gases from the turbine exhaust are used in a supplementary fired boiler to produce superheated steam at high temperatures for a steam turbine.

The computations of the gas turbine are the same as shown for the simple cycle. The steam turbine calculations are:

Steam generator heat

$$_4Q_1 = h_{1S} - h_{4S} \tag{2-33}$$

Turbine work

$$W_{tS} = \dot{m}_S(h_{1S} - h_{2S}) \tag{2-34}$$

Pump work

$$W_P = \dot{m}_S(h_{4S} - H_{3S})/\eta_p \tag{2-35}$$

The combined cycle work is equal to the sum of the net gas turbine work and the steam turbine work. About one-third to one-half of the design output is

84 • COGENERATION AND COMBINED CYCLE POWER PLANTS

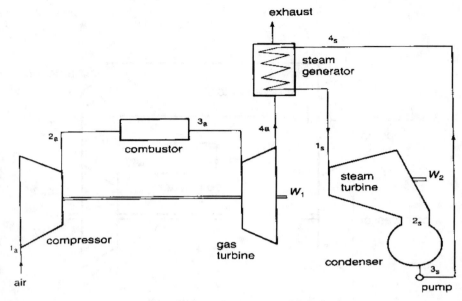

The Brayton-Rankine cycle.

Figure 2-35. The Combined Cycle

available as energy in the exhaust gases. The exhaust gas from the turbine is used to provide heat to the recovery boiler. Thus, this heat must be credited to the overall cycle. The following equations show the overall cycle work and thermal efficiency:

Overall cycle work

$$W_{cyc} = W_{ta} + W_{ts} - W_c - W_p \quad (2-36)$$

Overall cycle efficiency

$$\eta = \frac{W_{cyc}}{\dot{m}_f (LHV)} \quad (2-37)$$

This system, as can be seen from Figure 2-36, indicates that the net work is about the same as one would expect in a steam injection cycle, but the efficiencies are much higher. The disadvantages of this system are its high initial costs. However, just as in the steam injection cycle, the NO_x content of its exhaust remains the same and is dependent on the gas turbine used. This system is being used widely because of its high efficiency.

Cycles • 85

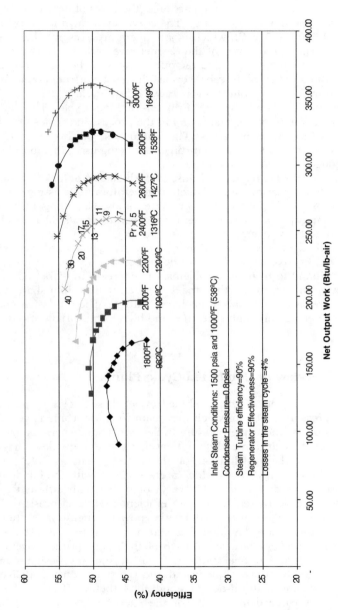

Figure 2-36. The Performance Map of a Typical Combined Cycle Power Plant

Summation of Cycle Analysis

Figures 2-37 and 2-38 give a good comparison of the effect of the various cycles on the output work and thermal efficiency. The curves are drawn for a turbine inlet temperature of 2400° (1316°C), which is a temperature presently being used by manufacturers. The output work of the regenerative cycle is very similar to the output work of the simple cycle, and the output work of the regenerative reheat cycle is very similar to that of the reheat cycle. The most work per pound of air can be expected from the intercooling, regenerative reheat cycle.

The most effective cycle is the Brayton-Rankine Cycle. This cycle has tremendous potential in power plants and in the process industries where steam turbines are in use in many areas. The initial cost of this system is high; however, in most cases where steam turbines are being used, this initial cost can be greatly reduced.

Regenerative cycles are popular because of the high cost of fuel. Care should be observed not to indiscriminately attach regenerators to existing units. The regenerator is most efficient at low pressure ratios. Cleansing turbines with abrasive agents may prove to be a problem in regenerative units since the cleansers can get lodged in the regenerator and cause hot spots.

Water injection, or steam injection systems are being used extensively to augment power. Corrosion problems in the compressor diffuser and combustor have not been found to be major problems. The increase in work and efficiency with a reduction in NO_x makes the process very attractive. Split-shaft cycles are attractive for use in variable-speed mechanical drives. The off-design characteristics of such an engine are high efficiency and high torque at low speeds.

A General Overview of Combined Cycle Plants

There are many concepts of the combined cycle, these cycles range from the simple single pressure cycle, in which the steam for the turbine is generated at only one pressure, to the triple pressure cycles, where the steam generated for the steam turbine is at three different levels. The energy flow diagram in Figure 2-39 shows the distribution of the entering energy into its useful component and the energy losses which are associated with the condenser and the stack losses. This distribution will vary somewhat with different cycles as the stack losses are decreased with more efficient multilevel pressure HRSGs. Distribution in the energy produced by the power generation sections as a function of the total energy produced is shown in Figure 2-40. This diagram shows that the load characteristics of each of the major prime-movers change drastically with off-design operation. The gas turbine at design conditions supplies 60% of the total energy delivered and the steam turbine delivers 40% of the energy, whereas at off-design conditions (below 50% of the design energy), the gas turbine delivers 40% of the energy while the steam turbine delivers 40% of the energy.

To fully understand the various cycles, it is important to define a few major parameters of the combined cycle. In most combined cycle applications, the gas

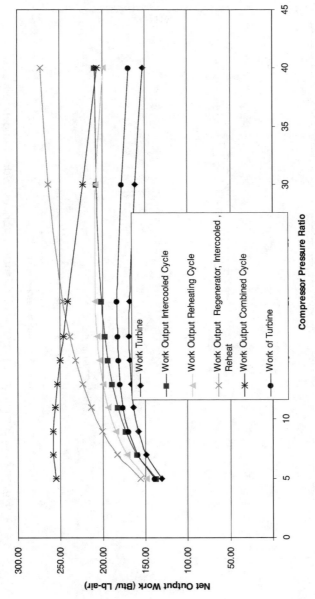

Figure 2-37. Comparison of Net Work Output of Various Cycles Temperature

88 • COGENERATION AND COMBINED CYCLE POWER PLANTS

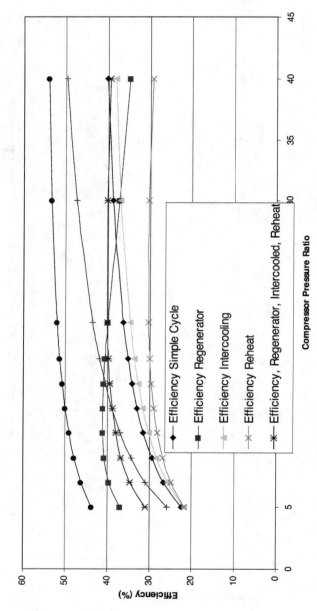

Figure 2-38. Comparison of Thermal Efficiency of Various Cycles Temperature

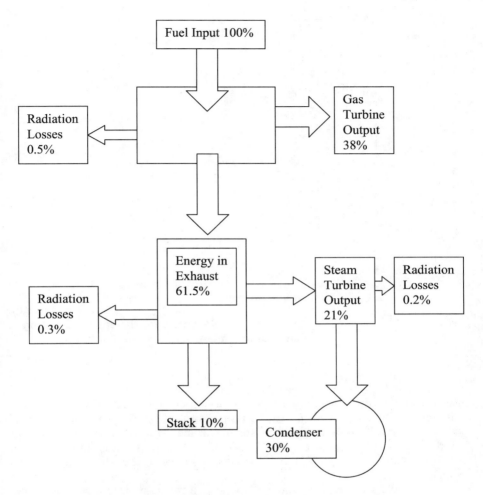

Figure 2-39. Energy Distribution in a Combined Cycle Power Plant

turbine is the topping cycle and the steam turbine is the bottoming cycle. Thermal efficiencies of the combined cycles can reach as high as 60%. In the typical combination, the gas turbine produces about 60% of the power and the steam turbine about 40%. Individual unit thermal efficiencies of the gas turbine and the steam turbine are between 30% and 40%. The steam turbine utilizes the energy in the exhaust gas of the gas turbine as its input energy. The energy transferred to the heat recovery steam generator (HRSG) by the gas turbine is usually equivalent to about the rated output of the gas turbine at design conditions. At off design conditions, the inlet guide vanes (IGV) are used to regulate the air so as to maintain a high temperature to the HRSG.

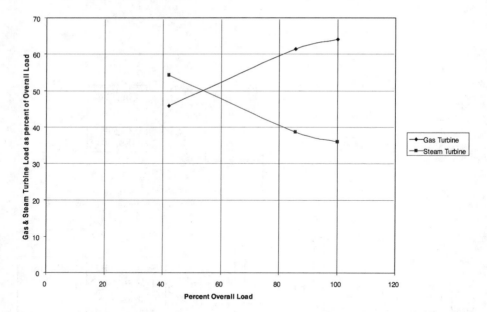

Figure 2-40. Load Sharing Between Prime Movers Over the Entire Operating Range of a Combine Cycle Power Plant

The heat recovery steam generating (HRSG) is where the energy from the gas turbine is transferred to the water to produce steam. There are many different configurations of the HRSG units. Most HRSG units are divided into the same amount of sections as the steam turbine. In most cases, each section of the HRSG has a pre-heater or economizer, an evaporator, and then one or two stages of Superheaters. The steam entering the steam turbine is superheated.

The condensate entering the HRSG goes through a deaerator where the gases from the water or steam are removed. This is important because a high oxygen content can cause corrosion of the piping and the components which would come into contact with the water/steam medium. An oxygen content of about 7 to 10 parts per billion (ppb) is recommended. The condensate is sprayed into the top of the deaerator, which is normally placed on the top of the feedwater tank. Deaeration takes place when the water is sprayed and then heated, thus releasing the gases that are absorbed in the water/steam medium. Dearation must be done on a continuous basis because air is introduced into the system at the pump seals and piping flanges since they are under vacuum.

Dearation can be either vacuum or over pressure dearation. Most systems use vacuum dearation because all the feedwater heating can be done in the feedwater tank and there is no need for additional heat exchangers. The heating steam in the vacuum dearation process is a lower quality steam, thus leaving the steam in the steam cycle for expansion work through the steam turbine; this increases the output of the steam turbine and therefore the efficiency of the combined cycle. In the case of the overpressure dearation, the gases can be exhausted directly to atmosphere independently of the condenser evacuation system.

Dearation also takes place in the condenser. The process is similar to that in the dearator, the turbine exhaust steam condenses and collects in the condenser hotwell while the incondensable hot gases are extracted by means of evacuation equipment. A steam cushion separates the air and water so re-absorption of the air cannot take place. Condenser dearation can be as effective as the one in a dearator. This could lead to not utilizing a separate dearator/feedwater tank, and the condensate being fed directly into the HRSG from the condenser. The amount of make-up water added to the system is a factor since make-up water is fully saturated with oxygen. If the amount of makeup water is less than 25% of the steam turbine exhaust flow, condenser dearation may be employed but in cases where there is steam extraction for process use and therefore the make-up water is large, a separate deaerator is needed.

The economizer in the system are used to heat the water close to its saturation point. If they are not carefully designed economizers can generate steam thus blocking the flow. To prevent this from occurring, the feedwater at the outlet is slightly subcooled. The difference between the saturation temperature and the water temperature at the economizer exit is known as the approach temperature. The approach temperature is kept as small as possible between 10°F and 20°F (5.5°C to 11°C). To prevent steaming in the evaporator, it is also useful to install a feedwater control valve downstream of the economizer, which keeps the pressure high and steaming is prevented. Proper routing of the tubes to the drum also prevents blockage if it occurs in the economizer.

Another important parameter is the temperature difference between the evaporator outlet temperature on the steam side and on the exhaust gas side. This difference is known as the pinch point. Ideally, the lower the pinch point, the more heat recovered, but this calls for more surface area and, consequently, increases the back pressure and cost. Also, excessively low pinch points can mean inadequate steam production if the exhaust gas is low in energy (low mass flow or low exhaust gas temperature). General guidelines call for a pinch point of 15 to 40°F (8 to 22°C). The final choice is obviously based on economic considerations.

The steam turbines in most of the large power plants are at a minimum divided into two major sections the high-pressure (HP) section and the low-pressure (LP) section. In some plants, the high-pressure section is further divided into a high-pressure section and an intermediate pressure (IP) section. The heat recovery steam generating (HRSG) is also divided into sections corresponding with the steam turbine. The LP steam turbine's performance is further dictated by the condenser backpressure, which is a function of the cooling and the fouling.

The efficiency of the steam section in many of these plants varies from 30% to 40%. To ensure that the steam turbine is operating in an efficient mode, the gas turbine exhaust temperature is maintained over a wide range of operating

92 • COGENERATION AND COMBINED CYCLE POWER PLANTS

conditions. This enables the HRSG to maintain a high degree of effectiveness over this wide range of operation.

The major components that make up a combined cycle are the gas turbine, the HRSG and the steam turbine. Figure 2-41 shows a typical combined cycle power plant with a single pressure HRSG. In a combined cycle plant, high steam pressures does not necessarily convert to a high thermal efficiency for a combined cycle power plant. Expanding the steam at higher steam pressure causes an increase in the moisture content at the exit of the steam turbine. The increase in moisture content creates major erosion and corrosion problems in

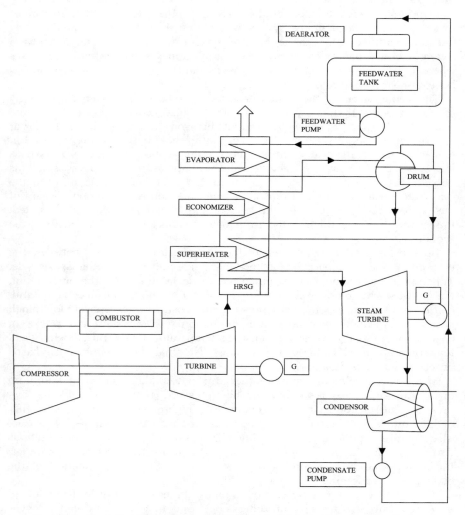

Figure 2-41. A Typical Pressure HRSG in a Combined Cycle Power Plant

the later stages of the turbine. A limit is set at about 10% (90% steam quality) moisture content.

The advantages for a high steam pressure are that the mass flow of the steam is reduced and that the turbine output is also reduced. The lower steam flow reduces the size of the exhaust steam section of the turbine thus reducing the size of the exhaust stage blades. The smaller steam flow also reduces the size of the condenser and the amount of water required for cooling. It also reduces the size of the steam piping and the valve dimensions. This all accounts for lower costs especially for power plants which use the expensive and high energy consuming air cooled condensers.

Increasing the steam temperature at a given steam pressure lowers the steam output of the steam turbine slightly. This occurs because of two contradictory effects: first, the increase in enthalpy drop, which increases the output; but second, the decrease in flow, which causes a loss in steam turbine output. The second effect is more predominant, which accounts for the lower steam turbine amount. Lowering the temperature of the steam also increase the moisture content.

Figure 2-42 is a schematic of a typical combined cycle power plant with a dual pressure HRSG. Understanding the design characteristics of the dual pressure HRSG and its corresponding dual section steam turbine (HP and LP turbines) is important. Increasing pressure of either section will increase the work output of the section for the same mass flow. However, at higher pressure, the mass flow of the steam generated is reduced. This effect is more significant for the LP Turbine. The pressure in the LP evaporator should not be below about 45 psia (3.1 Bar) because the enthalpy drop in the LP steam turbine becomes very small, and the volume flow of the steam becomes very large, thus the size of the LP section becomes large, with long expensive blading. Increase in the steam temperature brings substantial improvement in the output. In the dual pressure cycle, more energy is made available to the LP section if the steam team to the HP section is raised. Figure 2-43 shows an energy/temperature diagram for a dual-pressure HRSG.

An examination of the diagram shows that most of the heat exchange takes place in the HP portion of the HRSG. Examining this diagram with the single pressure HRSG, one notes that how energy utilization in the cold end of the HRSG has been improved.

Figure 2-44 is a schematic of a typical combined cycle power plant with triple pressure HRSG. There is a very small increase in the overall cycle efficiency between a dual pressure cycle and a triple pressure cycle. To maximize their efficiency, these cycles are operated at high temperatures, and extracting most heat from the system, thus creating relatively low stack temperatures. This means that in most cases, they must be only operated with natural gas as the fuel, as this fuel contains a very low to no sulfur content. Users have found that in the presence of even low levels of sulfur, such as when firing diesel fuel (No. 2 Fuel oil), stack temperatures must be kept above 300°F (149°C) to avoid acid gas corrosion. The increase in efficiency between the dual and triple pressure cycle is due to the steam being generated at the IP level than the LP level. The HP flow is slightly less than in the dual pressure cycle because the IP superheater is at a higher level than

94 • COGENERATION AND COMBINED CYCLE POWER PLANTS

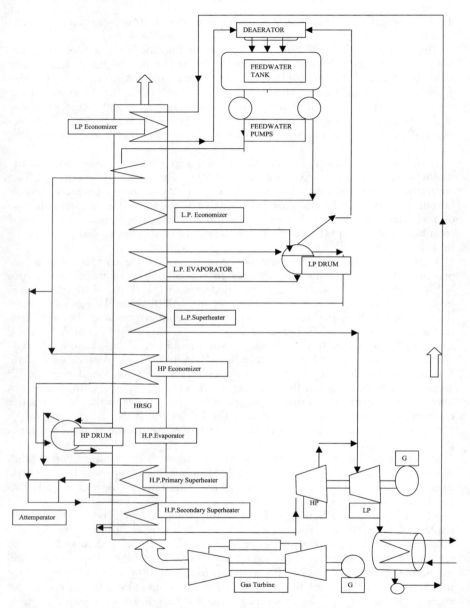

Figure 2-42. A Typical Dual Pressure HRSG in a Combined Cycle Power Plant

the LP superheater, thus removing energy from the HP section of the HRSG. In a triple pressure cycle, the HP and IP section pressure must be increased together. Moisture at the steam turbine LP section exhaust plays a governing

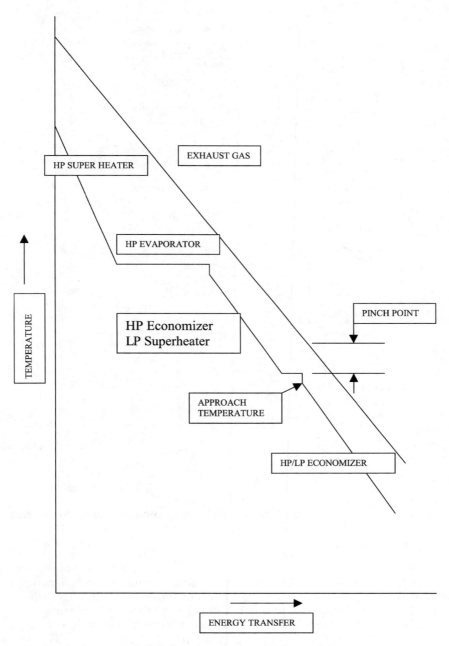

Figure 2-43. Energy Temperature Dual Pressure Diagram

96 • COGENERATION AND COMBINED CYCLE POWER PLANTS

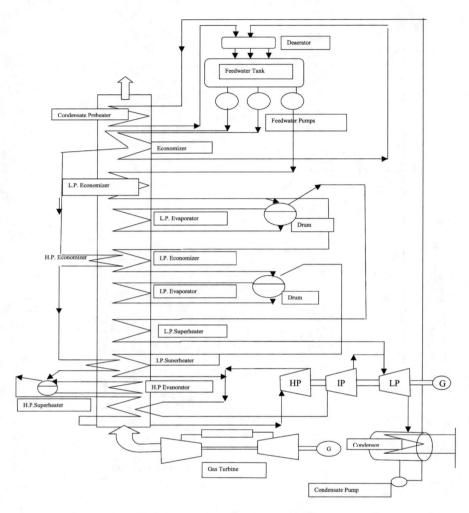

Figure 2-44. A Typical Triple Pressure HRSG in a Combined Cycle Power Plant

role. At inlet pressure of about 1500 psia (103.4 Bar), the optimum pressure of the IP section is about 250 psia (17.2 Bar). The maximum steam turbine output is clearly definable with the LP steam turbine pressure. The effect of the LP pressure also affects the HRSG surface area, as the surface area increases with the decrease in LP steam pressure, because less heat exchange increases at the low temperature end of the HRSG. Figure 2-45 is the energy/temperature diagram of the triple pressure HRSG. The IP and LP flows are much smaller than the HP steam turbine flow. The ratio is in the neighborhood of 25:1.

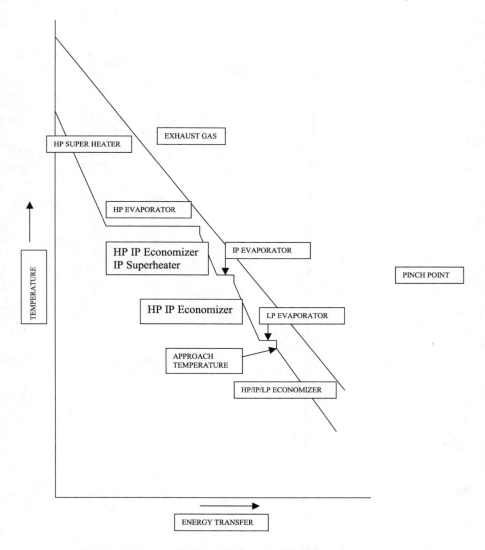

Figure 2-45. Energy Temperature Triple Pressure Diagram

Chapter 3

PERFORMANCE AND MECHANICAL EQUIPMENT STANDARDS

Combined cycle power plants and cogeneration plants all over the world require detailed specifications both from a performance as well as a mechanical point of view. The first decision is what type of combined cycle power plant should be built. Combined cycle plants are generally configured with each gas turbine and steam turbine driving a separate electrical generator. Since the two generators are on separate shafts, this configuration is called a multi-shaft combined cycle power plant. The term single-shaft combined cycle power plant refers to a configuration in which the gas turbine and steam turbine drive a single, common generator, with the various rotors connected by couplings to form a single shaft. These types of turbines are getting very popular in large plants around the world. The single-shaft combined cycle can be arranged in one of two ways: without a clutch or with a clutch.

The first step in designing a cogeneration or combined cycle power plant is the determination of the customers' requirements. Table 3-1 is a typical cogeneration questionnaire, and Table 3-2 is a typical questionnaire for the independent power industry, which should be answered so that a preliminary analysis can be made.

Once the power plant need is defined and the preliminary analysis is completed the following are the requirements that need to be provided to the engineering contractor:

1. The Overall Power Requirements
2. The Overall Process Steam Required
    2.1. Flow
    2.2. Pressure
    2.3. Temperature

3. Type of Plant
    3.1. Multi-Shaft Plant Configuration
    3.2. Single-Shaft Configured

4. Type of Gas Turbine
    4.1. Aero-derivative Type Gas Turbine
    4.2. Frame Type Unit
5. Type of Fuel

5.1. Natural gas
5.2. Liquid Fuel
    5.2.1. Distillate Oil
    5.2.2. Residual Fuel
    5.2.3. Crude Oil

6. Type of HRSG
    6.1. Drum Type HRSG
        6.1.1. Single Pressure
        6.1.2. Dual Pressure
        6.1.3. Triple Pressure
    6.2. Once through Steam Generators
        6.2.1. Single Pressure
        6.2.2. Dual Pressure
    6.3. Fired HRSG

7. Boiler Feed Water Pumps
    7.1. Single
    7.2. Spared

8. Condensate Pumps
    8.1. Single
    8.2. Spared

9. Type of Steam Turbines
    9.1. Condensing Steam Turbine
    9.2. Extraction Condensing Steam Turbine
    9.3. Back Pressure Steam Turbine

10. Water Treatment Systems
11. Type of Condenser
    11.1. Fresh Water Cooled Condenser
    11.2. Salt Water Cooled Condensers
    11.3. Air Cooled Condenser

12. Cooling Water
    12.1. Temperature: maximum and minimum
    12.2. Pressure at inlet and back pressure, if any
    12.3. Whether open or closed cooling system desired
    12.4. Source of water
    12.5. Fresh, salt, or brackish
    12.6. Silty or corrosive

13. Electrical Equipment
    13.1. Step-up Transformer
    13.2. Link to HighVoltage Grid
    13.3. Switch Gear

14. Distributed Control System
15. Plant Start-up System
    15.1. Black-Start Capability
    15.2. Batteries
    15.3. Battery Charger
16. General
    16.1. Mineral or Synthetic Lubricants
    16.2. Indoor or outdoor installation
    16.3. Floor space, special shape. Provide a sketch
    16.4. Soil characteristics
    16.5. List accessories desired and advise which are to be spared
    16.6. Pulsation dampeners or intake or discharge silencers to be supplied
17. Specifications
    17.1. Provide each bidder with three copies of any specification for the particular project
    17.2. Complete information enables all manufacturers to bid competitively on the same basis and assists the purchaser in evaluating bids.

Major Variables for a Combined Cycle Power Plant

The major variables that affect the location and selection of the type of combined cycle power plant and the gas turbines associated with them are:

- Plant Location and Site Configuration
- Plant Type; Aero-derivative, or Frame Type
- Plant size and efficiency
- Type of fuel
- Types of Condensers
- Enclosures
- Plant operation mode; Base or peaking
- Start-up techniques.

Each of the above points is discussed below.

Plant Location and Site Configuration

The location of the plant is the principal determination of the type of plant best configured to meet its needs. Is the plant located near transmission lines; the location from fuel port or pipe lines, type of fuel availability. Site configuration is generally not a constraint. Periodically, sites are encountered where one plant configuration, multi-shaft combined cycle power plant or single-shaft combined cycle power plant, is more suitable than the other is. As a general rule, long, narrow sites are ideal for single-shaft combined cycle power plant plants.

Table 3-1. Cogeneration Questionnaire

Name of Party:..
Address :..
....
....
Location To : Major Highway:......km Railroad.........km Pipe Line........km
Contact Person Management:..
Title:..................................Phone Number..............................
Fax..e-mail..
Technical:..
Title:..................................Phone Number..............................
Fax..e-mail..
Type of Industry:..
Market Share in Country:..
Type of Corporation: Public....................Private............................
Annual Report of the Company for the Past Three years.
Present Power Consumption: Max.........kW   Total Usage........kW·hr/year
Last Three Years Power Consumption:1999.........kW·hr/year   2000..........kW·hr/year   2001.........kW·hr/year
Power Usage Distribution:
Per Day: Morning.......kW Afternoon...... kW Night........kW
Per Week: Daily ....kW·hr 7 days........kW·hr
Per Month : Jan.........kW·hr Feb........kW·hr March......kW·hr April.......kW·hr
   May........kW·hr   June.......kW·hr   July.........kW·hr   Aug........kW·hr
   Sept........kW·hr   Oct........kW·hr   Nov........kW·hr   Dec........kW·hr

Monthly Average Readings:

Month	Day Temperature	Day Barometer	Day Relative Humidity	Night Temperature	Night Barometer	Night Relative Humidity
January						
February						
March						
April						
May						
June						
July						
August						
September						
October						
November						
December						

Other Sources of Energy Usage: Steam.........kg/hr Air-Conditioning...........Tons
   Heating..
   Other..

Availability of Sale To National Grid: ....MW.........$/kW
Present Cost of Electrical Power:....................$/kW·hr
Present Cost of Steam Generation:...................$/kg/hr
Availability of Land at Plant Site....................Acres
Availability of Cooling Water................ kW·hr gal/hr
Present Connection:
   Transformer Size......kVa/.......kVa Incoming Line......Volts .....Amps
Type of Power: Volts......................Cycle ....Hz
Type of Fuel Available:
Natural gas: Heating Value:...........kJ/kg Cost........$/m^3 Quantity............m^3/hr
Diesel Fuel: Heating Value:...........kJ/kg Cost.......$/tonne Quantity............ton/year
Furnace Oil: Heating Value:...........kJ/kg Cost.......$/tonne Quantity..... ....ton/year
Heavy Crude: Heating Value:...........kJ/kg Cost.......$/tonne Quantity..... ....ton/year
Environmental Regulations:
   NOx.......ppm UHC......ppm   CO.......ppm Partciculate ....ppm
   Noise at Boundary...... dba

Table 3-2. Independent Power Producer

Name of Party:..

Address :...
....
....

Location To : Major Highway:.......km Railroad..........km Pipe Line.........km

Contact Person Management:...
Title:...Phone Number...............................
Fax...e-mail...

Technical:...
Title:...Phone Number...............................
Fax...e-mail...

Monthly Average Readings:

Month	Day Temperature	Day Barometer	Day Relative Humidity	Night Temperature	Night Barometer	Night Relative Humidity
January						
February						
March						
April						
May						
June						
July						
August						
September						
October						
November						
December						

Availability of Sale To National Grid: ....MW.........$/kW

Availability of Land at Plant Site.........Acres Cooling Water.............gal/hr

Type of Power: Volts......................Cycle ....Hz

Type of Fuel Available:

Natural gas: Heating Value:............kJ/kg Cost........$/m^3 Quantity..............m^3/hr
Diesel Fuel: Heating Value:............kJ/kg Cost........$/tonne Quantity............ton/year
Furnace Oil: Heating Value:............kJ/kg Cost........$/tonne Quantity...... ....ton/year
Heavy Crude: Heating Value:............kJ/kg Cost........$/tonne Quantity...... ....ton/year

Environmental Regulations:

    NOx.......ppm UHC......ppm CO.......ppm Partciculate ....ppm
    Noise at Boundary....... dba

Financial:
    Local State Bank Guarantees Available:..
    Government Guarantees Available:...
    Local Investors..
    Local Interest Rates..

Business:
    Local Partner..
    Percentage Local Partner..

Investment:
    Bid Bond...
    Investment Bond..
    Local Investment...
    Foreign Investment..

Plant Type

The determination to have an aero-derivative type gas turbine or a frame type gas turbine is the plant location. In most cases, if the plant is located off-shore on a platform, then an aero-derivative plant is required. On most on-shore applications, if the size of the plant exceeds 100 MW, then the frame type is best suited for the gas turbine. In smaller plants between 2 and 20 MW, the Industrial type small turbines best suit the application, and in plants between 20 and 100 MW, both aero-derivative and frame types can apply. Aero-derivatives have lower maintenance and have high heat-recovery capabilities. In many cases, the type of fuel and service facilities may be the determination. Natural gas or diesel no. 2 would be suited for aero-derivative gas turbines, but heavy fuels would require a frame type gas turbine.

Plant Size and Efficiency

Plant size is important in the cost of the plant. The larger the plant, the less the initial cost per kilowatt, and the efficiency increases with plant size. Figure 3-1 shows typical plant cost and efficiency as a function of plant load.

Large combined cycle power plants are combinations of single-shaft combined cycle power plants or a multiple shaft combined cycle power plants. The plant arrangements shown in Figure 3-2a are categorized as multi-shaft 1:1:1, i.e., one gas turbine, one heat recovery steam generator (HRSG), and one steam turbine. The plant arrangements shown in Figures 3-2b and 3-2c are categorized as single-shaft 1:1:1, and single shaft without and with clutch, respectively. If the desired plant output is higher than can be produced by a single gas turbine plant, other possible arrangements in multiple shaft combined cycle power plants configurations are 2:2:1 (two gas turbines, two HRSGs, and one steam turbine), which is the most common in plants above 300 MW, as shown in Figure 3-3a, or a 3:3:1 (three gas turbines, three HRSGs, and one steam turbine), as shown in Figure 3-4a. Corresponding arrangements for the single-shaft combined cycle power plants configurations are $2 \times (1:1:1)$ or $3 \times (1:1:1)$, i.e., two or three trains of single-shaft arrangement, as shown in Figures 3-3b and 3-3c or Figures 3-4b and 3-4c, respectively. Arrangements for the single-shaft combined cycle power plants configurations are just multiple of trains. For example, two 1500 MW plants in Europe have five single-shaft combined cycle power plant trains, and a major 2800 MW plant in China has eight single-shaft combined cycle power plant trains. The total plant gross output and the individual gross outputs of the gas turbine generator and steam turbine generator, the overall plant efficiency and the heat rate for typical plants are shown in Table 3-3 for both the configurations. The numbers shown in these tables are for F technology gas turbines and represent the best of the F technology gas turbines available from the major gas turbine suppliers. The main point here is that from a performance point of view, it does not matter whether the turbine configuration is a single-shaft combined cycle power plant, or a multiple shaft combined cycle power plant. The most important effect on performance is whether the system is a three-pressure or dual-pressure steam

Performance and Mechanical Equipment Standards • 105

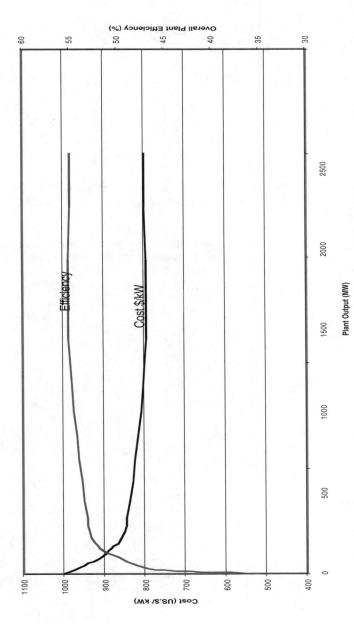

Figure 3-1. A Typical Plant Cost and Efficiency as a Function of a Plant Load

106 • COGENERATION AND COMBINED CYCLE POWER PLANTS

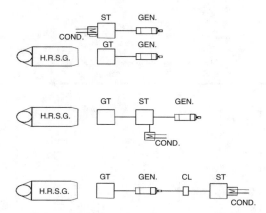

Figure 3-2. The Single-Shaft Combined Cycle Myth By Ram G. Narula (Courtesy ASME Paper No. 2000-GT-0594 IGTI – Munich 2000)

cycle, whether or not reheat is used. Based on manufacturer data of the F turbines, we note that the dual pressure is about 1% to 2% lower than the triple pressure and about 5% to 7% for reheat systems as seen in Figure 3-5. However

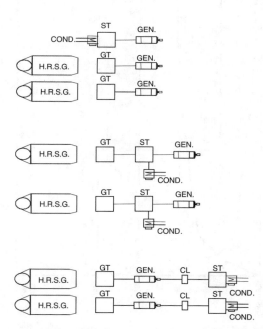

Figure 3-3. The Single-Shaft Combined Cycle Myth By Ram G. Narula (Courtesy ASME Paper No. 2000-GT-0594 IGTI – Munich 2000)

Performance and Mechanical Equipment Standards • 107

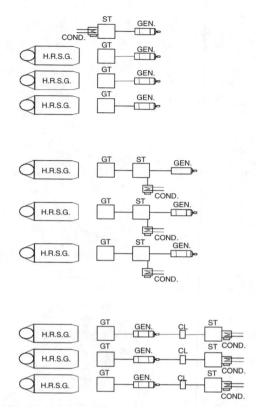

Figure 3-4. The Single-Shaft Combined Cycle Myth By Ram G. Narula
(Courtesy ASME Paper No. 2000-GT-0594 IGTI – Munich 2000)

in some cases, much higher pressure, dual pressure HRSGs are used making up for the lost of production in the Intermediate pressure steam turbine, and actually resulting in a higher output and efficiency. Figure 3-5 also introduces the once through heat recovery steam generator (OSTG). These steam generators have a once through flow path on the water/steam side, therefore all water that enters leaves as steam without any recirculation, thus getting rid of the traditional steam drums. These OSTGs are in many cases faster, cheaper, and have smaller foot prints than the traditional HRSGs. A very important point to remember is that an increase in cycle efficiency of 1% to 2% can justify an increase in initial cost of about 3% to 5%.

The primary advantage of a single-shaft combined cycle power plant over the multiple shaft combined cycle power plant is that combining the gas turbine and steam turbine generators into a single larger generator results in one less generator, main transformer, and associated electrical sub-systems. However, as the overall plant size increases, requiring multiple gas turbines, the advantage of one less generator/transformer grows proportionally smaller. As an example, the

Table 3-3. Data on Combined Cycle Power Plants

Plant Arrangement	50 Hz					60 Hz				
	Total Plant Output, MW	Single Gas Turbine Output, MW	Steam Turbine Output, MW	Overall Plant Eff., %	Overall Plant Heat Rate, Btu/kW·h (kJ/kW·h)	Total Plant Output, MW	Single Gas Turbine Output, MW	Steam Turbine Output, MW	Overall Plant Eff., %	Overall Plant Heat Rate, Btu/kW·h (kJ/kW·h)
Multiple Shaft Combined Cycle Power Plants (1×1×1)	189.2	121.6	67.6	52	6561 (6935)	262	170	92	56	6093 (6425)
Multiple Shaft Combined Cycle Power Plants (2×2×1)	480	156	164	52.9	6450 (6805)	529	170	189	56.5	6038 (6375)
Multiple Shaft Combined Cycle Power Plants (3×3×1)	309	67.3	107	54.7	6238 (6580)	309	67.3	107	54.7	6238 (6580)
Single-Shaft Combined Cycle Power Plants (1×1×1)	384	243	143	52.7	6474 (6840)	271	172	99	57.6	5923 (6250)

Performance and Mechanical Equipment Standards • 109

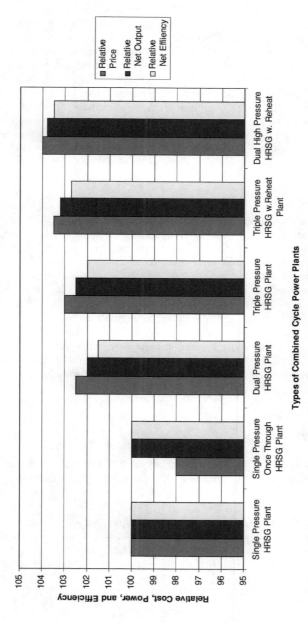

Figure 3-5. Comparison of Cost, Output Power, and Cycle Efficiency of Various Types of Combine Cycle Power Plants

advantage in electrical costs (on a US$ per kilowatt basis) of going from four to three generators is not as great as going from two generators to one.

The primary disadvantage of a multiple shaft combined cycle power plant is that the number of steam turbines, condensers, and condensate systems - and perhaps the cooling towers and circulating water systems - increases to match the number of gas turbines. For multiple shaft combined cycle power plant, there is only one steam turbine, condenser and the rest of the heat sink for up to three gas turbines; only their size increases. Having only one larger steam turbine and heat sink results in lower cost because of economies of scale. Further, a larger steam turbine also allows the use of a higher pressure and more efficient steam cycle. Thus, the overall plant size and the associated number of gas turbines required have a major impact on whether single-shaft combined cycle power plant or multiple shaft combined cycle power plant is more economical.

Type of Fuel

The type of fuel is one of the most important aspects that govern the selection of a plant. Natural gas would be the choice of most operators if natural gas was available since its effects on pollution is minimal and maintenance cost would also be the lowest. Table 3-4 shows how the maintenance cost would increase from natural gas to the heavy oils.

Aero-derivative gas turbines cannot operate on heavy fuels, thus if heavy fuels was a criterion, then the Frame type turbines would have to be used. With heavy fuels, the power delivered would be reduced after about two weeks of operation by about 10%. On-line turbine wash is recommended for turbines with high vanadium content in their fuel since magnesium salts have to be added to counteract vanadium. These salts cause the vanadium when combusted in the turbine to be turned to ash. This ash settles on the turbine blades and reduces the cross-sectional area, thus reducing the turbine power.

Types of HRSG

The selection of the HRSG depends on many factors but essentially the choice of the HRSG depends on initial cost and the efficiency of the overall plant. A 1%

Table 3-4. Typical Gas Turbine Maintenance Cost Based on Type of Fuel

Type of Fuel	Expected Actual Maintenance Cost ($/MWh)	Relative Maintenance Cost Factor
Natural gas	0.35	1.0
No. 2 Distillate Oil	0.49	1.4
Typical Crude Oil	0.77	2.2
No. 6 Residual Oil	1.23	3.5

increase in efficiency can support a 3% to 4% increase in cost; this is due to the fact that in a life cycle cost, the initial cost contributes 10% to the life cycle cost, the maintenance cost, 15%, and the fuel cost, about 75%.

The typical HRSG units used are the drum-type HRSG. The drum type HRSG is composed of drums, separate economizers, separate superheaters, circulation systems, generating tubes, and blowdown systems. In the past one decade once through heat recovery steam generators (OTSG) have evolved into a cost competitive and technologically advanced heat recovery steam generator. Water tube drum units were developed to prevent scaling, corrosion and control of the steam generating process. Today's technology in the area of materials, water treatment systems, and modern control technology do not require the traditional boiler components. This enables the OTSG to be more competitive in initial costs and also erection and installation costs.

The increase in cost and efficiency between a single pressure and triple pressure systems as well as the increase cost is shown in Table 3-5.

Types of Condensers

Large steam turbines are usually condensing type steam turbines. Condensers can be water cooled or air cooled units. Water cooled condensers are the most prevalent and have an higher degree of effectiveness. The cooling effectiveness of the condenser is very important to the performance of the low pressure steam turbine. If the cooling effectiveness is reduced, then the back pressure of the low pressure steam turbine is increased, and the power output of the steam turbine is reduced.

Steam Turbines traditionally have down (or bottom) exhaust, where the condenser is located below the steam turbine. In such an arrangement, the steam pedestal is raised about 30 ft (10 m) above the base slab where the condenser sits. Care must be taken that none of the condensate splashes onto the turbine casing, otherwise casing distortion and cracks can emit.

For smaller steam turbines up to 250 MW, axial exhaust is available from some steam turbine suppliers. Side exhaust is also available from some suppliers for steam turbines up to 450 MW. In the axial or side exhaust arrangement, the condenser is located side by side with the steam turbine. As a result, the steam turbine pedestal can be lowered by 15 to 21 ft (5 to 7 m); this reduces civil/

Table 3-5. Typical Cost and Efficiency Increases in Combined Cycle Power Plants

Type of Process	Increase in cost	Increase in Efficiency	Increase in Power
Single Pressure HRSG	1.0	1	1
Single pressure OTSG	0.98	1	1
Dual Pressure HRSG	1.025	1.015	1.015
Triple Pressure HRSG	1.03	1.02	1.02
Dual Pressure Reheat HRSG	1.04	1.035	1.035
Triple Pressure Reheat HRSG	1.035	1.027	1.027

structural costs. Further, if the steam turbine is enclosed in a building, then the building height is also reduced, further lowering civil/structural costs. It should be noted that some of these savings are partially offset by the increased cost of main steam and cold/hot reheat piping since lowering the pedestal increases the vertical length of this piping. With axial or side exhaust machines, proper attention must be paid to the moisture drainage and bearing lubrication aspects.

The use of multi-shaft combined cycle power plants have one major benefit and that is the gas turbine, and the steam turbine can be from two different manufacturers, thus a major limitation of a single-shaft combined cycle power plant is that it does not allow the flexibility of choosing the best available steam turbine for a given plant.

Enclosures

Gas turbines usually come packaged in their own enclosures. These enclosures are designed so that they limit the noise to 70 dB at 100 ft (30 m) from the gas turbine. The entire plant consisting of the gas turbine, HRSG, and the steam turbine can be either inside or outside. While open plants are less expensive than enclosed plants, some owners prefer to enclose their steam turbines in a building and use permanent cranes for maintenance, thus leaving the gas turbine and the HRSG in the open environment. In severe climate areas, the entire plant is enclosed in a building. Single-shaft combined cycle power plant with the generator in the middle require a wider building to allow the generator to be moved to facilitate rotor removal and inspection. Plant arrangements that do not use axial or side exhaust steam turbines result in a taller building and higher building costs.

Plant Operation Mode: Base or Peaking

Combined cycle power plants are not as were originally planned base loaded plants. It is not uncommon for the plant to be cycled from 40% to 100% load in a single day, every day of the year. The intended plant operation mode has some bearing on the selection between multi-shaft combined cycle power plant and single-shaft combined cycle power plant, especially for the larger plants with multiple gas turbines. A larger multi-shaft combined cycle power plant is inherently more efficient than the corresponding single-shaft combined cycle power plant because of the larger steam turbines and, as such, is more suitable as a base load plant. However, if the plant has to be operated as a cycling plant with multiple starts and with extended hours at part loads as low as 40%, then a single-shaft combined cycle power plant with multiple smaller small turbines may be more desirable. Further, if the plant must operate in simple cycle mode because of phased construction or when the steam turbine is down, then a multi-shaft combined cycle power plant or single-shaft combined cycle power plant with a clutch is more desirable than a single-shaft combined cycle power plant without a clutch.

Start-up Techniques

The startup of a gas turbine is done by the use of electrical motors, diesel motors, and in plants where there is an independent source of steam by a steam turbine. New turbines use the generator as a motor for startup. After combustion occurs and the turbine reaches a certain speed, the motor declutches and becomes a generator. Use of a synchronous clutch between two rotating pieces of equipment is not new. It is very common in use with start-up equipment. In the case of single-shaft combined cycle power plants, a synchronous clutch can be used to connect the steam turbine to the gas turbine. However, use of a clutch in transmitting over 100 MW of power is new and has not found unequivocal customer acceptance. While use of a synchronous clutch leads to additional space requirements, additional capital and O&M costs, and potentially reduced availability, it does offer the tangible benefit of easy and fast plant startup. A major drawback of an single-shaft combined cycle power plant with a clutch is that the generator installation and maintenance and power evacuation are more complex and costly because the generator is located in the middle.

Performance Standards

The purpose of the ASME Performance Test Codes is to provide standard directions and rules for the conduct and report of tests of specific equipment and the measurement of related phenomena. These codes provide explicit test procedures with accuracies consistent with current engineering knowledge and practice. The codes are applicable to the determination of performance of specific equipment. They are suitable for incorporation as part of commercial agreements to serve as a means to determine fulfillment of contract obligations. The parties to the test should agree to accept the code results as determined or, alternatively, agree to mutually acceptable limits of uncertainty established by prior agreement of the principal parties concerned.

The performance tests must be run as much as possible to meet the ASME performance codes. These codes are very well written and fully delineate the tests required. Meetings should be held in advance with the vendors to decide which part of the code would not be valid and what assumptions and correction factors must be under taken to meet the various power and efficiency guarantees. The determination of special data or verification of particular guarantees, which are outside of the scope of the codes, should be made only after written agreement of both parties to the test, especially regarding methods of measurement and computation, which should be completely described in the test report.

ASME, Performance Test Code on Overall Plant Performance, ASME PTC 46 1996

This code is written to establish the overall plant performance. Power plants, which produce secondary energy output such as cogeneration facilities are included within the scope of this code. For cogeneration facilities, there is no requirement for a minimum percentage of the facility output to be in the form of electricity; however, the guiding principles, measurement methods, and calculation procedures are predicated on electricity being the primary output. As a result, a test of a facility with a low proportion of electric output may not be capable of meeting the expected test uncertainties of this code. This code provides explicit procedures for the determination of power plant thermal performance and electrical output. Test results provide a measure of the performance of a power plant or thermal island at a specified cycle configuration, operating disposition and/or fixed power level, and at a unique set of base reference conditions. Test results can then be used as defined by a contract for the basis of determination of fulfillment of contract guarantees. Test results can also be used by a plant owner for either comparison to a design number or to trend performance changes over time of the overall plant. The results of a test conducted in accordance with this Code will not provide a basis for comparing the thermoeconomic effectiveness of different plant design.

Power plants are comprised of many equipment components. Test data required by this code may also provide limited performance information for some of this equipment; however, this code was not designed to facilitate simultaneous code level testing of individual equipment. ASME PTCs, which address testing of major power plant equipment, provide a determination of the individual equipment isolated from the rest of the system. PTC 46 has been designed to determine the performance of the entire heat-cycle as an integrated system. Where the performance of individual equipment operating within the constraints of their design-specified conditions are of interest, ASME PTCs developed for the testing of specific components should be used. Likewise, determining overall thermal performance by combining the results of ASME code tests conducted on each plant component is not an acceptable alternative to a PTC 46 test.

ASME, Performance Test Code on Test Uncertainty: Instruments and Apparatus PTC 19.1, 1988

This test code specifies procedures for evaluation of uncertainties in individual test measurements, arising from both random errors and systematic errors, and for the propagation of random and systematic uncertainties into the uncertainty of a test results. The various statistical terms involved are defined. The end result of a measurement uncertainty analysis is to provide numerical estimates of systematic uncertainties, random uncertainties, and the combination of these into a total uncertainty with an approximate confidence level. This is especially very important when computing guarantees in plant output, and plant efficiency.

ASME, Performance Test Code on Gas Turbines, ASME PTC 22 1997

The object of the code is to detail the test to determine the power output and thermal efficiency of the gas turbine when operating at the test conditions, and correcting these test results to standard or specified operating and control conditions. Procedures for conducting the test, calculating the results, and making the corrections are defined.

The code provides for the testing of gas turbines supplied with gaseous or liquid fuels (or solid fuels converted to liquid or gas prior to entrance to the gas turbine). Test of gas turbines with water or steam injection for emission control and/or power augmentation are included. The tests can be applied to gas turbines in combined-cycle power plants or with other heat recovery systems.

Meetings should be held with all parties concerned as to how the test will be conducted and an uncertainty analysis should be performed prior to the test. The overall test uncertainty will vary because of the differences in the scope of supply, fuel(s) used, and driven equipment characteristics. The code establishes a limit for the uncertainty of each measurement required; the overall uncertainty is then calculated in accordance with the procedures defined in the code and by ASME PTC 19.1.

ASME, Performance Test Code on Gas Turbine Heat Recovery Steam Generators, ASME PTC 4.4 1981, American Society of Mechanical Engineers Reaffirmed 1992

The purpose of this code is to establish procedures for the conduct and report of tests of heat recovery steam generators (HRSG) employed in combined cycle installations. Combined cycle, as defined in the code, is a gas turbine exhausting into an HRSG, which may or may not be arranged for supplemental firing. This code provides standard test procedures, which will yield results having the highest level of accuracy consistent with current engineering knowledge and practice.

The purposes of testing under this code are the determination of (a) efficiency, or effectiveness, at specified operating conditions; (b) capacity at specified operating conditions; (c) other related operating characteristics such as steam temperature and control range, inlet gas flow and temperature; pressure drops in combustion air, gas, steam and water circuits; quality and/or purity of steam; and air and bypass stack gas leakage.

Further testing can be carried out to check the effectiveness, capacity, temperatures and flows for (a) checking the actual performance against guarantee; (b) comparing these items with a standard of operation; (c) comparing different conditions or methods of operation; (d) determining the specific performance of individual parts or sections of the HRSG unit; (e) comparing performance when firing different fuels; (f) determining the effects of changes to equipment.

The code applies to HRSG units employed in combined cycle installations. Units operating that is units operating with approximately 40% excess air. Units operating with less than approximately 20% excess air should be tested in accordance with Performance Test Code PTC-4.1, Steam Generating Units. For units operating between 20% and 40% excess air, guidance as to the preferred test method is also given in the appendix of the code.

ASME, Performance Test Code on Steam Turbines, ASME PTC 6 1996

The code provides procedures for the accurate testing of steam turbines. It is recommended for use in conducting acceptance tests of steam turbines and for any other situation in which performance levels must be determined with minimum uncertainty.

The performance parameters, which are addressed are:

1. heat rate
2. generator output
3. steam flow
4. steam rate
5. feedwater flow.

This code applies to steam turbines operating either with a significant amount of superheat in the initial steam such as the steam turbines in a combined cycle power plant. It also contains procedures and techniques required to determine enthalpy values within the moisture region and modifications necessary to permit testing.

This code contains rules and procedures for the conduct and reporting of steam turbine testing, including mandatory requirements for pretest arrangements, instruments to be employed, their application and methods of measurement, testing techniques, and methods of calculation of test results. This then establishes a limit for the uncertainty of each measurement required; the overall uncertainty is then calculated in accordance with the procedures defined in the code and by ASME PTC 19.1.

ASME, Performance Test Code on Steam Condensing Apparatus, ASME PTC 12.2 1983

This Code provides rules for determining the performance of a condenser with regard to one or more of the following:

1. The absolute pressure the apparatus will maintain at the steam inlet nozzle when transferring heat rejected by the prime mover at a given rate in Btu/hr kJ/hr with a given flow and temperature of circulating water, and a given tube cleanliness factor.
2. The thermal transmittance of surface condensers for given operating conditions. A test method for determining, concurrent with other measurements, the degree of tube fouling, which is expressed as a cleanliness factor or fouling resistance, is described in the code.
3. The amount of undercooling of the condensate.
4. The amount of dissolved oxygen in the condensate.

Testing of the condenser, as per the code, does not require testing for the cleanliness factor. Parties to the test could agree to some assumed degree of cleanliness after cleaning by a method agreed upon by all the parties involved.

Failing an agreement on this point prior to the test will require determination of the cleanliness factor.

ASME, Performance Test Code on Atmospheric Water Cooling Equipment PTC 23, 1997

The objectives of this code is to outline procedures for the accurate testing and evaluation of the performance of mechanical and natural draft cooling towers. Under good, stable testing conditions and following the rules established in PTC 19.1, uncertainties of less 2.5% can be obtained.

The code includes procedures to be used and the instrumentation required to obtain the following measurements: (1) circulating water flow entering the tower, (2) hot water temperature to the tower, (3) cold water temperature leaving the tower, (4) wet bulb temperature of the air entering the tower, (5) dry bulb temperature of the air entering the tower, (6) wet bulb temperature of ambient air, (7) dry bulb temperature of ambient air, (8) makeup water flow, (9) makeup water temperature, (10) blowdown water flow, (11) blowdown water temperature, (12) fan-driver power, (13) wind velocity, (14) atmospheric pressure, (15) water analysis, (16) tower pumping head, (17) air flow through the tower, (18) exhaust air wet bulb temperature, (19) exhaust air dry bulb temperature, (20) measurement of drift losses and (21) sound level.

ISO, Natural Gas — Calculation of Calorific Value, Density and Relative Density International Organization for Standardization ISO 6976-1983(E)

This international standard specifies methods for the calculation of calorific value, both higher and lower heating values, density, and relative density of natural gas when values for the physical properties of the pure components and the composition of the gas by mole fraction are known. The standard also describes the determination of the precision of the calculated calorific value from the precision of the method of analysis.

Table of Physical Constants of Paraffin Hydrocarbons and other Components of Natural Gas — Gas Producers Association Standard 2145-94

The objective of the standard is to provide the gas processing industry and natural gas users a convenient compilation of authoritative numerical values for the paraffin hydrocarbons and other compounds occurring in natural gas and natural gas liquids. The physical properties selected are those considered to be the most needed by engineers in analytical computations in gas plants and engines.

Mechanical Parameters

Some of the best standards from a mechanical point of view have been written by the American Petroleum Institute (API), as part of their Mechanical Equipment Standards. These Mechanical Equipment Standards are an aid in specifying and selecting equipment for general petrochemical use; however, they are general in nature and can be used as guide lines for power plants. The intent of these specifications is to facilitate the development of high-quality equipment with a high degree of safety and standardization. The user's problems and experience in the field are considered in writing these specifications. The task force, which writes the specifications, consists of members from the user, the contractor, and the manufacturers. Thus, the task-force team brings together both experience and know-how, especially from the operation and maintenance point of view. The petroleum industry is one of the largest users of cogeneration power. Thus, the specifications written are well suited for this industry, and the tips of operation and maintenance apply for all industries. This section deals with some of the applicable API and ASME standards for the gas turbine, steam turbine, HRSG and other various associated pieces of equipment as used in a combined cycle or cogeneration mode for power plants.

The combination of the extensive ASME and API codes covers the gas turbine, which is the center piece of the combined cycle power plant.

API Std 616, Gas Turbines for the Petroleum, Chemical and Gas Industry Services, Fourth Edition, August 1998

This standard covers the minimum requirements for open, simple and regenerative-cycle combustion gas turbine units for services of mechanical drive, generator drive, or process gas generation. All auxiliary equipment required for starting and controlling gas turbine units and for turbine protection is either discussed directly in this standard or referred to in this standard through references to other publications. Specifically, gas turbine units that are capable of continuous service firing gas or liquid fuel or both are covered by this standard. In conjunction with the API specifications, the following ASME codes also supply significant data in the proper selection of the gas turbine.

ASME Basic Gas Turbines B 133.2 Published: 1977 (Reaffirmed year: 1997). This standard presents and describes features that are desirable for the user to specify in order to select a gas turbine that will yield satisfactory performance, availability and reliability. The standard is limited to a consideration of the basic gas turbine including the compressor, combustion system and turbine.

ASME Gas Turbine Fuels B 133.7M Published: 1985 (Reaffirmed year: 1992). Gas turbines may be designed to burn either gaseous or liquid fuels, or both with or without changeover while under load. This standard covers both types of fuel.

ASME Gas Turbine Control and Protection Systems B133.4 Published: 1978 (Reaffirmed year: 1997). The intent of this standard is to cover the normal requirements of the majority of applications, recognizing that economic trade-offs and reliability implications may differ in some applications. The user may desire to add, delete or modify the requirements in this standard to meet his specific needs, and he has the option of doing so in his own bid specification. The gas turbine control system shall include sequencing, control, protection and operator information which shall provide for orderly and safe startup of gas turbine, control of proper loading and an orderly shutdown procedure. It shall include an emergency shutdown capability which can be operated automatically by suitable failure detectors or which can be operated manually. Coordination between gas turbine control and driven equipment must be provided for startup, operation and shutdown.

ASME Gas Turbine Installation Sound Emissions B133.8 Published: 1977 (Reaffirmed: 1989). This standard gives methods and procedures for specifying the sound emissions of gas turbine installations for industrial, pipeline, and utility applications. Included are practices for making field sound measurements and for reporting field data. This standard can be used by users and manufacturers to write specifications for procurement, and to determine compliance with specification after installation. Information is included, for guidance, to determine expected community reaction to noise.

ASME Measurement of Exhaust Emissions from Stationary Gas Turbine Engines B133.9 Published: 1994. This standard provides guidance in the measurement of exhaust emissions for the emissions performance testing (source testing) of stationary gas turbines. Source testing is required to meet federal state, and local environmental regulations. The standard is not intended for use in continuous emissions monitoring although many of the online measurement methods defined may be used in both applications. This standard applies to engines that operate on natural gas and liquid distillate fuels. Much of this standard also will apply to engines operated on special fuels such as alcohol, coal gas, residual oil, or process gas or liquid. However, these methods may require modification or be supplemented to account for the measurement of exhaust components resulting from the use of a special fuel.

ASME Procurement Standard for Gas Turbine Electrical Equipment B133.5 Published: 1978 (Reaffirmed year: 1997). The aim of this standard is to provide guidelines and criteria for specifying electrical equipment, other than controls, which may be supplied with a gas turbine. Much of the electrical equipment will apply only to larger generator drive installations, but where applicable, this standard can be used for other gas turbine drives. Electrical equipment described here, in almost all cases, is covered by standards, guidelines, or recommended practices documented elsewhere. This standard is intended to supplement those references and point out the specific areas of interest for a gas turbine application. For a few of the individual items, no other standard is referenced for the entire subject, but where applicable, a standard is referenced for a sub-item. A user is advised to employ this and other more-detailed standards to improve his specification for a gas turbine installation. In addition, regulatory requirements such as OSHA and local codes should be considered in completing the final specification. Gas turbine electrical equipment covered by this standard is divided into four major areas: main power system, auxiliary power system, direct current system, and relaying. The main power system includes all electrical equipment from the generator neutral grounding connection up to the main power transformer or bus but not including a main transformer or bus. The auxiliary power system is the gas turbine section AC supply and includes all equipment necessary to provide such station power as well as motors utilizing electrical power. The DC system includes the battery and charger only. Relaying is confined to electric system protective relaying that is used for protection of the gas turbine station itself.

ASME Procurement Standard for Gas Turbine Auxiliary Equipment B133.3 Published: 1981 (Reaffirmed year: 1994). The purpose of this Standard is to provide guidance to facilitate the preparation of gas turbine procurement specifications. It is intended for use with gas turbines for industrial, marine, and electric power applications. The standard also covers auxiliary systems such as lubrication, cooling, fuel (but not its control), atomizing, starting, heating-ventilating, fire protection, cleaning, inlet, exhaust, enclosures, couplings, gears, piping, mounting, painting, and water and steam injection.

API Std 618, Reciprocating Compressors for Petroleum, Chemical and Gas Industry Services, Fourth Edition, June 1995

This standard could be adapted to the fuel compressor for the natural gas to be brought up to the injection pressure required for the gas turbine. It covers the minimum requirements for reciprocating compressors and their drivers used in petroleum, chemical, and gas industry services for handling process air or gas with either lubricated or non-lubricated cylinders. Compressors covered by this standard are of moderate-to-low speed and in critical services. The non-lubricated cylinders types of compressors are used for injecting of the fuel in gas turbines at the high pressure needed. Also covered are related lubricating systems, controls, instrumentation, intercoolers, after-coolers, pulsation suppression devices, and other auxiliary equipment.

API Std 619, Rotary-Type Positive Displacement Compressors for Petroleum, Chemical, and Gas Industry Services, Third Edition, June 1997

The dry helical lobe rotary compressors non-lubricated cylinders types of compressors are used for injecting the fuel in gas turbines at the high pressure needed. The gas turbine application requires that the compressor be dry. This standard is primarily intended for compressors that are in special purpose applications, and covers the minimum requirements for dry helical lobe rotary compressors used for vacuum or pressure or both in petroleum, chemical, and gas industry services. This edition also includes a new inspector's checklist and new schematics for general purpose and typical oil systems.

API Std 613 Special Purpose Gear Units for Petroleum, Chemical and Gas Industry Services, Fourth Edition, June 1995

Gears wherever used can be a major source of problem and downtime. This standard specifies the minimum requirements for special-purpose, enclosed, precision, single- and double-helical one- and two-stage speed increasers and reducers of parallel-shaft design for refinery services. This standard is primarily intended for gears that are in continuous service without installed spare equipment. This standard applies for gears used in the power industry.

API Std 677, General-Purpose Gear Units for Petroleum, Chemical and Gas Industry Services, Second Edition, July 1997, Reaffirmed March 2000

This standard covers the minimum requirements for general-purpose, enclosed single- and multi-stage gear units incorporating parallel-shaft helical and right angle spiral bevel gears for the petroleum, chemical, and gas industries. Gears manufactured according to this standard are limited to the following pitchline velocities: helical gears shall not exceed 12,000 ft/min (60 m/sec) and spiral bevel gears shall not exceed 8000 ft/min (40 m/sec). This standard includes related lubricating systems, instrumentation, and other auxiliary equipment. Also included in this edition is new material related to gear inspection.

API Std 614, Lubrication, Shaft-Sealing, and Control-Oil Systems and Auxiliaries for Petroleum, Chemical and Gas Industry Services, Fourth Edition, April 1999

Lubrication besides providing lubrication also provides cooling for various components of the turbine. This standard covers the minimum requirements for lubrication systems, oil-type shaft-sealing systems, and control-oil systems for special-purpose applications. Such systems may serve compressors, gears, pumps, and drivers. The standard includes the systems' components, along with the required controls and instrumentation. Data sheets and typical schematics of both system components and complete systems are also provided. Chapters include General Requirements, Special Purpose Oil Systems, General Purpose Oil Systems and Dry Gas Seal Module Systems. This standard is well written and the tips detailed are good practices for all type of systems.

ANSI/API Std 610 Centrifugal Pumps for Petroleum, Heavy Duty Chemical and Gas Industry Services, Eight Edition, August 1995 (-1995)

The centrifugal pumps are used for boiler feed water pumps, and condensate pumps in a combined cycle power plant. The vertical pumps are also used for sump pumps and for water for cooling purpose in the condenser. The pump types covered by this standard can be broadly classified as overhung, between bearings, and vertically suspended.

API Publication 534 Heat Recovery Steam Generators, First Edition, January 1995

Provides the guidelines for the selection or evaluation of heat recovery steam generator (HRSG) systems. Details of related equipment designs are considered only where they interact with the HRSG system design. The document does not provide rules for design but indicates areas that need attention and offers information and description of HRSG types available to the designer/user for purposes of selecting the appropriate HRSG.

API RP 556 Fired Heaters & Steam Generators, First Edition, May 1997

RP 556 was written to aid in the installation of the more generally used measuring, control, and analytical instruments, transmission systems, and related accessories to achieve safe, continuous, accurate, and efficient operation with minimum maintenance. Although the information has been prepared primarily for petroleum refineries, much of it is applicable without change in combined cycle power plants.

ISO 10436:1993 Petroleum and Natural Gas Industries — General Purpose Steam Turbine for Refinery Service, First Edition

Specifies the minimum requirements for general-purpose steam turbines for use in petroleum refinery service. The requirements include basic design, materials, related lubrication systems, controls, auxiliary equipment and accessories.

API Std 671, Special Purpose Couplings for Petroleum Chemical and Gas Industry Services, Third Edition, October 1998

This standard covers the minimum requirements for special purpose couplings intended to transmit power between the rotating shafts of two pieces of refinery equipment. These couplings are designed to accommodate parallel offset, angular misalignment, and axial displacement of the shafts without imposing excessive mechanical loading on the coupled equipment.

ANSI/API Std 670 Vibration, Axial-Position, and Bearing-Temperature Monitoring Systems, Third Edition, November 1993

Provides a purchase specification to facilitate the manufacture, procurement, installation, and testing of vibration, axial position, and bearing temperature monitoring systems for petroleum, chemical, and gas industry services. Covers the minimum requirements for monitoring radial shaft vibration, casing vibration, shaft axial position, and bearing temperatures. It outlines a standardized monitoring system and covers requirements for hardware (sensors and instruments), installation, testing, and arrangement. Standard 678 has been incorporated into this edition of Standard 670. This is a well-documented and standard and widely used in all industries.

API Std 672, Packaged, Integrally Geared Centrifugal Air Compressors for Petroleum, Chemical, and Gas Industry Services, Third Edition, September 1996

This specification covers the air compressors for instrument air for the plant. The standard establishes the minimum requirements for constant-speed, packaged, integrally geared centrifugal air compressors, including their accessories. It may be applied for gas services other than air that are non-hazardous and non-toxic. This standard is not applicable to machines that develop a pressure rise of less than 5.0 psi (0.35 Bar) above atmospheric pressure.

API Std 681, Liquid Ring Vacuum Pumps and Compressors, First Edition, February 1996

Defines the minimum requirements for the basic design, inspection, testing, and preparation for shipment of liquid ring vacuum pump and compressor systems for service in the petroleum, chemical, and gas industries. It includes both vacuum pump and compressor design and system design.

It is not the intent here to detail the API or ASME standards but to discuss some of the pertinent points of these standards and other available options. It is strongly recommended that the reader obtain from the API all mechanical equipment standards. A more in-depth study of the major components and the standards outlined above that affect the combined cycle power plant's operation are presented next.

Gas Turbine

The API Standard 616, Gas Turbines for the Petroleum, Chemical and Gas Industry Services is intended to cover the minimum specifications necessary to maintain a high degree of reliability in an open-cycle gas turbine used in refinery service for mechanical drive, generator drive, or hot-gas generation. The standard also covers the necessary auxiliary requirements directly or indirectly by referring to other listed standards.

The standard defines terms used in the industry and describes the basic design of the unit. It deals with the casing, rotors and shafts, wheels and blades, combustors, seals, bearings, critical speeds, pipe connections and auxiliary piping, mounting plates, weather-proofing, and acoustical treatment.

The specifications call preferably for a two-bearing construction. Two-bearing construction is desirable in single-shaft units, as a three-bearing configuration can cause considerable trouble, especially when the center bearing in the hot zone develops alignment problems. The preferable casing is a horizontally split unit with easy visual access to the compressor and turbine, permitting field balancing planes without removal of the major casing components. The stationary blades should be easily removable without removing the rotor.

A requirement of the standards is that the fundamental natural frequency of the blade should be at least two times the maximum continuous speed and at least 10% away from the passing frequencies of any stationary parts. Experience has shown that the natural frequency should be at least four times the maximum continuous speed. Care should be exercised on units where there is a great change in the number of blades between stages.

A controversial requirement of the specifications is that rotating blades or labyrinths for shrouded rotating blades be designed for slight rubbing. A slight rubbing of the labyrinths is usually acceptable, but excessive rubbing can lead to major problems. New gas turbines use "squealer blades"; some manufacturers suggest using ceramic tips, but whatever is done, great care should be exercised or blade failure and housing damage may occur.

Labyrinth seals should be used at all external points, and sealing pressures should be kept close to atmosphere. The bearings can be either rolling element bearings usually used in aero-derivative gas turbines and hydrodynamic bearings used in the heavier frame type gas turbines. In the area of hydrodynamics bearings, tilting pad bearings are recommended since they are less susceptible to oil whirl and can better handle misalignment problems.

Critical speeds of a turbine operating below its first critical should be at least 20% above the operating speed range. The term commonly used for units operating below their first critical is that the unit has a "stiff shaft", while units operating above their first critical are said to have a "flexible shaft". There are many exciting frequencies that need to be considered in a turbine. Some of the sources that provide excitation in a turbine system are:

1. Rotor Unbalance
2. Whirling mechanisms such as:
    2.1. Oil Whirl
    2.2. Coulomb Whirl
    2.3. Aero-dynamic Cross Coupling Whirl
    2.4. Hydrodynamic Whirl
    2.5. Hysteretic Whirl
3. Blade and Vane Passing Frequencies
4. Gear Mesh frequencies
5. Misalignment
6. Flow separation in boundary layer exciting blades
7. Ball/race frequencies in anti-friction bearings usually used in Aero-derivative gas turbines

Torsional criticals should be at least 10% away from the first or second harmonics of the rotating frequency. Torsional excitations can be excited by some of the following:

1. Start-up conditions such as speed detents
2. Gear problems such as unbalance and pitch line runout
3. Fuel pulsation especially in low NO_x combustors.

The maximum unbalance is not to exceed 2.0 mils on rotors with speeds below 4000 rpm, 1.5 mils for speeds between 4000 and 8000 rpm, 1.0 mil for speeds between 8000 and 12,000 rpm, and 0.5 mils for speeds above 12,000 rpm. These requirements are to be met in any plane and also include shaft runout. The following relationship is specified by the API standard:

$$L_v = \sqrt{\frac{12000}{N}} \qquad (3-1)$$

where
 L_v = Vibration limit, mils (thousandth of an inch) or mm (mils × 25.4)
 N = Operating speed, rpm

The maximum unbalance per plane (journal) shall be given by the following relationships:

$$U_{max} = 4W/N \qquad (3-2)$$

where
 U_{max} = Residual unbalance, oz-in. (g-mm)
 W = Journal static weight, lb (kg)

A computation of the force on the bearings should be calculated to determine whether or not the maximum unbalance is an excessive force.

The concept of an amplification factor (AF) is introduced in the new API 616 standard. Amplification factor is defined as the ratio of the critical speed to the speed change at the root mean square of the critical amplitudes.

$$AF = \frac{N_{cl}}{(N_2 - N_1)} \qquad (3-3)$$

Figure 3-6 is an amplitude-speed curve showing the location of the running speed to the critical speed, and the amplitude increase near the critical speed. The amplification factor should be less than eight, preferably about five. This same concept can be applied to gas turbines.

Balancing requirement in the specifications require that the rotor with blades assembled must be dynamically balanced without the coupling but with the half key, if any, in place. The specifications do not discuss whether this balancing is to be done at high-speeds or low-speeds. The balancing conducted in most shops is at low-speed. A high-speed balancing should be used on problem shafts, and any units, which operate above the second critical. Field balancing requirements should be specified.

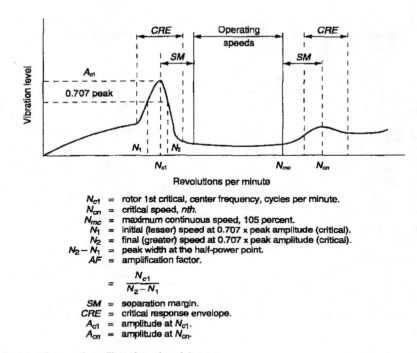

**Figure 3–6. Rotor Response Plot Chapter 3 [8]
(Courtesy of the American Petroleum Institute)**

The lubrication system for the turbine is described. This system closely follows the outline in API Standard 614, which is discussed in detail in Chapter 11. Separate lubrication systems for various sections of the turbine and driven equipment may be supplied. Many vendors and some manufacturers provide two separate lubrication systems: one for hot bearings and another for cool bearings. These and other lubrication systems should be detailed in the specifications.

The inlet and exhaust systems in gas turbines are described. The inlet and exhaust systems consist of an inlet filter, silencers, ducting, and expansion joints. The design of these systems can be critical to the overall design of a gas turbine. Proper filtration is a must, otherwise problems of blade contamination and erosion ensue. The standards are minimal for specifications, calling for a coarse metal screen to prevent debris from entering, a rain or snow shield for protection from the elements, and a differential pressure alarm. Most manufacturers are now suggesting so-called high-efficiency filters that have two stages of filtration, an inertia stage to remove particles above 5 μm followed by one or more filter screens, self cleaning filters, pad type pre filters, or a combination of them, to remove particles below 5 μm. Differential pressure alarms are provided by manufacturers, but the trend among users has been to ignore them. It is suggested that more attention be paid to differential pressure than in the past to assure high-efficiency operation.

Silencers are also minimally specified. Work in this area has progressed dramatically in the past few years with the NASA quiet engine program. There are some good silencers now available on the market, and inlets can be acoustically treated.

Starting equipment will vary, depending on the location of the unit, starting drives include electric motors, steam turbines, diesel engines, expansion turbines, and hydraulic motors. The sizing of a starting unit will depend on whether the unit is a single-shaft turbine or a multiple shaft turbine with a free-power turbine. The vendor is required to produce speed-torque curves of the turbine and driven equipment with the starting unit torque superimposed. In a free-power turbine design, the starting unit has to overcome only the torque to start the gas generator system. In a single-shaft turbine, the starting unit has to overcome the total torque. Turning gears are recommended in the specifications, especially on large units to avoid shaft bowing. They should always be turned on after the unit has been "brought down" and should be kept operational until the rotor is cooled.

The gears should meet API Standard 613. Gear units should be double-helical gears provided with thrust bearings. Load gears should be provided with a shaft extension to permit torsional vibration measurements. On high-speed gears, proper use of the lubricant as a coolant should be provided. Spraying oil as a coolant on the teeth and face of the units is recommended to prevent distortion.

Couplings should be designed to take the necessary casing and shaft expansion. Expansion is one reason for the wide acceptance of the dry flexible coupling. A flexible diaphragm coupling is more forgiving in angular alignment; however, a gear-type coupling is better for axial movement. Access for hot alignment checks must be provided. The couplings should be dynamically balanced independently of the rotor system.

Controls, instrumentation, and electrical systems in a gas turbine are defined. The outline in the standard is the least a user needs for safe operation of a unit. More details of the instrumentation and controls are given in Chapter 12.

The starting system can be manual, semi-automatic, or automatic, but in all cases should provide controlled acceleration to minimum governor speed and then, although not called for in the standards, to full-speed. Units, which do not have controlled acceleration to full-speed, have burned out first- and second-stage nozzles when combustion occurred in those areas instead of in the combustor. Purging the system of the fuel after a failed start is mandatory, even in the manual operation mode. Sufficient time for the purging of the system should be provided so that the volume of the entire exhaust system has been displaced at least five times.

Alarms should be provided on a gas turbine. The standards call for alarms to be provided to indicate malfunction of oil and fuel pressure, high exhaust temperature, high differential pressure across the air filter, excessive vibration levels, low oil reservoir levels, high differential pressure across oil filters, and high oil drain temperatures from the gearings shutdown occurs with low oil pressure, high exhaust temperature, and combustor flameout. It is recommended that shutdown also occur with high thrust bearing temperatures and with a temperature differential in the exhaust temperature. Vibration detectors suggested in the standards are non-contacting probes. Presently, most manufacturers provide velocity transducers mounted on the casing, but these are inadequate. A combination of non-contacting probes and accelerometers are needed to ensure the smooth operation and diagnostic capabilities of the unit.

Fuel systems can cause many problems, and fuel nozzles are especially susceptible to trouble. A gaseous fuel system consists of fuel filters, regulators, and gauges. Fuel is injected at a pressure of about 60 psi (4 Bar) above the compressor discharge pressure for which a gas compression system is needed. Knockout drums or centrifuges are recommended and should be implemented to ensure no liquid carry-overs in the gaseous system.

Liquid fuels require atomization and treatment to inhibit sodium and vanadium content. Liquid fuels can drastically reduce the life of a unit if not properly treated. More details are given in Chapter 10.

Recommended materials are outlined in the standards. Some of the recommendations in the standard are carbon steel for base plates, heat-treated forged steel for compressor wheels, heat-treated forged alloy steel for turbine wheels, and forged steel for couplings. The growth of materials technology has been so rapid especially in the area of high temperature materials, which the standard does not deal with it. Details of some of the materials technology of the high temperature alloys and single crystal blades are dealt with in Chapter 4.

The turbine undergoes three basic tests; these are hydrostatic, mechanical, and performance. Hydrostatic tests are to be conducted on pressure-containing parts with water at least one-and-a-half times the maximum operating pressure. The mechanical run tests are to be conducted for at least a period of 4 hours at maximum continuous speed. This test is usually done at no-load conditions. It checks out the bearing performance and vibration levels as well as overall mechanical operability. It is suggested that the user have a representative at this test to tape record as much of the data as possible. The data is helpful in further evaluation of the unit or can be used as base-line data. Performance tests should be conducted at maximum power with normal fuel composition. The tests should be conducted in accordance with **ASME PTC-22**, which is described in more detail in Chapter 13.

Gears

This standard API Standard 613 covers special purpose gears. They are defined as gears, which have either or both actual pinion speeds of more than 2900 rpm and pitchline velocities of more than 5000 ft/min (27 m/sec). The standard applies to helical gears employed in speed-reducer or speed-increaser units.

The scope and terms used are well defined and includes a listing of standards and codes for reference. The purchaser is required to make decisions regarding gear-rated horsepower and rated input and output speeds.

This standard includes basic design information and is related to AGMA Standard 421. Specifications for cooling water systems are given as well as information about shaft assembly designation and shaft rotation. Gear-rated power is the maximum power capability of the driver. Normally, the horsepower rating for gear units between a driver and a driven unit would be 110% of the maximum power required by the driven unit or 110% of the maximum power of the driver, whichever is greater.

The tooth pitting index or K factor is defined as

$$K = \frac{W_t}{Fd} \times \frac{(R+1)}{R} \qquad (3-4)$$

where
W_t = transmitted tangential load, lbs (at the operating pitch diameter)
$W_t = \dfrac{12600 \times \text{Gear rated horse power}}{\text{Pinion rpm} \times d}$
F = net face width, in.
d = pinion pitch diameter, in.
R = ratio (number teeth in gear divided by number teeth in pinion)

The allowable K factor is given by

$$\text{Allowable } K = \text{Material index number/Service Factor} \qquad (3-5)$$

Service factors and material index number tabulation are provided for various typical applications, allowing the determination of the K factor. Gear tooth size and geometry are selected so that bending stresses do not exceed certain limits. The bending stress number is given by:

$$S_t = \text{bending Stress Number} = \frac{W_t \times P_{nd}}{F} \times (SF) \times \frac{1.8 \cos \varphi}{J} = \qquad (3-6)$$

where
W_t = as defined in Eq. (4–4)
P_{nd} = normal diametral pitch
F = net face width, in.
φ = helix angle
J = geometry factor (from AGMA 226)
SF = service factor

Design parameters on casings, joint supports, and bolting methods. Some service and size criteria are included.

Critical speeds correspond to the natural frequencies of the gears and the rotor bearings support system. A determination of the critical speed is made by knowing the natural frequency of the system and the forcing function. Typical forcing functions are caused by rotor unbalance, oil filters, misalignment, and a synchronous whirl.

Gear elements must be multiplane and dynamically balanced. Where keys are used in couplings, half keys must be in place. The maximum allowable unbalanced force at maximum continuous speed should not exceed 10% of static weight load on the journal. The maximum allowable residual unbalance in the plane of each journal is calculated using the following relationship:

$$F = \dot{m}rw^2 \qquad (3-7)$$

Since the force must not exceed 10% of the static journal load,

$$\dot{m}r = \frac{0.1W}{\varpi^2} \qquad (3-8)$$

Taking the correction constants, the equation can be written as:

$$\text{Max. unbalanced forced} = \frac{56,347 \times \text{Journal static weight load}}{(\text{rpm})^2} \qquad (3-9)$$

The double amplitude of unfiltered vibration in any plane measured on the shaft adjacent to each radial bearing is not to exceed 2.0 mils (0.05 mm) or the value given by:

$$\text{Amplitude} = \sqrt{\frac{12000}{\text{rpm}}} \qquad (3-10)$$

where rpm is the maximum continuous speed. It is more meaningful for gears to be instrumented using accelerometers. Design specifications for bearings, seals, and lubrication are also given.

Accessories such as couplings, coupling guards, mounting plates, piping, instrumentation, and controls are described. Inspection and testing procedures are detailed. The purchaser is allowed to inspect the equipment during manufacture after notifying the vendor. All welds in rotating parts must receive 100% inspection. To conduct a mechanical run test, the unit must be operated at maximum continuous speed until bearing and lube oil temperatures have stabilized. Then, the speed is increased to 110% of maximum continuous speed and run for 4 hours.

Lubrication Systems

This API Standard 614 standard covers the minimum requirements for lubrication systems, oil shaft sealing systems, and related control systems for special purpose applications. The terms are fully defined, references are well documented and basic design is described.

Lubrication systems should be designed to meet continuously all conditions for a nonstop operation of three years. Typical lubricants should be hydrocarbon oils with approximate viscosities of 150 SUS at 100°F (37.8°C). Oil reservoirs should be sealed to prevent the entrance of dirt and water and sloped at the bottom to facilitate drainage. The reservoir working capacity should be sufficient for at least a 5-min flow. The oil system should include a main oil pump and a standby oil pump. Each pump must have its own driver sized according to API Standard 610. Pump capacities should be based on the systems' maximum usage plus a minimum of 15%. For seal oil systems, the pump capacity should be maximum capacity plus 20% or 10 gpm, whichever is greater. The standby oil pump should have an automatic startup control to maintain safe operation if the main pump fails. Twin oil coolers should be provided, and each should be sized to accommodate the total cooling load. Full-flow twin oil filters should be furnished downstream of the coolers. Filtration should be 10 μm nominal. The pressure drop for clean filters should not exceed 5 psi at 100°F operating temperature during normal flow.

Overhead tanks, purifiers, and degasing drums are covered. All pipe welding is to be done according to Section IX of the ASME code, and all piping must be seamless carbon steel, minimum schedule 80 for sizes 1 1/2 in. (38.1 mm) and smaller, and a minimum of schedule 40 for pipe sizes 2 in. (50.8 mm) or greater.

The lubrication control system should enable orderly startup, stable operation, warning of abnormal conditions, and shutdown of main equipment in the event of impending damage. A list of required alarm and shutdown devices is provided. The purchaser has the right to inspect the work and testing of sub-components if he informs the vendor in advance. Each cooler, filter, accumulator, and other pressure vessels should be hydrostatically tested at one and one-half times design pressure. Cooling water jackets and other water-handling components should be tested at one and one-half times design pressure. The test pressure should not be less than 115 psig (7.9 Bar). Tests should be maintained for durations of at least 30 minutes.

Operational tests should:

1. Detect and correct all leaks
2. Determine relief pressures and check for proper operation of each relief valve
3. Accomplish a filter cooler changeover without causing startup of the standby pump
4. Demonstrate the control valves have suitable capacity, response, and stability
5. Demonstrate the oil pressure control valve can control oil pressure.

Vibration Measurements

The API Standard 670 covers the minimum requirements for non-contacting vibration in an axial-position monitoring system.

The accuracy for the vibration channels should meet a linearity of ±5% of 200 mV/mil sensitivity over a minimum operating range of 80 mils. For the axial position, the channel linearity must be ±5% of 200 mV/mil sensitivity and a ±1.0 mil of a straight line over a minimum operating range of 80 mils. Temperature should not affect the linearity of the system by more than 5% over a temperature range of -30°F to 350°F (-34.4°C to 176.7°C) for the probe and extension cable. The oscillator demodulator is a signal-conditioning device powered by -24 V of DC. It sends a radio frequency signal to the probe and demodulates the probe output. It should maintain linearity over the temperature range of -30°F to 150°F (-34.4°C to 65.6°C). The monitors and power supply should maintain their linearity over a temperature range of -20°F to 150°F (-28.9°C to 65.6°C). The probes, cables, oscillator demodulators, and power supplies installed on a single train should be physically and electrically interchangeable.

The non-contacting vibration and axial position monitoring system consists of probe, cables, connectors, oscillator demodulator, power supply, and monitors. The probe tip diameters should be 0.190 to 0.195 in. (4.8 to 4.95 mm) with body diameters of $1/4$ (6.35 mm) - 28 UNF - 2A threaded, or 0.3 to 0.312 in. (7.62 to 7.92 mm) with a body diameter of 3/8 (9.52 mm) - 24 UNF - 24A threaded. The probe length is about 1-in. long. Tests conducted on various manufacturer's probes indicate that the 0.3 to 0.312-in. (7.62 to 7.92 mm) probe has a better linearity in most cases. The integral probe cables have a cover of tetra-flouroethylene, a flexible stainless steel armoring which extends to within 4 in. of the connector. The overall physical length should be approximately 36 in. (914.4 mm) measured from probe tip to the end of the connector. The electrical length of the probe and integral cable should be 6 ft. The extension cables should be coaxial with electrical and physical lengths of 108 in. (2743.2 mm). The oscillator demodulator will operate with a standard supply voltage of -24 V DC and will be calibrated for a standard electrical length of 15 ft (5 m). This length corresponds to the probe integral cable and extension. Monitors should operate from a power supply of 117 V ±5% with the linearity requirements specified. False shutdown from power interruption will be prevented regardless of mode or duration. Power supply failure should actuate an alarm.

The radial transducers should be placed within 3 in. of the bearing, and there should be two radial transducers at each bearing. Care should be taken not to place the probe at the nodal points. The two probes should be mounted 90 deg. apart (±5 deg.) at a 45 deg. (±5 deg.) angle from each side of the vertical center. Viewed from the drive end of the machine train, the x probe will be on the right side of the vertical, and the y probe will be on the left side of the vertical.

The axial transducers should have one probe sensing the shaft itself within 12 in. (305 mm) of the active surface of the thrust collar with the other probe sensing the machined surface of the thrust collar. The probes should be mounted facing in opposite directions. Temperature probes embedded in the bearings are often more useful in preventing thrust-bearing failures than the proximity probe. This is because of expansion of the shaft casing and the probability that the probe is located far from the thrust collar.

When designing a system for thrust bearing protection, it is necessary to monitor small changes in rotor axial movement equal to oil film thickness. Probe system accuracy and probe mounting must be carefully analyzed to minimize temperature drift. Drift from temperature changes can be unacceptably high.

A functional alternative to the use of proximity probes for bearing protection is bearing temperature, bearing temperature rise (bearing temperature minus bearing oil temperature), and rate of change in bearing temperature. A matrix combining these functions can produce a positive indication of bearing distress.

A phase angle transducer should also be supplied with each train. This transducer should record one event per revolution. Where intervening gearboxes are used, a mark and phase angle transducer should be provided for each different rotational speed.

Specifications

The previous API standards are guidelines to information regarding machine train applications. The more pertinent the information obtained during the evaluation of the proposal, the better the selection for the problem. The following list contains items the user should consider in his attempt to properly evaluate the bid. Some of these points are covered in the API standards.

Table 3-6 indicates the main points an engineer must consider in evaluating different gas turbine units, while the important points to consider in evaluating a steam turbine are listed in Table 3-7. These tables will enable the engineer to make a proper evaluation of each critical point and ensure that he is purchasing units of high reliability and efficiency.

Table 3-6. Point to Consider in a Gas Turbine

1.	Type of Turbine	
	• Aero-derivative	
	• Frame Type	
2.	Type of Fuel	
3.	Type of Compressor	
4.	No. of stages and pressure ratio	
5.	Types of blades, blade attachment, and wheel attachment	
6.	No. of bearings	
7.	Type of bearings	
8.	Type of thrust bearings	
9.	Critical speed	
10.	Torsional criticals	
11.	Campbell diagrams	
12.	Balance planes	
13.	Balance pistons	
14.	Type of combustor	
15.	Wet and dry combustors	
16.	Types of Fuel nozzles	
17.	Transition pieces	
18.	Type of turbine	
19.	Power Transmission curvic coupling	
20.	No. of stages	
21.	Free-power turbine	
22.	Turbine inlet temperature	
23.	Type of fuels	
24.	Fuel additives	
25.	Types of couplings	
26.	Alignment data	
27.	Exhaust diffuser	
28.	Performance map of turbine and compressor	
29.	Gearing	
30.	Drawings	
Accessories		
1.	Lubrication Systems	
2.	Intercoolers	
3.	Inlet filtration system	
4.	Control system	
5.	Protection system	

Table 3-7. Points to Consider in a Steam Turbine

1. Number of Sections
   - HP Section
   - IP Section
   - LP Section
2. Steam Characteristics (per casing)
   - Mass Flow
   - Temperature
   - Pressure
3. Type of seals (inner seal) and oil seals
4. Type of bearings (radial)
5. Bearing stiffness coefficients
6. Types of thrust bearings (Tapered land, non-equalizing tilting pad and Kingsbury)
7. Thrust float
8. Temperature for journal and thrust bearings (operating temperature)
9. Critical speed diagram (Speed vs. bearing stiffness curve)
10. Types of Rotors
    10.1. Shrouded or unshrouded Blades
    10.2. Blading Type (Impulse or Reaction)
    10.3. Attachment of blades to hub and shroud
11. Attachment of Rotor to shaft
    11.1. Shrink fit
    11.2. Key fit
    11.3. Curvic Couplings
12. Campbell diagrams of Rotors and Nozzles.
    12.1. No. of blades (impellers)
    12.2. No. of blades (diffuser)
    12.3. No. of blades (guide vanes)
13. Balance planes (location)
    13.1. How is it balanced (detail)
14. Weight of rotor (assembled)
15. Type of casing
    15.1. Split casing
16. Data on torsional vibration (Bending criticals)
17. Alignment data
18. Type of coupling between Units
19. Performance curves (separate casings)
    19.1. Temperature
    19.2. Pressure
    19.3. Efficiency
20. Condenser
    20.1. Back Pressure
    20.2. Pressure drop
    20.3. Effectiveness

Chapter 4

AN OVERVIEW OF GAS TURBINES

The gas turbine is a power plant that produces a great amount of energy for its size and weight. The gas turbine has found increasing service in the past 15 years in the power and petrochemical industry throughout the world. Its compactness, low weight and multiple fuel application make it a natural power plant in all applications, from power plants to offshore platforms. Today, there are gas turbines which run on natural gas, diesel fuel, naphtha, methane, crude, low-Btu gases, vaporized fuel oils, and even waste.

The last 20 years has seen a large growth in gas gurbine technology. The growth is spearheaded by the growth of materials technology, new coatings, and new cooling schemes. This, with the conjunction of increase in compressor pressure ratio, has increased the gas turbine thermal efficiency from about 15% to over 45%.

The advanced gas turbines are operating at very high pressure ratios and very high firing temperatures, ensuring high performance of power and efficiency. These turbines are pushing the envelope of technology in the areas of material science and aerodynamics to their limit. The new gas turbines are the basis of the growth of combined cycle power plant and will be the power for most of the first half of the new millennium. Exceeding efficiencies of over 45%, these turbines in a combined cycle mode reach plant efficiencies of nearly 60%. Since fuel costs are nearly 75% of the life cycle cost of a plant, these new advanced gas turbines are here to stay and will be in large demand.

The aerospace engines have been the leaders in most of the technology in the gas turbine. The design criteria for these engines was high reliability, high performance, with many starts and flexible operation throughout the flight envelope. The engine life of about 3500 hours between major overhauls was considered good. The aerospace engine performance has always been rated primarily on its thrust/weight ratio. Increase in engine thrust/weight ratio is achieved by the development of high-aspect ratio blades in the compressor, as well as optimizing the pressure ratio and firing temperature of the turbine for maximum work output per unit flow.

The industrial gas turbine has always emphasized long life, and this conservative approach has resulted in the industrial gas turbine in many aspects, giving up high performance for rugged operation. The industrial gas turbine has been conservative in the pressure ratio and the firing temperatures. This has all changed in the last 10 years; spurred on by the introduction of the "aero-derivative gas turbine", the industrial gas turbine has dramatically improved its performance in all operational aspects. This has resulted in dramatically reducing the performance gap between these two types of gas turbines. The gas turbine, to date in the combined cycle mode, is fast replacing the steam turbine as the base load provider of electrical power throughout the world. This is even true in Europe and the United States of America, where the large steam turbine is the only type of base load power in the fossil energy sector. The gas turbine from the 1960s to the late 1980s was used only as peaking power in those countries; it was used as base load mainly in the "developing countries", where the need of power was increasing rapidly, and that the wait of 3 to 6 years for a steam plant was unacceptable.

Figures 4-1 and 4-2 show the growth of the pressure ratio and firing temperature. The growth of both the pressure ratio and firing temperature parallel each other, as both growths are necessary in achieving the optimum thermal efficiency.

Gas turbines in the power industry can be classified into five broad groups:

1. Industrial heavy-duty gas turbines
2. Aircraft-derivative gas turbines
3. Medium-range gas turbines
4. Small gas turbines
5. Micro-turbines.

In the past, the gas turbine was perceived as a relatively inefficient power source when compared to other power sources. Its efficiencies were as low as 15% in the early 1950s; today, its efficiencies are in the 45% to 50% range. The limiting factor for most gas turbines has been the turbine inlet temperature. With new schemes of air-cooling and breakthroughs in blade metallurgy, higher turbine temperatures have been achieved. The new gas turbines have fired inlet temperatures as high as 2300°F (1260°C), and pressure ratios of 40:1 with efficiencies of 45% and above.

Some factors one must consider in deciding what type of power plant is best suited for the needs at hand are capital cost, time from planning to completion, maintenance costs and fuel costs. The gas turbine has the lowest maintenance and capital cost. It also has the fastest completion time to full operation of any other plant. Its disadvantage was its high heat rate, but this has been addressed and the new turbines are among the most efficient types of prime movers. The combination of plant cycles further increases the efficiencies to the low 60s.

The design of any gas turbine must meet essential criteria based on operational considerations. Chief among these criteria are:

1. High efficiency
2. High reliability and, thus, high availability
3. Ease of service
4. Ease of installation and commission
5. Conformance with environmental standards
6. Incorporation of auxiliary and control systems, which have a high degree of reliability
7. Flexibility to meet various service and fuel needs.

A look at each of these criteria will enable the user to get a better understanding of the requirements.

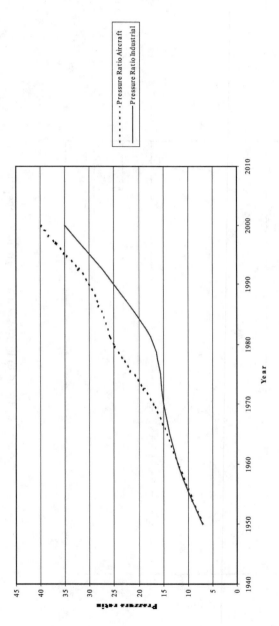

Figure 4-1. Development of Engine Pressure Ratio Over the Years

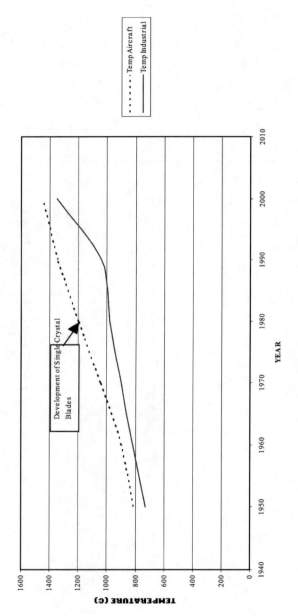

Figure 4-2. Trend in Improvement in Firing Temperature

The two factors which most affect high turbine efficiencies, are pressure ratios and temperature. The axial flow compressor, which produces the high-pressure gas in the turbine, has seen dramatic change as the gas turbine pressure ratio has increased from 7:1 to 40:1. The increase in pressure ratio increases the gas turbine thermal efficiency when accompanied with the increase in turbine firing temperature. Figure 4-3 shows the effect on the overall cycle efficiency of the increasing Pressure Ratio and the Firing Temperature. The increase in the pressure ratio increases the overall efficiency at a given temperature; however, increasing the pressure ratio beyond a certain value at any given firing temperature can actually result in lowering the overall cycle efficiency. It should also be noted that the very high pressure ratios tend to reduce the operating range of the turbine compressor. This causes the turbine compressor to be much more intolerant to dirt buildup in the inlet air filter and on the compressor blades, and creates large drops in cycle efficiency and performance. In some cases, it can lead to compressor surge, which in turn can lead to a flameout, or even serious damage and failure of the compressor blades and the radial and thrust bearings of the Gas Turbine.

The effect of temperature is very predominant — for every 100°F (55.5°C) increase in temperature, the work output increases approximately 10% and gives about a $1^1/_2\%$ increase in efficiency. Higher pressure ratios and turbine inlet temperatures improve efficiencies on the simple-cycle gas turbine. Figure 4-4 shows a simple cycle gas turbine performance map as a function of pressure ratio and turbine inlet temperature.

Another way to achieve higher efficiencies is with regenerators. Figure 4-5 shows the effects of pressure ratio and temperatures on efficiencies and work for a regenerative cycle. The effect of pressure ratio for this cycle is opposite to that experienced in the simple cycle. Regenerators can increase efficiency as much as 15% to 20% at today's operating temperatures. The optimum pressure ratios are about 20:1 for a regenerative system, compared to 40:1 for the simple-cycle at today's higher turbine inlet temperatures that approach 3000°F (1649°C).

The Availability of a gas turbine is the percent of time the gas turbine is available to generate power in any given period at its acceptance load. The Acceptance Load or the Net Established Capacity would be the net electric power generating capacity of the gas turbine at design or reference conditions established as result of the Performance Tests conducted for acceptance of the plant. The actual power produced by the gas turbine would be corrected to the design or reference conditions and is the actual net available capacity of the gas turbine. Thus it is necessary to calculate the effective forced outage hours which are based on the maximum load the plant can produce in a given time interval when the plant is unable to produce the power required of it. The effective forced outage hours is based on the following relationship:

$$EFH = HO \times \frac{(MW_d - MW_a)}{MW_d} \qquad (4-1a)$$

where:
$\quad MW_d$ = Desired Output corrected to the design or reference conditions. This must be equal to or less than the gas turbine load measured and

142 • COGENERATION AND COMBINED CYCLE POWER PLANTS

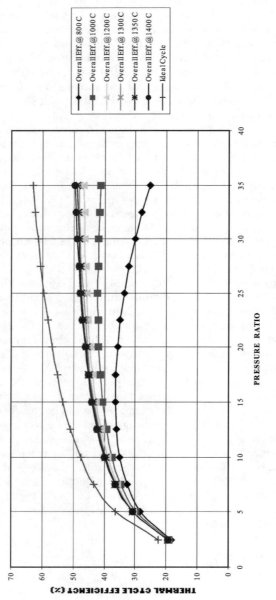

Figure 4-3. Overall Cycle Efficiency

An Overview of Gas Turbines • 143

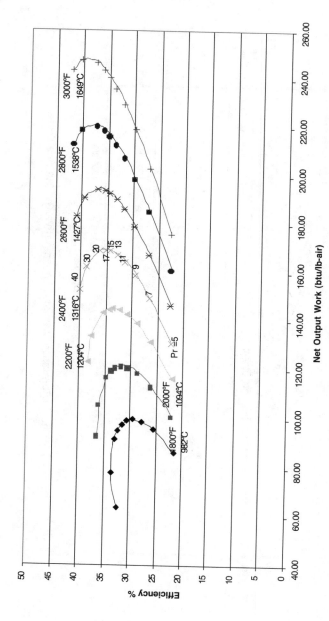

Figure 4-4. Performance Map of a Simple Cycle Gas Turbine

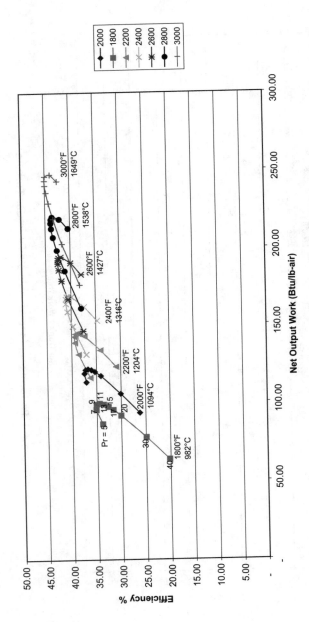

Figure 4-5. The Performance Map of a Regenerative Gas Turbine Cycle

corrected to the design or reference conditions at the acceptance test.
MW$_a$ = Actual maximum acceptance test produced and corrected to the design or reference conditions.
HO = Hours of operation at reduced load.

The Availability of a gas turbine can now be calculated by the following relationship, which takes into account the stoppage due to both forced and planed outages, as well as the forced effective outage hours:

$$A = \frac{(PT - PM - FO - EFH)}{PT} \quad (4-1b)$$

where:
PT = Time Period (8760 hrs/year)
PM = Planned Maintenance Hours
FO = Forced Outage Hours
EFH = Equivalent forced outage hours

The reliability of the gas turbine is the percentage of time between planed overhauls and is defined as:

$$R = \frac{(PT - FO - EFH)}{PT} \quad (4-2a)$$

Availability and Reliability have a very major impact on the plant economy. Reliability is essential in that when the power is needed it must be there. When the power is not available it must be generated or purchased and can be very costly in the operation of a plant. Planned outages are scheduled for non peak periods. Peak periods is when the majority of the income is generated as usually there are various tiers of pricing depending on the demand. Many power purchase agreements have clauses, which contain capacity payments, thus making plant availability critical in the economics of the plant. A 1% reduction in plant Availability could cost $500,000 in income on a 100MW plant.

Reliability of a plant depends on many parameters, such as the type of fuel, the preventive maintenance programs, the operating mode, the control systems, and the firing temperatures. Another very important factor in a gas turbine is the Starting Reliability (SR). This reliability is a clear understanding of the successful starts that have taken place and is given by the following relationship:

$$SR = \frac{number\ of\ starting\ successes}{(number\ of\ starting\ successes + number\ of\ starting\ failures)} \quad (4-2b)$$

To achieve a high availability and reliability factor, the designer must keep in mind many factors. Some of the more important considerations, which govern the design, are blade and shaft stresses, blade loadings, material integrity, auxiliary systems, and control systems. The high temperatures required for high efficiencies have a disastrous effect on turbine blade life. Proper cooling must be provided to achieve blade metal temperatures of about 1000°F (537°C), 1300°F (704°C) below the levels of the onset of hot corrosion. Thus, the right type of cooling systems with proper blade coatings and materials are needed to ensure the high reliability of a turbine.

Serviceability is an important part of any design, since fast turnarounds result in high availability to a turbine and reduces maintenance and operations costs. Service can be accomplished by providing proper checks, such as exhaust temperature monitoring, shaft vibration monitoring, and surge monitoring. Also, the designer should incorporate borescope ports for fast visual checks of hot parts in the system. Split casings for fast disassembly, field balancing ports for easy access to the balance planes, and combustor cans which can be easily disassembled without removing the entire hot section are some of the many ways that afford ease of service.

Ease of installation and commissioning is another reason for gas turbine use. A gas turbine unit can be tested and packaged at the factory. Use of a unit should be carefully planned so as to cause as few start cycles as possible. Frequent startups and shutdowns at commissioning greatly reduce the life of a unit.

Environmental considerations are critical in the design of any system. The system's impact on the environment must be within legal limits and, thus, must be addressed by the designer carefully. Combustors are the most critical component, and great care must be taken to design them to provide low smoke and low NO_x output. The high temperatures result in increasing the NO_x emissions from the gas turbines. This resulted in initially attacking the NO_x problem by injecting water or steam in the combustor producing a wet combustor. The next stage was the development of Dry Low NO_x Combustors. The development of new Dry Low NO_x Combustors has been a very critical component in reducing the NO_x output as the gas turbine firing temperature is increased. The new low NO_x combustors increase the number of fuel nozzle and the complexity of the control algorithms.

Lowering the inlet velocities and providing proper inlet silencers can reduce air noise. Considerable work by NASA on compressor casings has greatly reduced noise.

Auxiliary systems and control systems must be designed carefully, since they are often responsible for the downtime in many units. Lubrication systems, one of the critical auxiliary systems, must be designed with a backup system and should be as close to failure-proof as possible. The advanced gas turbines are all digitally controlled and incorporate on-line condition monitoring to some extent. The addition of new on-line monitoring requires new instrumentation. Control systems provide acceleration-time and temperature-time controls for startups, as well as controls various anti-surge valves. At operating speeds, they must regulate fuel supply and monitor vibrations, temperatures, and pressures throughout the entire range.

Flexibility of service and fuels are criteria which enhance a turbine system, but they are not necessary for every application. The energy shortage makes closer to its operating point and, thus, operate at higher efficiencies. This flexibility may entail a two-shaft design incorporating a power turbine, which is separate and not connected to the Gasifier unit. Multiple fuel applications are now in greater demand, especially where various fuels may be in shortage at different times of the year.

The previous criteria are some of the many criteria that designers must meet to design successful units.

Industrial Heavy-Duty Gas Turbines

These gas turbines were designed shortly after World War II and introduced to the market in the early 1950s. The early heavy-duty gas turbine design was largely an extension of steam turbine design. Restrictions of weight and space were not important factors for these ground-based units, and so the design characteristics included heavy-wall casings split on horizontal centerlines, sleeve bearings, large-diameter combustors, thick airfoil sections for blades and stators, and large frontal areas. The overall pressure ratio of these units varied from 5:1 for the earlier units to 35:1 for the units in present-day service. Turbine inlet temperatures have been increased and run as high as 2500°F (1371°C) on some of these units; this makes the gas turbine one of the most efficient prime mover on the market today, reaching efficiencies of 50%. Projected temperatures approach 3000°F (1649°C) and, if achieved, would make the gas turbine an even more efficient unit. The Advanced Gas Turbine Programs sponsored by the U.S. Department of Energy has these high temperatures as one of its goals. To achieve these high temperatures, steam cooling is being used in the latest designs to achieve the goals of maintaining blade metal temperatures below 1300°F (704°C), and prevent hot corrosion problems.

The industrial heavy-duty gas turbines employ axial-flow compressors and turbines. The industrial turbine consists of a 15- to 17- stage axial flow compressor, with multiple can-annular combustors each connected to the other by cross-over tubes. The cross-over tubes help propagate the flames from one combustor can to all the other chambers, and also assures an equalization of the pressure between each combustor chamber. The earlier industrial European designs have single-stage side combustors. The new European designs do not use the side combustor in most of their newer designs. The newer European designs have can-annular or annular combustors since side (silo type) combustors had a tendency to distort the casing. Figure 4-6 is a cross-sectional representation of the GE Industrial Type Gas Turbine, with can-annular combustors, and Figure 4-7 is a cross-sectional representation of the Siemens Silo-Type Combustor Gas Turbine. The turbine expander consists of a two- to four-stage axial flow turbine, which drives both the axial flow compressor and the generator.

The large frontal areas of these units reduce the inlet velocities, thus reducing air noise. The pressure rise in each compressor stage is reduced, creating a large, stable operating zone.

The auxiliary modules used on most of these units have gone through considerable hours of testing and are heavy-duty pumps and motors.

The advantages of the heavy-duty gas turbines are their long life, high availability, and slightly higher overall efficiencies. The noise level from this type of turbine is considerably less than an aircraft-type turbine. The heavy-duty gas turbine's largest customers are the electrical utilities and independent power producers. Since the 1990s, the industrial turbines have been the bulwarks of most combined cycle power plants.

148 • COGENERATION AND COMBINED CYCLE POWER PLANTS

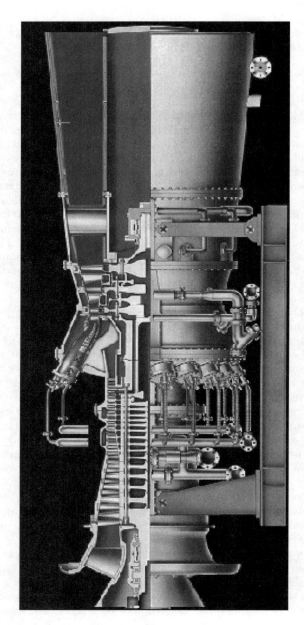

Figure 4-6. A Frame Type Gas Turbine with Can-annular Combustors (Courtesy of GE Turbines)

An Overview of Gas Turbines • 149

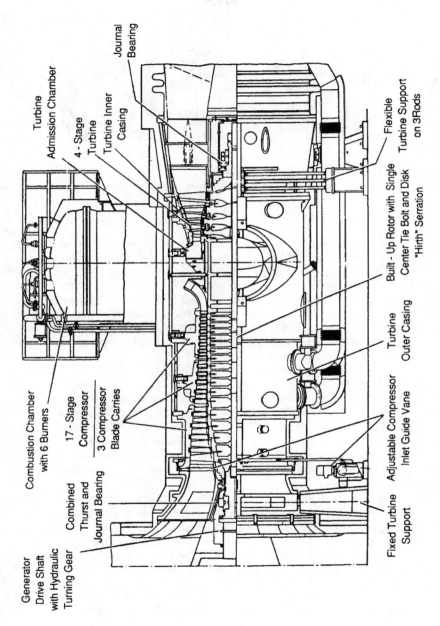

Figure 4-7. Frame Type Gas Turbine with Silo Type Combustors (Courtesy of Siemens Power Generation)

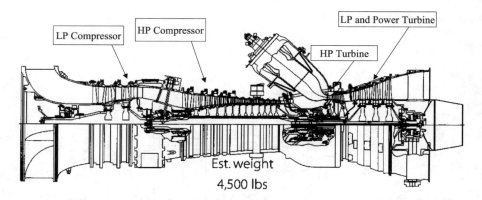

Figure 4-8. A Cross-section of an Aeroderivative Gas Turbine Engine

Aircraft-Derivative Gas Turbines

Jet gas turbines consist of two basic components: an aircraft-derivative gas generator and a free-power turbine. The gas generator serves as a producer of gas energy or gas horsepower. The gas generator is derived from an aircraft engine modified to burn industrial fuels. Design innovations are usually incorporated to ensure the required long-life characteristics in the ground-based environment. In case of fan jet designs, the fan is removed and a couple of stages of compression are added in front of the existing low-pressure compressor. The axial flow compressor, in many cases, is divided into two sections: a low-pressure compressor followed by a high-pressure compressor. In those cases, there are usually a high-pressure turbine and a low-pressure turbine, which drives the corresponding sections of the compressor. The shafts are usually concentric, thus the speeds of the high-pressure and low-pressure sections can be optimized. In this case, the power turbine is separate and is not mechanically coupled; the only connection is via an aerodynamic coupling. In these cases, the turbines have three shafts, all operating at independent speeds. The gas generator serves to raise combustion gas products to conditions of around 45 to 75 psi (3 to 5 bar) and temperatures of 1300°F to 1700°F (704°C to 927°C) at the exhaust flange. Figure 4-8 shows a cross-section of an aeroderivative engine.

Both the power industry and the petrochemical industries use the aircraft-type turbine. The power industry uses these units in a combined cycle mode for power generation, especially in remote areas where the power requirements are less than 100 MW. The petrochemical industry uses these types of turbines on off-shore platforms, especially for gas re-injection, and as power plants for these off-shore platforms, mostly due to their compactness and the ability to be easily replaced and then sent out to be repaired. The aero-derivative gas turbine also is used widely by gas transmission companies and petrochemical plants, especially for many variable speed mechanical drives. The benefits of the aeroderivative gas turbines are:

1. *Favorable installation cost.* The equipment involved is of a size and weight that can be packaged and tested as a complete unit within the

manufacturer's plant. Generally, the package will include either a generator or a driven pipeline compressor and all auxiliaries and control panels specified by the user. Immediate installation at the job site is facilitated by factory matching and debugging.

2. *Adaptation to remote control.* Users strive to reduce operating costs by automation of their systems. Many new offshore and pipeline applications today are designed for remote unattended operation of the compression equipment. Jet gas turbine equipment lends itself to automatic control, as auxiliary systems are not complex, water cooling is not required (cooling by oil-to-air exchanges), and the starting device (gas expansion motor) requires little energy and is reliable. Safety devices and instrumentation adapt readily for purposes of remote control and monitoring the performance of the equipment.

3. *Maintenance concept.* The off-site maintenance plan fits in well with these systems where minimum operating personnel and unattended stations are the objectives. Technicians conduct minor running adjustments and perform instrument calibrations. Otherwise, the aeroderivative gas turbine runs without inspection until monitoring equipment indicates distress or sudden performance change. This plan calls for the removal of the gasifier section (the aero-engine) and sending it back to the factory for repair while another unit is installed. The power turbine does not usually have problems since its inlet temperature is much lower. Downtime due to the removal and replacement of the gasifier turbine is about 8 hours.

Medium-Range Gas Turbines

Medium-range gas turbines are usually rated between 5 and 15 MW. These units are similar in design to the large heavy-duty gas turbines; their casing is thicker than the aeroderivative casing, but thinner than the industrial gas turbines. They usually are split-shaft designs that are efficient in part load operations. Efficiency is achieved by letting the gasifier section (the section which produces the hot gas) operate at maximum efficiency, while the power turbine operates over a great range of speeds. The compressor is usually a 10- to 16-stage subsonic axial compressor, which produces a pressure ratio from about 5:1 to 15:1. Most American designs use can-annular (about five to ten combustor cans mounted in a circular ring) or annular-type combustors. Most European designs use side combustors and have lower turbine inlet temperatures compared to their American counterparts. Figure 4-9 shows a medium-range gas turbine.

The gasifier turbine is usually a two- to three-stage axial turbine with an air-cooled first-stage nozzle and blade. The power turbine is usually a single- or two-stage axial-flow turbine. The medium-range turbines are used on offshore platforms and are finding increasing use in petrochemical plants. The straight simple-cycle turbine is low in efficiency, but by using regenerators to consume exhaust gases, these efficiencies can be greatly improved. In process plants, this exhaust gas is used to produce steam. The combined-cycle (air-steam) cogeneration plant has very high efficiencies and is the trend of the future.

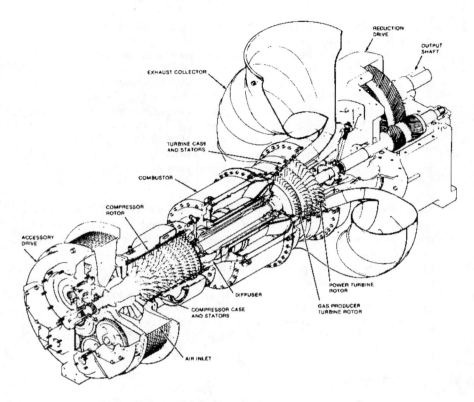

**Figure 4-9. A Medium Size Industrial Gas Turbine
(Courtesy of Solar Turbines Incorporated)**

These gas turbines have, in many cases, regenerators or recuperators to enhance the efficiency of these turbines. Figure 4-10 shows such a new recuperated gas turbine design, which has an efficiency of 38%.

The term "regenerative heat exchanger" is used for a system, in which the heat transfer between two streams is affected by the exposure of a third medium alternately to the two flows. (The heat flows successively into and out of the third medium, which undergoes a cyclic temperature.) In a "recuperative heat exchanger," each element of head-transferring surface has a constant temperature and, by arranging the gas paths in contraflow, the temperature distribution in the matrix in the direction of flow is that giving optimum performance for the given heat-transfer conditions. This optimum temperature distribution can be achieved ideally in a contraflow regenerator and approached very closely in a cross-flow regenerator.

Small Gas Turbines

Many small gas turbines which produce below 5 MW are designed similar to the larger turbines already discussed; however, there are many designs which incorporate centrifugal compressors or combinations of centrifugal and axial compressors, as well as radial-inflow turbines. A small turbine will often consist of a single-stage centrifugal compressor producing a pressure ratio as high as 6:1, a single side combustor, where temperatures of about 1800°F (982°C) are reached, and radial-inflow turbines. Figure 4-11 shows a schematic of such a typical turbine. Air is induced through an inlet duct to the centrifugal compressor, which rotating at high speed, imparts energy to the air. On leaving the impeller, air with increased pressure and velocity passes through a high-efficiency diffuser, which converts the velocity energy to static pressure. The compressed air, contained in a pressure casing, flows at low speed to the combustion chamber, which is a side combustor. A portion of the air enters the combustor head, mixes with the fuel and burns continuously. The remainder of the air enters through the wall of the combustor and mixes with the hot gases. Good fuel atomization and controlled mixing ensure an even temperature distribution in the hot gases, which pass through the volute (7) to enter the radial inflow turbine nozzles. High acceleration and expansion of the gases through the nozzle guide vane passages and turbine combine to impart rotational energy, which is used to drive the external load and auxiliaries on the cool side of the turbine. The efficiency of a small turbine is usually much lower than a larger unit

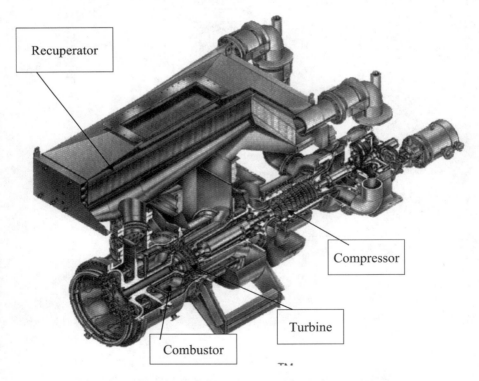

Figure 4-10. A Recuperative Medium-sized Industrial Gas Turbine (Courtesy of Solar Turbines Incorporated)

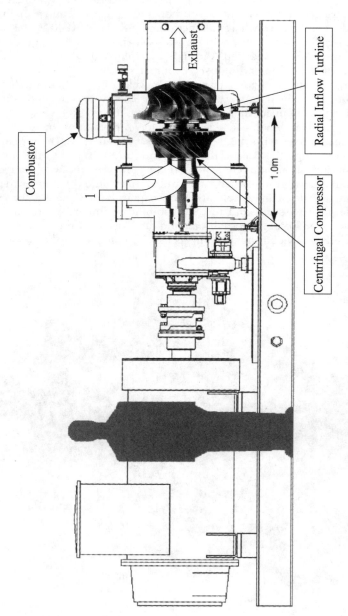

Figure 4-11. A Small Radial Flow Gas Turbine Cutaway Showing the Turbine Rotor

because of the limitation of the turbine inlet temperature and the lower component efficiencies. Turbine inlet temperature is limited because the turbine blades are not cooled. Radial-flow compressors and impellers inherently have lower efficiencies than their axial counterparts. These units are rugged and their simplicity in design assures many hours of trouble-free operation. A way to improve the lower overall cycle efficiencies, 18% to 23%, is to use the waste heat from the turbine unit. High thermal efficiencies (30% to 35%) can be obtained, since nearly all the heat not converted into mechanical energy is available in the exhaust, and most of this energy can be converted into useful work. These units when placed in a combined Heat power application can reach efficiencies of the total process as high as 60% to 70%.

Figure 4-12 shows an aeroderivative small gas turbine. This unit has three independent rotating assemblies mounted on three concentric shafts. This turbine has a three-stage axial flow compressor followed by a centrifugal compressor, each driven by a single stage axial flow compressor. Power is extracted by a two-stage axial flow turbine and delivered to the inlet end of the machine by one of the concentric shafts. The combustion system comprises of a reverse flow annular combustion chamber with multiple fuel nozzles and a spark igniter. This aeroderivative engine produces 4.9 MW and has an efficiency of 32%.

Major Gas Turbine Components

Compressors

A compressor is a device that pressurizes a working fluid. The types of compressors fall into three categories as shown in Figure 4-13. The positive displacement

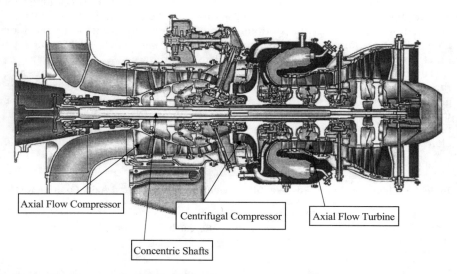

Figure 4-12. A Small Aeroderivative Gas Turbine, ST30 Marine and Industrial Gas Turbine (Courtesy of Pratt & Whitney Canada Corp.)

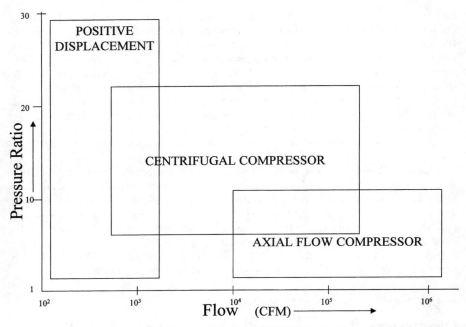

Figure 4-13. Performance Characteristics of Different Types of Compressors

compressors are used for low flow and high pressure (head), centrifugal compressors are for medium flow and medium head, and axial flow compressors are for high flow and low pressure. In gas turbines, the centrifugal flow and axial flow compressors, which are continuous flow compressors, are the ones used for compressing the air. Positive displacement compressors, such as the gear type units, are used for lubrication systems in the gas turbines.

The characteristics of these compressors are given in Table 4-1. The pressure ratio of the axial and centrifugal compressors have been classified into three groups: industrial, aerospace and research. The industrial pressure ratio is low for the reasons that the operating range needs to be large. The operating range is the range between the surge point and the choke point. Figure 4-14 shows the operating characteristics of a compressor. The surge point is the point when the flow is

Table 4-1. Characteristics of Positive, Displacement, Centrifugal and Axial Compressors

Types of Compressors	Pressure Ratio			Efficency	Operating Range
	Industrial	Aerospace	Research		
Positive Displacement	Up to 30	–	–	75–82%	–
Centrifugal	1.2–1.9	2.0–7.0	13	75–87%	Large, 25%
Axial	1.05–1.3	1.1–1.45	2.1	80–93%	Narrow, 3–10%

reversed in the compressor. The choke point is the point when the flow has reached a Mach = 1.0, the point where no more flow can get through the unit, a "stone wall". When surge occurs, the flow is reversed, and so are all the forces acting on the compressor, especially the thrust forces, which can lead to total destruction of the compressor. Thus, surge is a region that must be avoided. Choke conditions cause a large drop in efficiency, but does not lead to destruction of the unit.

It is important to note that with the increase in pressure ratio and the number of stages, the operating range is narrowed.

The turbocompressors discussed in this section transfers energy by dynamic means from a rotating member to the continuously flowing fluid. The two types of compressors used in gas turbines are axial and centrifugal. Nearly all gas turbines producing over 5 MW have axial flow compressors. Some small gas turbines employ a combination of an axial compressor followed by a centrifugal unit. Figure 4-15 shows a schematic of an axial-flow compressor followed by a centrifugal compressor, an annular combustor, and an axial-flow turbine, very similar to the actual engine depicted in Figure 4-12.

Thermodynamic and fluid mechanic relationships:

1. Equation of state

$$P/\rho^\gamma = \text{Const.} \qquad (4-3)$$

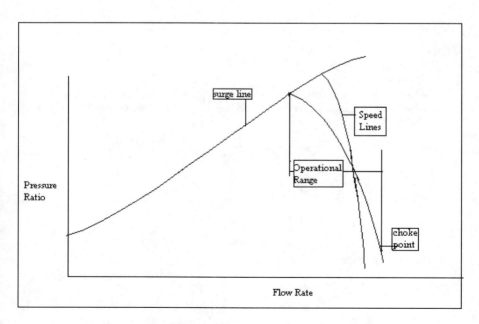

Figure 4-14. Operating Characteristics of a Compressor

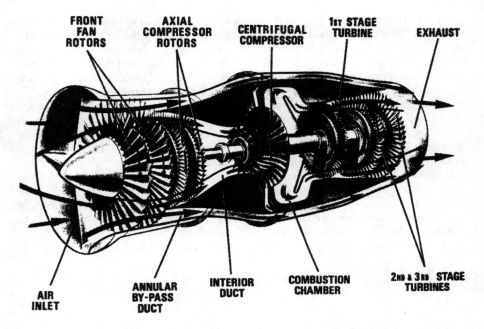

Figure 4-15. A Schematic of a Cutaway of a Small Gas Turbine Used in Helicopter or Vehicular Applications

where
P = pressure
ρ = density
γ = for an isentropic adiabatic process $\gamma = \frac{c_p}{c_v}$; where c_p and c_v are the specific heats of the gas at constant pressure and volume respectively and can be written as:

$$c_p - c_v = R \qquad (4-4)$$

where

$$c_p = \frac{\gamma R}{\gamma - 1} \quad \text{and} \quad c_v = \frac{R}{\gamma - 1} \qquad (4-5)$$

2. The energy equation:

$$h_1 + \frac{V_1^2}{2g_c J} + {}_1Q_2 = h_2 + \frac{V_2^2}{2g_c J} + {}_1W_2 \qquad (4-6)$$

where
 h = enthalpy
 V = absolute velocity
 W = work
 Q = heat rejection

and the continuity equation

$$\dot{m} = A_1 V_1 \rho_1 = A_2 V_2 \rho_2 \qquad (4-7)$$

where
 A = area
 V = velocity
 ρ = density
 J = mech. equiv. of heat
 g_c = gravitational constant
 h = enthalpy
 $\dot{m}$ = massflow

the flow per unit area can be written as follows:

$$\frac{\dot{m}}{A} = \sqrt{\frac{\gamma}{R}} \frac{P}{\sqrt{T}} \frac{M}{\left(1 + \frac{\gamma-1}{2}M^2\right)^{\frac{\gamma+1}{2(\gamma-1)}}} \qquad (4-8)$$

where the Mach number, M, is defined as:

$$M = \frac{V}{a} \qquad (4-9)$$

it is important to note that the Mach number is based on static temperature. The acoustic velocity, a, in a gas is given by the following relationship

$$a^2 = \left(\frac{\partial P}{\partial \rho}\right)_{s=c} \qquad (4-10)$$

for an adiabatic process (s = entropy = constant) the acoustic speed can be written as follows:

$$a = \sqrt{\frac{\gamma g_c R T_s}{MW}} \qquad (4-11)$$

where

T_s = static temperature

Total conditions occur when the flow is brought to rest in a reversible adiabatic manner, and the static conditions are the conditions of flow in a moving stream.

It is important to note that the pressure measured can be either total or static; however, only total temperature can be measured. For the total conditions of pressure and temperature to change, the energy must be added or extracted to the fluid stream. The relationship between total and static conditions for pressure and temperature are as follows:

$$T_o = T_s + \frac{V^2}{2c_p} \qquad (4-12)$$

where

T_o = total temperature
T_s = static temperature
V = gas stream velocity

and

$$P_o = P_s + \rho \frac{V^2}{2g_c} \qquad (4-13)$$

where

P_o = total pressure
P_s = static pressure

Equations 4-12 and 4-13 can be written in terms of the Mach number as follows:

$$\frac{T_o}{T_s} = \left(1 + \frac{\gamma - 1}{2} M^2\right) \qquad (4-14)$$

and

$$\frac{P_o}{P_s} = \left[1 + \frac{\gamma - 1}{2} M^2\right]^{\frac{\gamma}{\gamma - 1}} \qquad (4-15)$$

Axial-Flow Compressors. An axial-Flow compressor compresses its working fluid by first accelerating the fluid and then diffusing it to obtain a pressure increase. The fluid is accelerated by a row of rotating airfoils or blades (the rotor) and diffused by a row of stationary blades (the stator). The diffusion in the stator converts the velocity increase gained in the rotor to a pressure increase. One rotor and one stator make up a stage in a compressor. A compressor usually consists of multiple stages. One additional row of fixed blades (inlet guide vanes) is frequently used at the compressor inlet to ensure that air enters the first-stage rotors at the desired angle. In addition to the stators, additional diffuser at the exit of the compressor further diffuses the fluid and controls its velocity when entering the combustors.

In an axial compressor, air passes from one stage to the next with each stage raising the pressure slightly. By producing low-pressure increases on the order of 1.1:1 to 1.4:1, very high efficiencies can be obtained. The use of multiple stages permits overall pressure increases up to 40:1. The rule of thumb for a multiple-stage gas turbine compressor would be that the energy rise per stage would be constant rather than the pressure rise per stage there are however some units where the opposite is true. The energy rise per stage can be written as:

$$\Delta T_{\text{stage}} = T_{\text{in}} \left[\left(\frac{P_2}{P_1} \right)^{\frac{\gamma-1}{\gamma}} - 1 \right] \qquad (4-16)$$

Figure 4-16 shows multistage high-pressure axial flow turbine rotor. The turbine rotor depicted in this figure has a low-pressure compressor followed by a high-pressure compressor. There are also two turbine sections, the reason there is a large space between the two turbine sections is that this is a reheat turbine and the second set of combustors are located between the high-pressure and the low-pressure turbine sections. The compressor produces 30:1 pressure in 22 stages. The low-pressure increase per stage also simplifies calculations in the design of the compressor by justifying the air as incompressible in its flow through an individual

Figure 4-16. An High Pressure Ratio Turbine Rotor (Courtesy of ALSTOM)

stage. As with other types of rotating machinery, an axial compressor can be described by a cylindrical coordinate system. The Z-axis is taken as running the length of the compressor shaft, the radius, r, is measured outward from the shaft, and the angle of rotation θ is the angle turned by the blades in Figure 4-17. This coordinate system will be used throughout this discussion of axial-flow compressors.

Figure 4-18 shows the pressure, velocity, and total enthalpy variation for flow through several stages of an axial compressor. It is important to note here that the changes in the total conditions for pressure, temperature, and enthalpy occur only in the rotating component where energy is inputted into the system. As seen in Figure 4-16, the length of the blades and the annulus area, which is the area between the shaft and shroud, decreases through the length of the compressor. This reduction in flow area compensates for the increase in fluid density as it is compressed, permitting a constant axial velocity. In most preliminary calculations

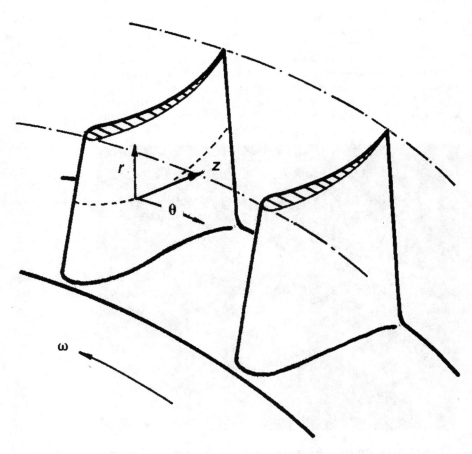

Figure 4-17. Coordinate System for Axial Flow Compressor

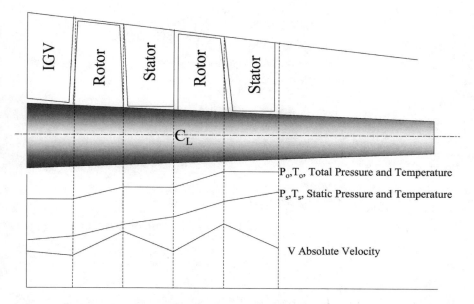

Figure 4-18. Variation of Flow and Thermodynamic Properties in an Axial Flow Compressor

used in the design of a compressor, the average blade height is used as the blade height for the stage.

To understand the flow in a turbomachine, the concepts of absolute and relative velocity must be grasped. Absolute velocity, V, is gas velocity with respect to a stationary coordinate system. Relative velocity, W, is the velocity relative to the rotor. In turbomachinery, the air entering the rotor will have a relative velocity component parallel to the rotor blade, and an absolute velocity component parallel to the stationary blades. Mathematically, this relationship is written as:

$$\vec{V} = \vec{U} \longmapsto \vec{W} \qquad (4-17)$$

where the absolute velocity, V, is the algebraic addition of the relative velocity, W, and the linear rotor velocity, U. The absolute velocity can be resolved into its components, the radial or meridional velocity, V_m, and the tangential component, V_θ. From Figure 4-19 the following relationships are obtained:

$$V_1^2 = V_{\theta 1}^2 + V_{m1}^2$$
$$V_2^2 = V_{\theta 2}^2 + V_{m2}^2$$
$$W_1^2 = (U_1 - V_{\theta 1})^2 + V_{m1}^2$$
$$W_2^2 = (U_2 - V_{\theta 2})^2 + V_{m2}^2 \qquad (4-18)$$

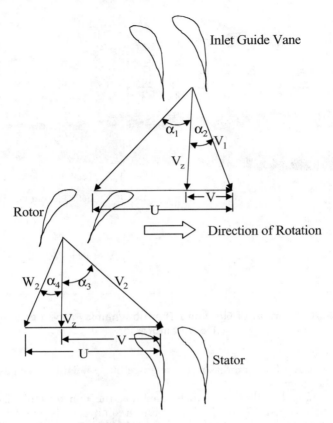

Figure 4-19. A Typical Velocity Diagram for an Axial Flow Compressor

The Euler turbine equation, a simplification of the momentum equation depicting the total energy transfer is given as follows:

$$E = \frac{\dot{m}}{g_c}(U_1 V_{\theta 1} - U_2 V_{\theta 2}) \qquad (4-19)$$

The total energy, E, when written for a unit mass flow is the head produced, H, per unit mass:

$$H = \frac{1}{g_c}(U_1 V_{\theta 1} - U_2 V_{\theta 2}) \qquad (4-20)$$

By placing these relationships into the Euler turbine equation, the following relationship is obtained:

$$E = \frac{1}{2g_c}\left[\left(V_1^2 - V_2^2\right) + \left(U_1^2 - U_2^2\right) + \left(W_1^2 - W_2^2\right)\right] \qquad (4-21)$$

As started earlier, an axial-flow compressor operates on the principle of putting work into the incoming air by acceleration and diffusion. Air enters the rotor as shown in Figure 4-19 with an absolute velocity V_1 and an angle α_1, which combines vector ally with the tangential velocity of the blade, U, to produce the resultant relative velocity W_1 at an angle α_2. Air flowing through the passages formed by the rotor blades is given a relative velocity W_2 at an angle α_4, which are less than α_2 because of the camber of the blades. Note that W_2 is less than W_1, resulting from an increase in the passage width as the blades become thinner toward the trailing edges. Therefore, some diffusion will take place in the rotor section of the stage. The combination of the relative exit velocity and blade velocity produce an absolute velocity V_2 at the exit of the rotor. The air then passes through the stator, where it is turned through an angle so that the air is directed into the rotor of the next stage with a minimum incidence angle. The air entering the rotor has an axial component at an absolute velocity V_{z1} and a tangential component $V_{\theta 1}$, and assuming that the blade speeds at the inlet and exit velocity of the compressor are the same and noting the relationships, the Euler turbine equation per unit mass combined with the energy rise across the stage can be re-written as:

$$H = \frac{UV_z}{g_c}(\tan \alpha_1 - \tan \alpha_2) \qquad (4-22)$$

The pressure ratio across the stage can be written as:

$$\frac{P_2}{P_1} = \left\{\frac{UV_z}{gC_p T_{in}}[\tan \alpha_2 - \tan \alpha_4] + 1\right\}^{\frac{\gamma}{\gamma+1}} \qquad (4-23)$$

Degree of Reaction. The degree of reaction in an axial-flow compressor is defined as the ratio of the change of static head in the rotor to the head generated in the stage.

$$R = \frac{H_{rotor}}{H_{stage}} \qquad (4-24)$$

The change in static head in the rotor is equal to the change in relative kinetic energy:

$$H_{rotor} = \frac{1}{2g}(W_2^2 - W_1^2) \qquad (4-25)$$

and

$$W_1^2 = V_{z1}^2 + (V_{z1} \tan \alpha_2)^2 \qquad (4-26)$$

$$W_1^2 = V_{z1}^2 + (V_{z1} \tan \alpha_2)^2 \qquad (4-27)$$

Thus, the reaction of the stage can be written as

$$R = \frac{V_z}{2U}(\tan \alpha_2 + \tan \alpha_4) \qquad (4-28)$$

The 50% reaction stage is widely used, since an adverse pressure rise on either the rotor or stator blade surfaces is minimized for a given stage pressure rise. When designing a compressor with this type of blading, the first stage must be preceded by inlet guide vanes to provide prewhirl, and the correct velocity entrance angle to the first-stage rotor. With a high tangential velocity component maintained by each succeeding stationary row, the magnitude of W_1 is decreased. Thus, higher blade speeds and axial-velocity components are possible without exceeding the limiting value of 0.70 to 0.75 for the inlet Mach number. Higher blade speeds result in compressors of smaller diameter and less weight.

Another advantage of the symmetrical stage comes from the equality of static pressure rises in the stationary and moving blades, resulting in a maximum static pressure rise for the stage. Therefore, a given pressure ratio can be achieved with a minimum number of stages, a factor in the lightness of this type of compressor. The serious disadvantage of the symmetrical stage is the high exit loss resulting from the high axial velocity component. However, the advantages are of such importance in aircraft applications that the symmetrical compressor is normally used. In stationary applications, where weight and frontal area are of lesser importance, one of the other stage types is used.

The term "asymmetrical stage" is applied to stages with reaction other than 50%. The axial-inflow stage is a special case of an asymmetrical stage where the entering absolute velocity is in the axial direction. The moving blades impart whirl to the velocity of the leaving flow, which is removed by the following stator. From this whirl, the major part of the stage pressure rise occurs in the moving row of blades with the degree of reaction varying from 60% to 90%. The stage is designed for constant energy transfer and axial velocity at all radii so that the vortex flow condition is maintained in the space between blade rows.

Many present-day axial flow compressors have blade profiles, which are based on the NACA 65 Series blades, and the double circular arc blades. The original work by NACA and NASA is the basis on which most modern axial-flow compressors are designed. Under NACA, a large number of blade profiles were tested. The test data on these blade profiles is published. The cascade data conducted by NACA is the most extensive work on its kind. In most commercial axial-flow compressors, NACA 65 series blades are used. These blades are usually specified by notation similar to the following: 65-(18) 10. This notation means that the blade has a lift coefficient of 1.8, a profile shape 65, and a thickness/chord ratio of 10%. The lift coefficient can be directly related to the blade camber angle by the following relationship for 65 series blades:

$$\Theta \approx 25 \, C_L. \qquad (4-29)$$

New blade profiles have also used diffusion type blading in the early stages of the compressor. The new high-pressure and high-loading compressor requires a higher Mach number to increase their capacity and efficiency. The diffusion bladings increase the loading at the tips and tend to distribute the loading equally between the rotor and the tip. Efficiencies in the later stages of multiple stage axial flow compressors are lower than the earlier stages due to the distortions of the radial flow.

Centrifugal Flow Compressors. Centrifugal compressors are used in small gas turbines and are the driven units in most gas turbine compressor trains. They are an integral part of the petrochemical industry, finding extensive use because of their smooth operation, large tolerance of process fluctuations, and their higher reliability compared to other types of compressors. Centrifugal compressors range in size from pressure ratios of 1:3 per stage to as high as 13:1 on experimental models. Discussions here are limited to the compressors used in small gas turbines. This means that the compressor pressure ratio must be between 3:1 and 7:1 per stage. This is considered a highly loaded compressor. With pressure ratios which exceed 5:1, flows entering the diffuser from the rotor supersonic in their Mach number ($M > 1.0$). This requires special design of the diffuser.

In a typical centrifugal compressor, the fluid is forced through the impeller by rapidly rotating impeller blades. The velocity of the fluid is converted to pressure, partially in the impeller and partially in the stationary diffusers. Most of the velocity leaving the impeller is converted into pressure energy in the diffuser as shown in Figure 4-20. It is normal practice to design the compressor so that half the pressure rise takes place in the impeller and the other half in the diffuser. The diffuser consists essentially of vanes, which are tangential to the impeller. These vane passages diverge to convert the velocity head into pressure energy. The inner edge of the vanes is in line with the direction of the resultant airflow from the impeller as shown in Figure 4-21.

In the centrifugal or mixed-flow compressor, the air enters the compressor in an axial direction and exists in a radial direction into a diffuser. This combination of rotor (or impeller) and diffuser comprises a single stage. The air initially enters a centrifugal compressor at the inducer as shown in Figure 4-20. The inducer, usually an integral part of the impeller, is very much like an axial-flow compressor rotor. Many earlier designs kept the inducer separate. The air then goes through a 90 deg. turn and exits into a diffuser, which usually consists of a vaneless space followed by a vaned diffuser. This is especially true if the compressor exit is supersonic, as is the case with high-pressure ratio compressors. The vaneless space is used to reduce the velocity leaving the rotor to a value lower than Mach number equal to 1 ($M < 1$). From the exit of the diffuser, the air enters a scroll or collector. The centrifugal compressor is slightly less efficient than the axial-flow compressor, but it has a higher stability. A higher stability means that its operating range is greater (surge-to-choke margin).

The fluid comes into the compressor through an intake duct and can be a given prewhirl by the IGVs. The inlet guide vanes give circumferential velocity to the fluid at the inducer inlet. IGVs are installed directly in front of the inducer or, where an axial entry is not possible, located radially in an intake duct. The purpose of installing the IGVs is usually to decrease the relative Mach number at the inducer-tip (impeller eye) inlet because the highest relative velocity at the inducer

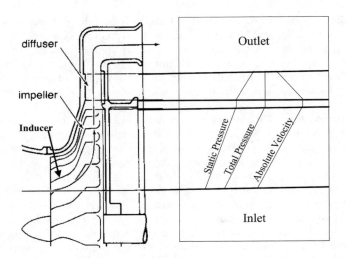

Figure 4-20. Aerodynamic and Thermodynamic Properties in a Centrifugal Compressor Stage

inlet is at the shroud. When the relative velocity is close to the sonic velocity or greater than it, a shock wave takes place in the inducer section. A shock wave produces shock loss and chokes the inducer.

The flow can enter the impeller axially, or with a positive rotation (rotation of the flow in the direction of rotation of the impeller), or with a negative rotation (rotation of the flow in the direction opposite to the rotation of the impeller). It then flows into an inducer with a minimal incidence angle, and its flow direction is changed from axial to radial.

For non-rotation (without IGVs or axial entry), $V_{\theta 1}$ is equal to zero. The disadvantage of positive rotation is that a positive inlet whirl velocity reduces the energy transfer from the Euler head:

$$H = \frac{1}{g}[U_1 V_{\theta 1} - U_2 V_{\theta 2}] \qquad (4-30)$$

to

$$H = \frac{1}{g}[-U_2 V_{\theta 2}] \qquad (4-31)$$

With positive prewhirl, the first term of the Euler equation remains; therefore, Euler work is reduced by the use of positive prewhirl. On the other hand, negative prewhirl increases the energy transfer by the amount $U_1 V_{\theta 1}$. This results in a larger pressure head being produced in the case of the negative prewhirl for the same impeller diameter and speed.

The positive rotation decreases the relative Mach number at the inducer inlet. However, negative prewhirl increases it. A relative Mach number is defined by

$$M_{\text{rel}} = W_1/a_1 \qquad (4-32)$$

where
    M_{rel} = relative Mach number
    W_1 = relative velocity at an inducer inlet
    a_1 = sonic velocity at inducer inlet conditions. It should be important to note that the sonic velocity is based on static temperature at the inlet.

Further, each of these rotations can be divided into three other pre-whirl categories. There are three kinds of prewhirl:

1. *Free-vortex prewhirl.* This type is represented by $r_1 V_{\theta 1}$ = constant with respect to the inducer inlet radius. This prewhirl distribution has the rotational component $V_{\theta 1}$ at a minimum at the inducer inlet shroud radius. Therefore, it is not effective in decreasing the relative Mach number in this manner.
2. *Forced-vortex prewhirl.* This type is shown as $V_{\theta 1}/r_1$ = constant. This prewhirl distribution has the rotational component $V_{\theta 1}$ at a maximum at

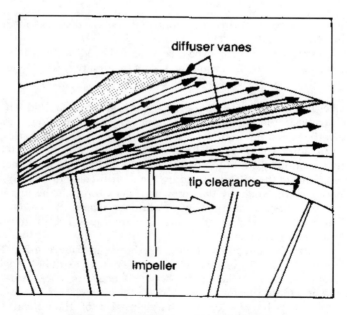

Figure 4-21. Flow in a Vaned Diffuser

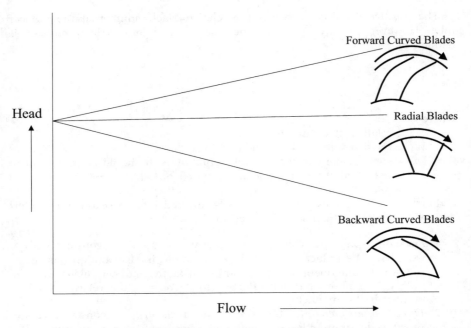

Figure 4-22. Theoretical Head Characteristics as a Function of the Flow in a Centrifugal Impeller

the inducer inlet shroud radius, contributing to a decrease in the inlet relative Mach number.
3. *Control-vortex prewhirl.* This type is represented by $V_{\theta 1} = AR_1 + B/r_1$, where A and B are constants. This equation shows the first type with $A = 0$, $B \neq 0$, and the second type with $B = 0$, $A \neq 0$.

An impeller in a centrifugal compressor imparts energy to a fluid. The impeller consists of two basic components: (1) an inducer like an axial-flow rotor, and (2) the blades in the radial direction where energy is imparted by centrifugal force. Flow enters the impeller in the axial direction and leaves in the radial direction. The velocity variations from hub to shroud resulting from these changes in flow directions complicate the design procedure for centrifugal compressors. The fluid is given energy at this stage by the rotor as it goes through the impeller while compressing. It is then discharged into a diffuser, where the kinetic energy is converted into static pressure. The flow enters the scroll from which the compressor discharge is taken.

There are three impeller vane types, as shown in Figure 4-22. These are defined according to the exit blade angles. Impellers with exit blade angle $\beta_2 = 90$ deg. are radial vanes. Impellers with $\beta_2 < 90$ deg. are backward-curved or backward-swept vanes, and for $\beta_2 > 90$ deg., the vanes are forward-curved or forward-swept. They have different characteristics of theoretical head-flow relationship to each other. In Figure 4-22, the forward-curved blade has the highest theoretical head. In actual practice, the head characteristics of all the impellers are similar to the backward-curved impeller. Most applications use backward-curved blades since they have the lowest velocity leaving the impeller, thus the diffuser has much smaller dynamic head to convert. Also, backward-curved blades have a much larger operational margin. Table 4-2 shows the advantages and disadvantages of various impellers.

Regenerators

Heavy-duty regenerators are designed for applications in large gas turbines in the 5000 to 100,000-kW range. The use of regenerators in conjunction with industrial gas turbines substantially increases cycle efficiency and provides an impetus to energy management by reducing fuel consumption up to 30%.

Figure 4-23 shows how a regenerator works. In most present-day regenerative gas turbines, ambient air enters the inlet filter and is compressed to a pressure ratio between 7:1 and 15:1, and a temperature of exiting the compressor between 530°F (277°C) and 800°F (427°C). The air is then piped to the regenerator, which heats the air to about 950°F (510°C). The heated air then enters the combustor, where it is further heated before entering the turbine. After the gas has undergone expansion in the turbine, it is about 1000°F (537°C) to 1100°F (590°C) and essentially at ambient pressure. The gas is ducted through the regenerator where the waste heat is transferred to the incoming air. The gas is then discharged into

Table 4-2. Advantages and Disadvantages of Various Impeller

Types of Impellers	Advantages	Disadvantages
Radial Blades	1. Reasonable compromise between low energy transfer and high absolute outlet velocity 2. No complex bending stress 3. Ease in manufacturing	Surge Margin is Narrow
Backward-Curved Blades	1. Low outlet kinetic energy 2. Low diffuser inlet Mach Number 3. Surge margin is widest of the three	1. Low Energy transfer 2. Complex Bending Stress 3. Difficulty in Manufacturing
Forward-Curved Blades	High Energy Transfer	1. High outlet Kinetic Energy 2. High Diffuser Inlet Mach Number 3. Complex Bending Stress 4. Difficulty in Manufacturing

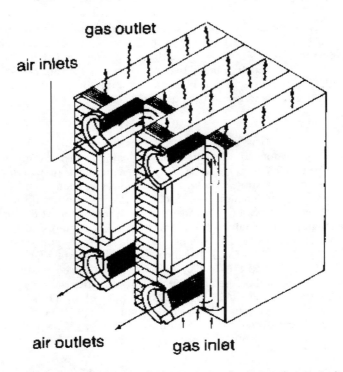

Figure 4-23. A Typical Plate and Fin Type Regenerator for an Industrial Gas Turbine

the ambient air through the exhaust stack. In effect, the heat that would otherwise be lost is transferred to the air, decreasing the amount of fuel that must be consumed to operate the turbine. For a 30,000-kW turbine, the regenerator heats 10 million lb of air per day.

Combustors

All gas turbine combustors perform the same function: they increase the temperature of the high-pressure gas. The gas turbine combustor uses very little of its air (10%) in the combustion process. The rest of the air is used for cooling and mixing. New combustors are also circulating steam for cooling purpose. The air from the compressor must be diffused before it enters the combustor. The velocity of the air leaving the compressor is about 400 to 600 fps (122 to 183 mps) and the velocity in the combustor must be maintained below 50 fps (15.2 mps). Even at these low velocities, care must be taken to avoid the flame to be carried on downstream. However, there are different methods to arrange combustors on a gas turbine. Designs fall into four categories:

1. Tubular (side combustors)
2. Can-annular

3. Annular
4. External (experimental).

The combustor is a direct-fired air heater, in which fuel is burned almost stoichiometrically with one-third or less of the compressor discharge air. Combustion products are then mixed with the remaining air to arrive at a suitable turbine inlet temperature. Despite the many design differences in combustors, all gas turbine combustion chambers have three features: (1) a recirculation zone, (2) a burning zone (with a recirculation zone which extends to the dilution region), and (3) a dilution zone, as seen in Figure 4-24. The air enters the combustor in a straight through flow, or reverse flow. Most aero-engines have straight-through flow type combustors. Most of the large frame type units have reverse flow. The function of the recirculation zone is to evaporate, partly burn, and prepare the fuel for rapid combustion within the remainder of the burning zone. Ideally, at the end of the burning zone, all fuel should be burnt so that the function of the dilution zone is solely to mix the hot gas with the dilution air. The mixture leaving the chamber should have a temperature and velocity distribution acceptable to the guide vanes and turbine. Generally, the addition of dilution air is so abrupt that if combustion is not complete at the end of the burning zone, chilling occurs, which prevents completion. However, there is evidence with some chambers that if the burning zone is run-over rich, some combustion does occur within the dilution region. Figure 4-25 shows the distribution of the air in the various regions of the combustor. The theoretical or reference velocity is the flow of combustor-inlet air through an area equal to the maximum cross-section of the combustor casing. The flow velocity is 25 fps (7.6 mps) in a reverse-flow combustor; and between 80 fps (24.4 mps) and 135 fps (41.1 mps) in a straight-through flow turbojet combustor.

Combustor inlet temperature depends on engine pressure ratio, load and engine type, and whether or not the turbine is regenerative or non-regenerative, especially at the low pressure ratios. The new industrial turbine pressure ratios are between 17:1 and 35:1, which means that the combustor inlet temperatures range from 850°F (454°C) to 1200°F (649°C). The new aircraft engines have pressure ratios, which are in excess of 40:1.

Combustor performance is measured by efficiency, the pressure decrease encountered in the combustor, and the evenness of the outlet temperature profile. Combustion efficiency is a measure of combustion completeness. Combustion completeness affects fuel consumption directly, since the heating value of any unburned fuel is not used to increase the turbine inlet temperature. To calculate combustion efficiency, the actual heat increase of the gas is ratioed to the theoretical heat input of the fuel (lower heating value).

$$\eta_{comb} = \frac{\Delta h_{actual}}{\Delta h_{theoretical}} = \frac{(\dot{m}_a + \dot{m}_f)h_3 - \dot{m}_a h_2}{\dot{m}_f LHV} \qquad (4-33)$$

The loss of pressure in a combustor is a major problem, since it affects both the fuel consumption and power output. A pressure loss occurs in a combustor

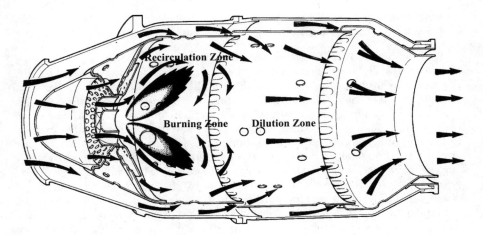

Figure 4-24. A Typical Combustor Can with Straight Through Flow

because of diffusion, friction, and momentum. Total pressure loss is usually in the range of 2% and 8% of static pressure (compressor outlet pressure). The efficiency of the engine will be reduced by an equal percent. The result is increased fuel consumption and lower power output that affects the size and weight of the engine.

The uniformity of the combustor outlet profile affects the useful level of turbine inlet temperature, since the average gas temperature is limited by the peak gas temperature. The profile factor is the ratio between the maximum exit temperature and the average exit temperature.

This uniformity assures adequate turbine nozzle life, which depends on operating temperature. The average inlet temperature to the turbine affects both fuel consumption and power output. A large combustor outlet gradient will work to reduce average gas temperature and, consequently, reduce power output and

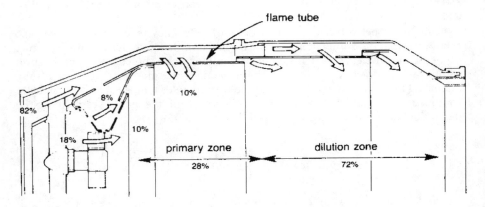

Figure 4-25. Air Distribution in a Typical Combustor

efficiency. The traverse number is defined as the peak gas temperature minus mean gas temperature divided by mean temperature rise in nozzle design. Thus, the traverse number must have a lower value — between 0.05 and 0.15 in the nozzle.

Equally important are the factors that affect satisfactory operation and life of the combustor. To achieve satisfactory operation, the flame must be self-sustaining, and combustion must be stable over a range of fuel-to-air ratios to avoid ignition loss during transient operation. Moderate metal temperatures are necessary to assure long life. Also, steep temperature gradients, which warps and cracks the liner, must be avoided. Carbon deposits can distort the liner and alter the flow patterns to cause pressure losses. Smoke is environmentally objectionable, as well as a fouler of heat exchangers. (5) Minimum carbon deposits and smoke emissions also help assure satisfactory operations.

The air entering a combustor is divided so that the flow is distributed between three major regions: (1) primary zone, (2) dilution zone, (3) annular space between the liner and casing.

The combustion in a combustor takes place in the primary zone. Combustion of natural gas is a chemical reaction that occurs between carbon, or hydrogen, and oxygen. Heat is given off as the reaction takes place. The products of combustion are carbon dioxide and water. The reaction is stoichiometric, which means that the proportions of the reactants are such that there are exactly enough oxidizer molecules to bring about a complete reaction to stable molecular forms in the products. The ratio of the oxygen content at stoichiometric conditions and actual conditions is called the equivalence ratio.

$$\Phi = (\text{oxygen/fuel at stoichiometric condition}) / (\text{oxygen/fuel at actual condition}) \quad (4-34)$$

Normal combustion temperatures range from 3400°F (1871°C) to 3500°F (1927°C). At this temperature, the volume of nitric oxide in the combustion gas is about 0.01%. If the combustion temperature is lowered, the amount of nitric oxide is substantially reduced.

The amount of oxygen in the combustion gas is regulated by controlling the ratio of air to fuel in the primary zone. The ideal volumetric ratio of air to methane is 10:1. If less than ten volumes of air are used with one volume of methane, the combustion gas will contain carbon monoxide. The reaction is as follows:

$$CH_4 + 1.5(O_2 \times 4N_2) \rightarrow 2H_2O + CO + 6N_2 + \text{Heat} \quad (4-35)$$

In gas turbines, there is plenty of air, so the carbon monoxide problem is not present.

Velocity is used as a criterion in combustor design, especially with respect to flame stabilization. The importance of air velocity in the primary zone is known. A transition zone is often included before the primary zone, so that the high-velocity air from the compressor is diffused to a lower velocity and higher pressure, and distributed around the combustion liner. In the primary zone, fuel-to-air ratios are about 60:1; the remaining air must be added somewhere. The secondary, or dilution, air should only be added after the primary reaction has reached completion. Dilution air should be added gradually so as not to quench the reaction. Flame tubes should be designed to produce a desirable outlet profile and to last a long time in the combustor environment. Adequate life is assured by film cooling of the liner.

The air enters the annular space between the liner and casing and is admitted into the space within the liner through holes and slots because of the pressure difference. The design of these holes and slots divides the liner into distinct zones for flame stabilization, combustion, dilution, and provides film cooling of the liner.

The liner experiences a high temperature because of heat radiated by the flame and combustion. To improve the life of the liner, it is necessary to lower the temperature of the liner and use a material, which has a high resistance to thermal stress and fatigue. The air-cooling method reduces the temperature both inside and outside the surface of the liner. This reduction is accomplished by fastening a metal ring inside the liner to leave a definite annular clearance. Air is admitted into this clearance space through rows of small holes in the liner and is directed by metal rings as a film of cooling air along the liner inside.

Combustor Design Considerations

Cross-Sectional Area. The combustor cross-section can be determined by dividing the volumetric flow at the combustor inlet by a reference velocity which has been selected as being appropriate for the particular turbine conditions on the basis of proven performance in a similar engine. Another basis for selecting a combustor cross-section comes from correlations of thermal loading per unit cross-section. Thermal loading is proportional to the primary zone airflow because fuel and air mixtures are near stoichiometric in all combustors.

Length. Combustor length must be sufficient to provide for flame stabilization, combustion, and mixing with dilution air. The typical value of the length-to-diameter ratio for liners ranges from three to six. Ratios for casing range from two to four.

Wobbe Number is an indicator of the characteristics and stability of the combustion process.

$$W_b = \frac{LHV}{\sqrt{Sp.Gr * T_{amb}}}$$

Increasing the Wobbe number can cause the flame to burn closer to the liner. Decreasing the Wobbe number can cause pulsations in the combustor.

Pressure Drop. The minimum practical pressure drop — excluding diffuser loss — is about 14 times the reference velocity pressure. Higher values are frequently used. Some values for this pressure loss are: 100 fps (30.5 mps), 4%; 80 fps (24.4 mps), 2.5%; 70 fps (21.3 mps), 2%; 50 fps (15.24 mps), 1%.

Volumetric Heat-release Rate. The heat-release rate is proportional to the fuel-to-air ratio and the combustor pressure, and it is a function of combustor capacity. Actual space required for combustion, as chemical limits are approached, varies with pressure to the 1.8 power.

Liner Holes. Liner area to casing area and liner hold area to casing area are important to the performance of combustors. For example, the pressure loss coefficient has a minimum value in the range of 0.6 of the liner area/casing area ratio with a temperature ratio of 4:1.

In practice, it has been found that the diameter of holes in the primary zone should be no larger than 0.1 of the liner diameter. Tubular liners with about ten rings of eight holes each give good efficiency. As discussed before, swirl vanes with holes yield better combustor performance. In the dilution zone, sizing of the holes can be used to provide a desired temperature profile.

Reliability of Combustors. The heat from combustion, pressure fluctuation, and vibration in the compressor may cause cracks in the liner and nozzle. Also, there are corrosion and distortion problems.

The edges of the holes in the liner are of great concern because the holes act as stress concentrators for any mechanical vibrations and, on rapid temperature fluctuations, high-temperature gradients are formed in the region of the hole edge, giving rise to a corresponding thermal fatigue.

It is necessary to modify the edge of the hole in various ways to reduce these stress concentrations. Some methods of modification are priming, plunging, and standard radiusing and polishing methods. For resistance against fatigue, Nimonic 75 is used with Nimonic 80 and Nimonic 90. Nimonic 75 is an 80 to 20 nickel-chromium alloy stiffened with a small amount of titanium carbide. Nimonic 75 has excellent oxidation and corrosion resistance at elevated temperatures, a reasonable creep strength, and good fatigue resistance. In addition, it is easy to press, draw, and mold. Many of today's combustors also have ceramic coating of liners to give further protection.

Typical Combustor Arrangements

Can-annular and Annular. In aircraft applications, where frontal area is important, either can-annular or annular designs are used to produce favorable radial and circumferential profiles because of the great number of fuel nozzles employed. The annular design is especially popular in new aircraft designs; however, the can-annular design is still used because of the developmental difficulties associated with annular designs. Annular combustor popularity increases with higher temperatures or low-Btu gases, since the amount of cooling air required is much less than in can-annular designs due to a much smaller surface area. The amount of cooling air required becomes an important consideration in low-BTu gas applications, since most of the air is used up in the primary zone and little is left for film cooling. Development of a can-annular design requires experiments with only one can, whereas the annular combustor must be treated as a unit and requires much more hardware and compressor flow. Can-annular combustors can be of the straight-through or reverse-flow design. If can-annular cans are used in aircraft, the straight-through design is used, while a reverse-flow design may be used on industrial engines. Annular combustors are almost always straight-through flow designs. Figure 4-26 shows a typical can-annular combustor used in frame type units, with reverse flow.

Tubular (side combustors). These designs are found on large industrial turbines, especially European designs, and some small vehicular gas turbines. They offer the advantages of simplicity of design, ease of maintenance, and long-life due to low heat release rates. These combustors may be of the "straight-through" or "reverse-flow" design. In the reverse-flow design, air enters the annulus between the combustor can and its housing, usually a hot-gas pipe to the turbine. Reverse-flow designs have minimal length. Figure 4-27 shows one such combustor design.

Air Pollution Problems

Smoke. In general, it has been found that much visible smoke is formed in small, local fuel-rich regions. The general approach to eliminating smoke is to develop leaner primary zones with an equivalence ratio between 0.9 and 1.5. Another supplementary way to eliminate smoke is to supply relatively small quantities of air to those exact, local, over-rich zones.

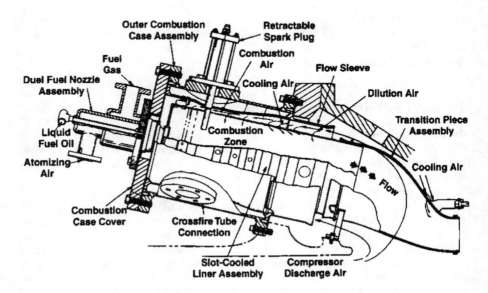

Figure 4-26. A Typical Reverse Flow Can Annular Combustor

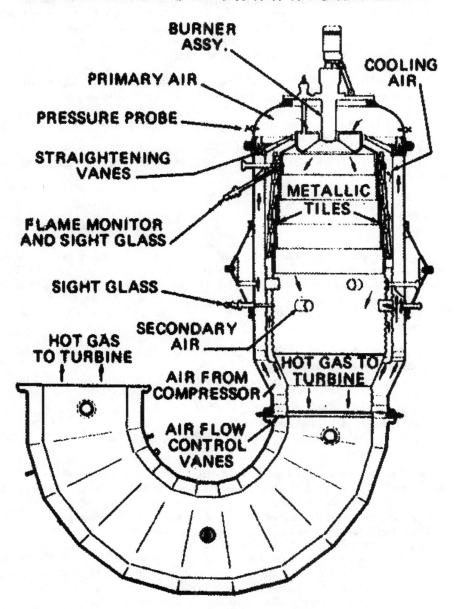

Figure 4-27. A Typical Single Can Side Combustor

Unburnt Hydrocarbons and Carbon Monoxide. Unburnt Hydrocarbon (UHC) and carbon monoxide (CO) are only produced in incomplete combustion typical of idle conditions. It appears probable that idling efficiency can be improved by detailed design to provide better atomization and higher local temperatures.

Oxides of Nitrogen. The main oxides of nitrogen produced in combustion are NO, with the remaining 10% as NO_2. These products are of great concern because of their poisonous character and abundance, especially at full-load conditions.

The formation mechanism of NO can be explained as follows:

1. Fixation of atmospheric oxygen and nitrogen at high-flame temperature.
2. Attack of carbon or hydrocarbon radicals of fuel on nitrogen molecules, resulting in NO formation.
3. Oxidation of the chemically bound nitrogen in fuel.

In 1977, the Environmental Protection Agency (EPA) in the U.S. issued Proposed Rules that limited the emissions of new, modified and reconstructed gas turbines to:

- 75 vppm NO_x at 15% oxygen (dry basis)
- 150 vppm SO_x at 15% oxygen (dry basis), controlled by limiting fuel sulfur content to less than 0.8% wt.

These standards applied to simple and regenerative cycle gas turbines, and to the gas turbine portion of combined cycle steam/electric generating systems. The 15% oxygen level was specified to prevent the NO_x ppm level being achieved by dilution of the exhaust with air.

In 1977, it was recognised that there were a number of ways to control oxides of nitrogen:

1. Use of a rich primary zone in which little NO is formed, followed by rapid dilution in the secondary zone.
2. Use of a very lean primary zone to minimize peak flame temperature by dilution.
3. Use of water or steam admitted with the fuel for cooling the small zone downstream from the fuel nozzle.
4. Use of inert exhaust gas recirculated into the reaction zone.
5. Catalytic exhaust cleanup.

"Wet" control became the preferred method in the 1980s and most of 1990s, since "dry" controls and catalytic cleanup were both at very early stages of development. The catalytic converters were used in the 1980s and are still being widely used; however, the cost of rejuvenating the catalyst is very high.

There has been a gradual tightening of the NO_x limits over the years from 75 down to 25 ppm, and now the new turbine goals are 9.

Advances in combustion technology now make it possible to control the levels of NO_x production at source, removing the need for "wet" controls. This of course opened up the market for the gas turbine to operate in areas with limited supplies of suitable quality water, e.g., deserts or marine platforms.

Although water injection is still used, "dry" control combustion technology has become the preferred method for the major players in the industrial power generation market. DLN (dry low NO_x) was the first acronym to be coined, but with the requirement to control NO_x without increasing carbon monoxide and unburned hydrocarbons, this has now become DLE (dry low emissions).

The majority of the NO_x produced in the combustion chamber is called "thermal NO_x". It is produced by a series of chemical reactions between the nitrogen (N_2) and the oxygen (O_2) in the air that occur at the elevated temperatures and pressures in gas turbine combustors. The reaction rates are highly temperature-dependent, and the NO_x production rate becomes significant above flame temperatures of about 3300°F (1815°C). Figure 4-28 shows, schematically, the flame temperatures and, therefore, NO_x production zones inside a "conventional" combustor. This design deliberately burned all of the fuel in a series of zones, going from fuel-rich to fuel-lean, to provide good stability and combustion efficiency over the entire power range.

The great dependence of NO_x formation on temperature reveals the direct effect of water or steam injection on NO_x reduction. Recent research showed an 85% reduction of NO_x by steam or water injection with optimizing combustor aerodynamics.

In a typical combustor as shown in Figure 4-28, the flow entering the primary zone is limited to about 10%. The rest of the flow is used for mixing the combusted air and cooling the combustor can. The maximum temperature is reached in the primary or stoichiometric zone of about 4040°F (2230°C), and after the mixing of the combustion process with the cooling air, the temperature drops down to a low of 2500°F (1370°C).

Basis for NO_x Prevention. Emissions from turbines are a function of temperature and, thus, a function of the F/A ratio. Figure 4-29 shows that as the temperature is increased, the amount of NO_x emissions are increased, and the CO and the unburnt hydrocarbons are decreased. The principal mechanism for NO_x formation is the oxidation of nitrogen in air when exposed to high temperatures in the combustion process; the amount of NO_x is thus dependent on the temperature of the combustion gases and also, to a lesser amount, on the time the nitrogen is exposed to these high temperatures.

The challenge in these designs is to lower the NO_x without degradation in unit stability. In the combustion of fuels that do not contain nitrogen compounds, NO_x compounds (primarily NO) are formed by two main mechanisms, thermal mechanism, and the prompt mechanism. In the thermal mechanism, NO is formed by the oxidation of molecular nitrogen through the following reactions.

NO_x is primarily formed through high-temperature reaction between nitrogen and oxygen from the air.

$$O + N_2 \leftrightarrow NO + N \qquad (4-36)$$

$$N + O_2 \leftrightarrow NO + O \qquad (4-37)$$

$$N + OH \leftrightarrow NO + H \qquad (4-38)$$

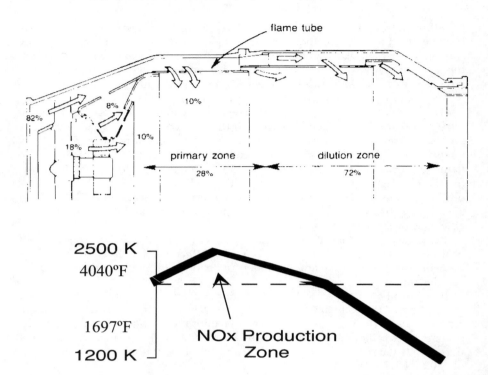

Figure 4-28. A Typical Combustor Showing the NO_x Production Zone

Hydrocarbon radicals, predominantly through the reaction, initiate the prompt mechanism

$$CH + N_2 \rightarrow HCN + N \qquad (4-39)$$

HCN and N are converted rapidly to NO by reaction with oxygen and hydrogen atoms in the flame.

The prompt mechanism predominates at low temperatures under fuel-rich conditions, whereas the thermal mechanism becomes important at temperatures above 2732°F (1500°C). Due to the onset of the thermal mechanism, the formation of NO_x in the combustion of fuel/air mixtures increases rapidly with temperature above 2732°F (1500°C) and also increases with residence time in the combustor.

The production rate of NO can be given as follows:

$$\frac{d(NO)}{dt} = \frac{K}{\sqrt{T}} e^{\frac{1}{T}} \sqrt{O_2}(N_2) \qquad (4-40)$$

The important parameters in the reduction of NO_x, as seen in the above equation, are the temperature of the flame, the nitrogen and oxygen content, and the resident time of the gases in the combustor. Figure 4-30 is a correlation between the adiabatic flame temperature and the emission of NO_x. Reduction of any and all these parameters will reduce the amount of NO_x emitted from the turbine.

Dry Low NO_x Combustor

The gas turbine combustors have seen considerable change in their design as most new turbines have progressed to dry low emission NO_x combustors from the wet combustors, which were injected by steam in the primary zone of the combustor. The DLE approach is to burn most (at least 75%) of the fuel at cool, fuel-lean conditions to avoid any significant production of NO_x. The principal features of such a combustion system is the pre-mixing of the fuel and air before the mixture enters the combustion chamber, and the leanness of the mixture strength in order to lower the flame temperature and reduce NO_x emission. This action brings the full-load operating point down on the flame temperature curve as seen in Figure 4-31 and closer to the lean limit. Controlling CO emissions thus can be difficult, and rapid engine off-loads bring the problem of avoiding flame

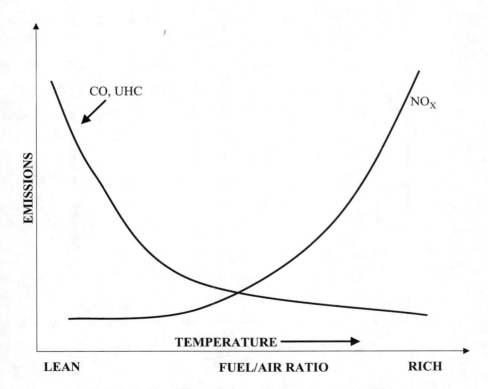

Figure 4-29. The Effect of Flame Temperature on Emissions

184 • COGENERATION AND COMBINED CYCLE POWER PLANTS

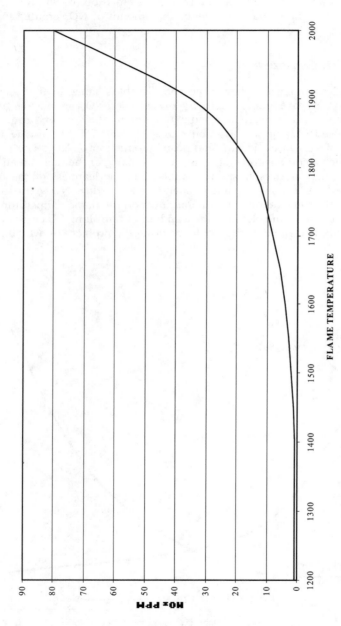

Figure 4-30. Correlation of Adiabatic Flame Temperature with NO_x Emissions

An Overview of Gas Turbines • 185

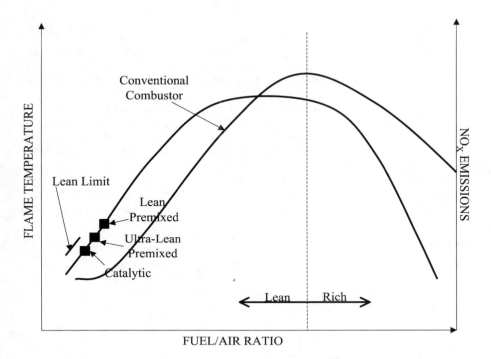

Figure 4-31. Effect of Fuel/Air Ratio on Flame Temperature and NO$_x$ Emissions

extinction, which if it occurs, cannot be safely reestablished without bringing the engine to rest and going through the restart procedure.

Figure 4-32 shows a schematic comparison of a typical dry low emission NO$_x$ combustor and conventional combustors. In both cases, a swirler is used to create the required flow conditions in the combustion chamber to stabilize the flame. The DLE fuel injector is much larger because it contains the fuel/air pre-mixing chamber and the quantity of air being mixed is large, approximately 50% to 60% of the combustion air flow.

The DLE injector has two fuel circuits. The main fuel, approximately 97% of the total, is injected into the air stream immediately downstream of the swirler at the inlet to the pre-mixing chamber. The pilot fuel is injected directly into the combustion chamber with little, if any, pre-mixing. With the flame temperature being much closer to the lean limit than in a conventional combustion system, some action has to be taken when the engine load is reduced to prevent flame out. If no action were taken, flameout would occur since the mixture strength would become too lean to burn. A small proportion of the fuel is always burned richer to provide a stable "piloting" zone, while the remainder is burned lean. In both cases, a swirler is used to create the required flow conditions in the combustion chamber to stabilize the flame. The LP fuel injector is much larger because it contains the fuel/air pre-mixing chamber and the quantity of air being mixed is large, approximately 50% to 60% of the combustion air flow.

186 • COGENERATION AND COMBINED CYCLE POWER PLANTS

Figure 4-33 shows a schematic of an actual dry low-emission NO_x combustor used by ALSTOM in their large turbines. With the flame temperature being much closer to the lean limit than in a conventional combustion system, some action has to be taken when the engine load is reduced to prevent flameout. If no action were taken, flameout would occur since the mixture strength would become too lean to burn.

One method is to close the compressor inlet guide vanes progressively as the load is lowered. This reduces the engine airflow and, hence, reduces the change in mixture strength that occurs in the combustion chamber. This method, on a single-shaft engine, generally provides sufficient control to allow low-emission operation to be maintained down to 50% engine load. Another method is to deliberately dump air overboard prior to or directly from the combustion section of the engine. This reduces the airflow and also increases the fuel flow required (for any given load) and, hence, the combustion fuel/air ratio can be held approximately constant

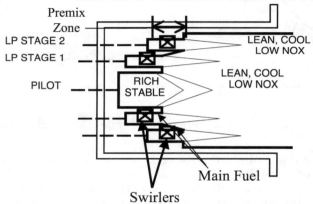

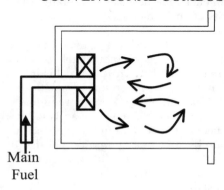

Figure 4-32. A Schematic Comparison of a Typical Dry Low Emission NO_x Combustor and a Conventional Combustors

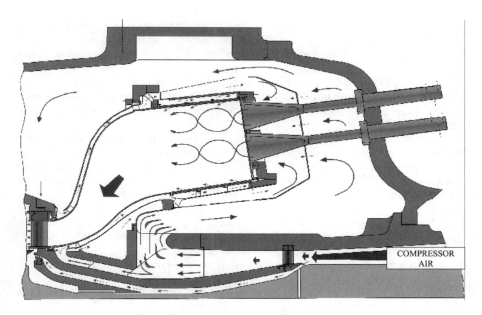

**Figure 4-33. Schematic of a Dry Low Emission NO$_x$ Combustor
(Courtesy of ALSTOM)**

at the full load value. This latter method causes the part load thermal efficiency of the engine to fall off by as much as 20%.

Even with these air management systems, lack of combustion stability range can be encountered particularly when load is rapidly reduced.

If the combustor does not feature variable geometry, then it is necessary to turn on the fuel in stages as the engine power is increased. The expected operating range of the engine will determine the number of stages, but typically, at least two or three stages are used as seen in Figure 4-34. Some units have very complex staging as the units are started or operated at off-design conditions.

Gas turbines often experience problems with these DLE combustors, some of the common problems experienced are:

- auto-ignition and flash-back
- combustion instability.

These problems can result in sudden loss of power, because a fault is sensed by the engine control system, and the engine is shutdown.

Auto-ignition is the spontaneous self-ignition of a combustible mixture. For a given fuel mixture at a particular temperature and pressure, there is a finite time before self-ignition will occur. Diesel engines (knocking) rely on it to work, but spark-ignition engines must avoid it.

DLE combustors have pre-mix modules on the head of the combustor to mix the fuel uniformly with air. To avoid auto-ignition, the residence time of the fuel in the pre-mix tube must be less than the auto-ignition delay time of the fuel. If auto-ignition does occur in the pre-mix module, then it is probable that the resulting damage will require repair and/or replacement of parts before the engine is run again at full load.

Some operators are experiencing engine shutdowns because of auto-ignition problems. The response of the engine suppliers to rectify the situation has not been encouraging, but the operators feel that the reduced reliability cannot be accepted as the "norm".

If auto-ignitions occur, then the design does not have sufficient safety margin between the auto-ignition delay time for the fuel and the residence time of the fuel in the pre-mix duct. Auto-ignition delay times for fuels do exist, but a literature search will reveal that there is considerable variability for a given fuel. Reasons for auto-ignition could be classified as follows:

- long fuel auto-ignition delay time assumed
- variations in fuel composition reducing auto-ignition delay time
- fuel residence time incorrectly calculated
- auto-ignition triggered "early" by ingestion of combustible particles.

Flashback into a pre-mix duct occurs when the local flame speed is faster than the velocity of the fuel/air mixture leaving the duct.

Flashback usually happens during unexpected engine transients, e.g., compressor surge. The resultant change of air velocity would almost certainly result in

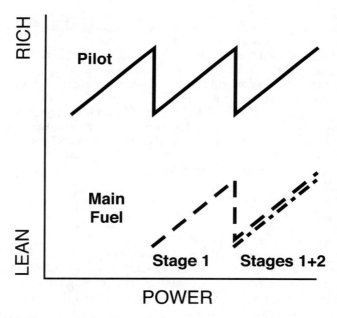

Figure 4-34. Shows the Staging of Dry Low Emissions Combustor as the Turbine is Brought to Full Power

flashback. Unfortunately, as soon as the flame-front approaches the exit of the pre-mix duct, the flame-front pressure drop will cause a reduction in the velocity of the mixture through the duct. This amplifies the effect of the original disturbance, thus prolonging the occurrence of the flashback.

Advanced cooling techniques could be offered to provide some degree of protection during a flashback event caused by engine surge. Flame detection systems coupled with fast-acting fuel control valves could also be designed to minimize the impact of a flashback. The new combustors also have steam cooling being provided.

High-Pressure burners for gas turbines use pre-mixing to enable combustion of lean mixtures. The stoichiometric mixture of air and fuel varies between 1.4 and 3.0 for gas turbines. The flames become unstable when the mixture exceeds a factor of 3.0, and below 1.4, the flame is too hot and NO_x emissions will rise rapidly. The new combustors are therefore shortened to reduce the time the gasses are in the combustor. The number of nozzles is increased to give better atomization and better mixing of the gases in the combustor. The number of nozzles in most cases increases by a factor of five to ten, which does lead to a more complex control system. The trend now is to an evolution towards the can-annular burners. For example, ABB GT9 turbine had one combustion chamber with one burner, the new ABB 13 E2 has 12 can-annular combustors and 72 burners.

Combustion instability only used to be a problem with conventional combustors at very low engine powers. The phenomenon was called "rumble". It was associated with the fuel-lean zones of a combustor, where the conditions for burning are less attractive. The complex 3D-flow structure that exists in a combustor will always have some zones that are susceptible to the oscillatory burning. In a conventional combustor, the heat release from these "oscillating" zones was only a significant percentage of the total combustor heat release at low power conditions.

With DLE combustors, the aim is to burn most of the fuel very lean to avoid the high combustion temperature zones that produce NO_x. So, these lean zones that are prone to oscillatory burning are now present from idle to 100% power. Resonance can occur (usually) within the combustor. The pressure amplitude at any given resonant frequency can rapidly build up and cause failure of the combustor. The modes of oscillation can be axial, radial or circumferential, or all three at the same time. The use of dynamic pressure transducer in the combustor section, especially in the low NO_x combustors ensures that each combustor can is burning evenly. This is achieved by controlling the flow in each combustor can till the spectrums obtained from each combustor can match. This technique has been used and found to be very effective and ensures combustor stability.

The calculation of the fuel residence time in the combustor or the pre-mixing tube is not easy. The mixing of the fuel and the air to produce a uniform fuel/air ratio at the exit of the mixing tube is often achieved by the interaction of flows. These flows are composed of swirl, shear layers and vortex. CFD modeling of the mixing tube aerodynamics is required to ensure the success of the mixing process and to establish that there is a sufficient safety margin for auto-ignition.

By limiting the flame temperature to a maximum of 2650°F (1454°C), single-digit NO_x emissions can be achieved. To operate at a maximum flame temperature of 2650°F (1454°C), which is up to 250°F (139°C) lower than the LP system previously described, requires pre-mixing 60% to 70% of the air flow with the fuel prior to admittance into the combustion chamber. With such a high amount of the available combustion air flow required for flame temperature control, insufficient air remains to be allocated solely for cooling the chamber wall or diluting the hot gases down to the turbine inlet temperature. Consequently, some of the air available has to do double duty, being used for both cooling and dilution. In engines using high turbine inlet temperatures, 2400°F to 2600°F (1316°C to 1427°C), although dilution is hardly necessary, there is not enough air left-over to cool the chamber walls. In this case, the air used in the combustion process itself has to do double duty and be used to cool the chamber walls before entering the injectors for pre-mixing with the fuel. This double duty requirement means that film or effusion cooling cannot be used for the major portion of the chamber walls. Some units are looking into steam cooling. Walls are also coated with thermal barrier coating (TBC), which has a low thermal conductivity and, hence, insulates the metal. This is a ceramic material that is plasma sprayed-on during combustion chamber manufacture. The temperature drop across the TBC, typically by 300°F (149°C), means the combustion gases are in contact with a surface that is operating at approximately 2000°F (1094°C), which also helps to prevent the quenching of the CO oxidation.

Catalytic Combustion

Catalytic combustion is a process in which a combustible compound and oxygen react on the surface of a catalyst, leading to complete oxidation of the compound. This process takes place without a flame and at much lower temperatures than those associated with conventional flame combustion. Due partly to the lower operating temperature, catalytic combustion produces lower emissions of nitrogen oxides (NO_x) than conventional combustion. Catalytic combustion is now widely used to remove pollutants from exhaust gases, and there is growing interest in applications in power generation, particularly in gas turbine combustors.

In catalytic combustion of a fuel/air mixture, the fuel reacts on the surface of the catalyst by a heterogeneous mechanism. The catalyst can stabilize the combustion of ultra-lean fuel/air mixtures with adiabatic combustion temperatures below 1500°C. Thus, the gas temperature will remain below 1500°C and very little thermal NO_x will be formed, as can be seen in Figure 4-30. However, the observed reduction in NO_x in catalytic combustors is much greater than that expected from the lower combustion temperature. The reaction on the catalytic surface apparently produces no NO_x directly, although some NO_x may be produced by homogeneous reactions in the gas phase initiated by the catalyst.

Features of Catalytic Combustion

Surface Temperatures. At low temperatures, the oxidation reactions on the catalyst are kinetically controlled and the catalyst activity is an important parameter. As the temperature increases, the buildup of heat on the catalyst surface due to the exothermic surface reactions produces ignition, and the catalyst surface temperature jumps rapidly to the adiabatic flame temperature of the fuel/air mixture on ignition. Figure 4-35 shows a schematic of the temperature profiles for catalyst and bulk gas in a traditional catalytic combustor. At the adiabatic flame temperature, oxidation reactions on the catalyst are very rapid, and the overall steady-state reaction rate is determined by the rate of mass transfer of fuel to the catalytic surface. The bulk gas temperature rises along the reactor because of heat transfer from the hot catalyst substrate and eventually approaches the catalyst surface temperature.

As the catalyst surface temperature is equal to the adiabatic flame temperature after ignition, it is independent of the overall conversion in the combustion reaction. It follows that the catalyst surface temperature cannot be reduced simply by limiting the conversion (by using a short reactor or a monolith with large cells, for example). Therefore, unless some other means of limiting the catalyst surface temperature is used, the catalyst materials must be able to withstand the adiabatic flame temperature of the fuel/air mixture during the combustion reaction. For the present generation of gas turbines, this temperature will be equal to the required turbine inlet temperature of 1300°C, which presents severe problems for existing combustion catalyst.

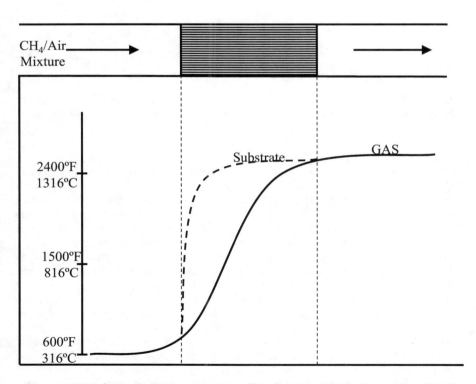

Figure 4-35. Schematic Temperature Profiles for Catalyst (Substrate) and Bulk Gas in the Traditional Catalytic Combustor

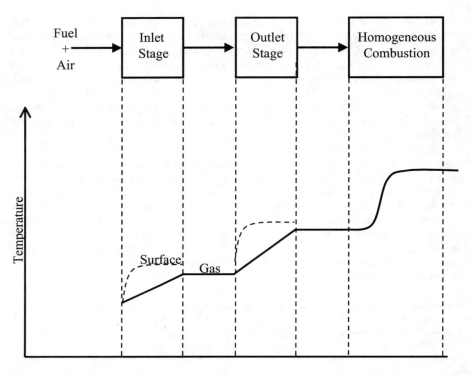

Figure 4-36. Schematic Temperature Profile for Catalytica Catalytic Combustion System in Which the Wall Temperature is Limited and Complete Combustion Occurs After the Catalyst

Catalytica has developed a new approach to catalytic combustion and Tanaka Kikinzoku Kogyo K.K. combines catalytic and homogeneous combustion in a multistage process. In this approach, shown schematically in Figure 4-36, the full fuel/air mixture required to obtain the desired combustor outlet temperature is reacted over a catalyst. However, a self-regulating chemical process described below limits the temperature rise over the catalyst. The catalyst temperature at the inlet stage therefore remains low, and the catalyst can maintain very high activity over long periods of time. Because of the high catalyst activity at the inlet stage, ignition temperatures are low enough to allow operation at, or close to, the compressor discharge temperature, which minimizes the use of a pre-burner. The outlet stage brings the partially combusted gases to the temperature required to attain homogeneous combustion. Because the outlet stage operates at a higher catalyst temperature, the stable catalyst in this stage will have a lower activity than the inlet stage catalyst. However, as the gas temperature in this stage is higher, the lower activity is adequate. In the final stage, homogeneous gas phase reactions complete the combustion of the fuel and bring the gases to the required combustor outlet temperature.

The temperature rise in the inlet stage is limited by taking advantage of the unique properties of palladium combustion catalysts. Under combustion conditions, palladium can be either in the form of the oxide or the metal. Palladium oxide is a highly active combustion catalyst, whereas palladium metal is much less active. Palladium oxide is formed under oxidizing conditions at temperatures higher than 400°F (200°C), but decomposes to the metal at temperatures between 1436°F (780°C) and 1690°F (920°C), depending on the pressure. So, when the catalyst temperature reaches about 1472°F (800°C), the catalytic activity will suddenly fall off, due to the formation of the less active palladium metal, preventing any further rise in temperature. The catalyst essentially acts as a kind of chemical thermostat that controls its own temperature.

Catalytic Combustor Design

Testing at full scale has been done in a catalytic combustor system developed by GE for its MS9001E gas turbine. The MS9001E combustor operates with a full-load firing temperature of 1105°C (2020°F) and a combustor exit temperature of about 1190°C (2170°F). The key components of the test stand at the GE Power Generation Engineering Laboratories in Schenectady, NY, are shown in Figure 4-37.

There are four major subassemblies in the overall combustion system: the pre-burner, the main fuel injector, the catalytic reactor, and the downstream liner leading to the transition piece. To summarize, their roles are:

- *Pre-burner* — the pre-burner carries the machine load at operating points where the conditions in the catalytic reactor are outside of the catalyst

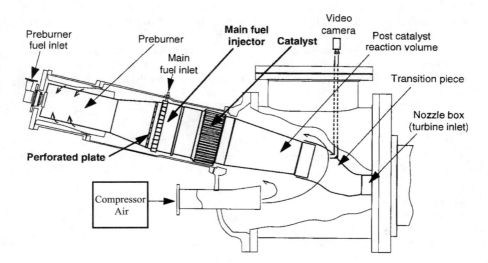

Figure 4-37. Schematic of a Full-Scale Catalytic Combustor (Courtesy of GE Power Systems and Catalytica Combustion System Inc.)

operating window. Most often, these are the low load points where the fuel required for turbine operation is insufficient for the catalyst to generate the necessary minimum exit gas temperature (cf. Figure 4-2). As the turbine load is increased, progressively more fuel is directed through the main injector, and progressively less goes to the pre-burner. Ultimately, the pre-burner receives only enough fuel to maintain the catalyst above its minimum inlet temperature.
- *Main fuel injector* — this unit is designed to deliver a fuel-air mixture to the catalyst that is uniform in composition, temperature, and velocity. A multi-venturi tube (MVT) fuel injection system was developed by GE specifically for this purpose. It consists of 93 individual venturi tubes arrayed across the flow path, with four fuel injection orifices at the throat of each venturi.
- *Catalytic reactor* — the role of the catalyst was described earlier; it must burn enough of the incoming fuel to generate an outlet gas temperature high enough to initiate rapid homogeneous combustion just past the catalyst exit.

The catalytic combustor has great potential in the application of gas turbines in new combined cycle power plants as the NO_x emissions in high attainment areas will have to be below 2 ppm.

Turbine Expander Section

There are two types of turbines used in gas turbines. These consist of the axial-flow type and the radial-inflow type. The axial-flow turbine is used in more than 95% of all applications.

The two types of turbines — axial-flow and radial-inflow turbines — can be divided further into impulse or reaction type units. Impulse turbines take their entire enthalpy drop though the nozzles, while the reaction turbine takes a partial drop through both the nozzles and the impeller blades.

The relative proportions of energy transfers obtained by a change of static and dynamic pressure are used to classify turbomachinery. The parameter used to describe this relationship is called the degree of reaction. Reaction is the energy transfer by means of a change in static pressure in a rotor to the total energy transfer in the rotor.

$$R = \frac{(h_2 - h_3)}{(H_{01} - H_{04})} = \frac{\left[\left(U_2^2 - U_2^2\right) + \left(W_1^2 - W_2^2\right)\right]}{\left[\left(V_1^2 - V_2^2\right) + \left(U_1^2 - U_2^2\right) + \left(W_1^2 - W_2^2\right)\right]} \qquad (4-41)$$

where
    H_{01} = total enthalpy entering the nozzle
    H_{04} = total enthalpy leaving the blade
    h_2 = static enthalpy at the leading edge of the blade
    h_4 = static enthalpy at the trailing edge of the blade
    U_1 = blade velocity at the leading edge of the blade
    U_2 = blade velocity at the trailing edge of the blade

V_1 = absolute velocity leaving the nozzle entering the blade (parallel to the exit angle of the nozzle)

V_2 = absolute velocity leaving the blade (parallel to the following nozzle entrance angle)

W_1 = relative velocity entering the blade (parallel to the blade entrance angle)

W_2 = relative velocity leaving the blade (parallel to the blade exit angle)

In a turbine, not all energy supplied can be converted into useful work even with an ideal fluid. There must be some kinetic energy at the exit velocity. The velocity exiting from a turbine is considered unrecoverable. The total head divided by the total head plus the absolute exit velocity head is known as the utilization factor.

Thus, the utilization factor is defined as the ratio of ideal work to the energy supplied

$$\epsilon = \frac{H_{ad}}{H_{ad} + \left(\frac{1}{2}V_4^2\right)} \qquad (4-42)$$

The effect of the utilization factor with speed is shown in Figure 4-38. The figure also shows the difference between an impulse and a 50% reaction turbine. An impulse turbine is a zero-reaction turbine.

In addition to the degree of reaction and the utilization factor, another parameter used to determine the blade loading is the work factor. The work factor is defined as the total head developed divided by the rotor head.

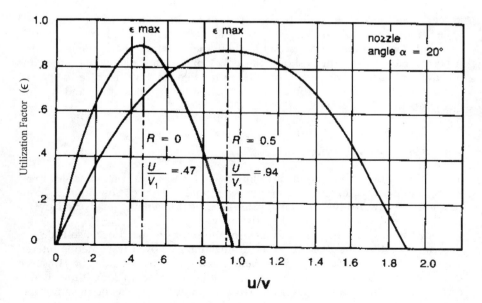

Figure 4-38. Variation of Utilization Factor with U/V

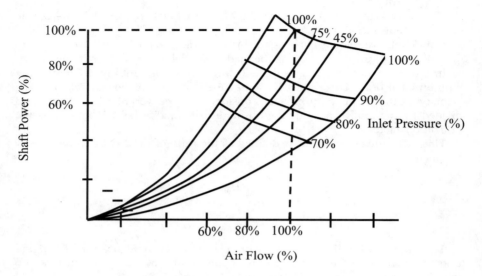

Figure 4-39. Turbine Performance Map

$$\Gamma = \frac{U_1 V_{\theta 1} - U_2 V_{\theta 2}}{U_2^2} \qquad (4-43)$$

The two conditions which vary the most in a turbine in the power generation mode are the inlet airflow and temperature, since speed is a constant. Two diagrams are needed to show their characteristics. Figure 4-39 is a performance map which shows the effect of turbine inlet temperature and airflow on the performance of the unit. Figure 4-40 is also a typical performance map of a turbine showing the effect of heat rate and exhaust gas temperature on power.

Radial-Inflow Turbine

The radial-inflow turbine, or inward-flow radial turbine, has been in use for many years. Basically, a centrifugal compressor with reversed flow and opposite rotation, the inward-flow radial turbine is used for smaller loads and over a smaller operational range than the axial turbine.

Radial-inflow turbines are only now beginning to be used because little was known about them heretofore. Axial turbines have enjoyed tremendous interest due to their low frontal area, making them suited to the aircraft industry. However, the axial machine is much longer than the radial machine, making it unsuited to certain applications. Radial turbines are used in turbochargers and in some types of expanders.

The inward-flow radial turbine has many components similar to a centrifugal compressor. There are two types of inward-flow radial turbines: the cantilever and the mixed-flow. The cantilever type in Figure 4-41 is similar to an axial-flow turbine, but it has radial blading. However, the cantilever turbine is not popular because of design and production difficulties.

Mixed-Flow Turbine

The turbine, as shown in Figure 4-42, is almost identical to a centrifugal compressor — except its components have different functions. The scroll is used to distribute the gas uniformly around the periphery of the turbine.

The nozzles, used to accelerate the flow toward the impeller tip, are usually straight vanes with no airfoil design. The vortex is a vaneless space and allows an equalization of the pressures. The flow enters the rotor radially at the tip with no appreciable axial velocity, and exits the rotor through the exducer axially with little radial velocity.

The nomenclature of the inward-flow radial turbine is shown in Figure 4-43. These turbines are used because of lower production costs, in part because the nozzle blading does not require any camber or airfoil design.

Axial-Flow Turbines

The axial-flow turbine, like its counterpart the axial-flow compressor, has flow which enters and leaves in the axial direction. There are two types of axial turbines: (1) impulse type and (2) reaction type. The impulse turbine has its entire enthalpy drop in the nozzle; therefore, it has a very high velocity entering the rotor. The reaction turbine divides the enthalpy drop in the nozzle and the rotor. Figure 4-44 is a schematic of an axial-flow turbine, also depicting the distribution of the pressure, temperature and the absolute velocity.

Most axial flow turbines consist of more than one stage, the front stages are usually impulse (zero reaction) and the later stages have about 50% reaction. The impulse stages produce about twice the output of a comparable 50% reaction stage, while the efficiency of an impulse stage is less than that of a 50% reaction stage.

The high temperatures that are now available in the turbine section are due to improvements of the metallurgy of the blades in the turbines. Development of directionally solidified blades, as well as the new single crystal blades, with the new coatings and the new cooling schemes, are responsible for the increase in firing temperatures. The high-pressure ratio in the compressor also causes the cooling air used in the first stages of the turbine to be very hot. The temperatures leaving the gas turbine compressor can reach as high as 1200°F (649°C). Thus, the present cooling schemes need revisiting, and also the cooling passages are, in many cases, also coated. The cooling schemes are limited in the amount of air they can use, before negating an effort in overall thermal efficiency due to an increase in the amount of air used in cooling. The rule of thumb in this area is that if you need

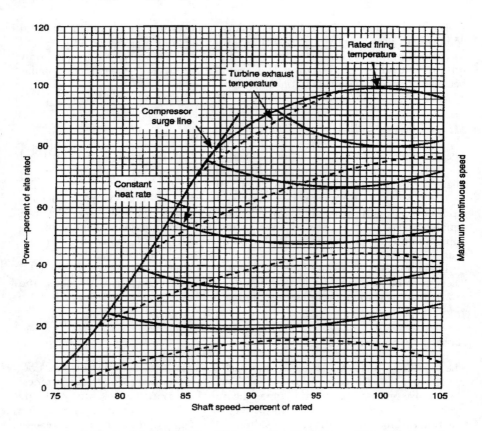

Figure 4-40. Performance Map Showing the Variation of Heat Rate and Power as a Function of Turbine Speed Chapter 3 [8] (Courtesy of American Petroleum Institute)

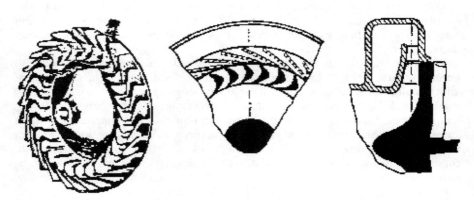

Figure 4-41. Cantilever Type Radial Inflow Turbine

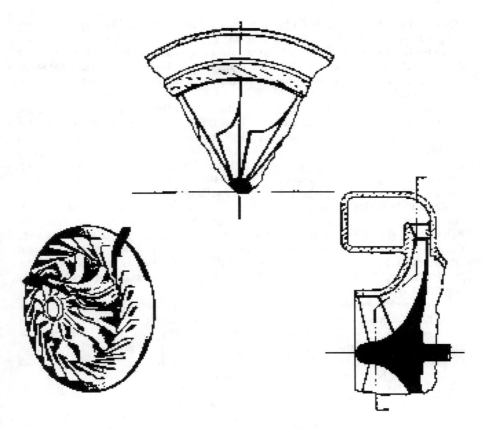

Figure 4-42. Mixed Flow Type Radial Inflow Turbine

more than 8% of the air for cooling, you are losing the advantage from the increase in the firing temperature.

The new gas turbines being designed for the new millennium are investigating the use of steam as a cooling agent for the first and second stages of the turbines. Steam cooling is possible in the new combined cycle power plants, which is the base of most of the new high performance gas turbines. Steam, as part of the cooling as well as part of the cycle power, will be used in the new gas turbines in the combined cycle mode. The extra power obtained by the use of steam is the cheapest MW/US$ available. The injection of about 5% of steam by weight of air amounts to about 12% more power. The pressure of the injected steam must be at least 40 Bar above the compressor discharge. The way steam is injected must be done very carefully so as to avoid compressor surge. These are not new concepts and have been used and demonstrated in the past. Steam cooling for example was the basis of the cooling schemes proposed by the team of United Technology and Stal-Laval in their conceptual study for the U.S. Department of Energy study on the

High Turbine Temperature Technology Program, which was investigating firing temperatures of 3000°F (1649°C), in the early 1980s.

Velocity Diagrams. An examination of various velocity diagrams for different degrees of reaction is shown in Figure 4-45. These types of blade arrangements with varying degrees of reaction are all possible; however, they are not all practical.

Examining the utilization factor, the discharge velocity, $V_4^2/2$, represents the kinetic energy loss or the unused energy part. For maximum utilization, the exit velocity should be at a minimum and, by examining the velocity diagrams; this minimum is achieved when the exit velocity is axial. This type of a velocity diagram is considered to have zero exit swirl. Figure 4-46 shows the various velocity diagrams as a function of the work factor and the turbine type. This diagram shows that zero exit swirl can exist for any type of turbine.

Zero Exit Swirl Diagram. In many cases, the tangential angle of the exit velocity, $V_{\theta 4}$, represents a loss in efficiency. A blade designed for zero exit swirl, $V_{\theta 4} = 0$, minimizes the exit loss. If the work parameter is less than two, this type of diagram produces the highest static efficiency. Also, the total efficiency is approximately the same as the other types of diagrams. If the Work Factor, Γ, is greater than 2.0, stage reaction is usually negative, a condition best avoided.

Impulse Diagram. For the impulse rotor, the reaction is zero, so the relative velocity of the gas is constant, or $W_3 = W_4$. if the work factor is less than 2.0, the exit swirl is positive, which reduces the stage work. For this reason, an impulse diagram should be used only if the work factor is 2.0 or greater. This type of diagram is a good choice for the last stage because for Γ greater than 2.0, an impulse rotor has the highest static efficiency.

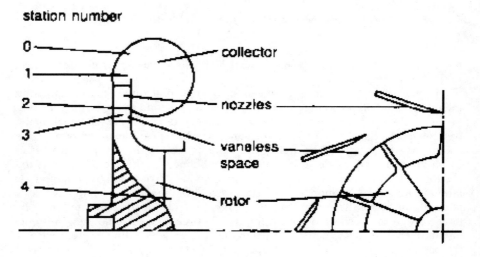

Figure 4-43. Components of a Radial Inflow Turbine

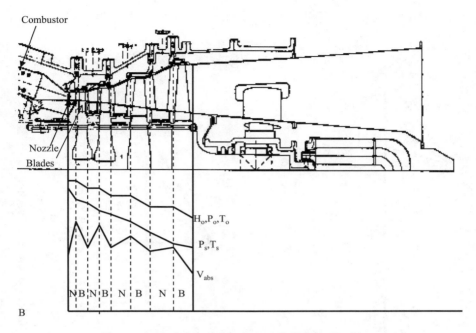

Figure 4-44. Schematic of an Axial Flow Turbine

Symmetrical Diagram. The symmetrical-type diagram is constructed so that the entrance and exit diagrams have the same shape, $V_3 = W_4$ and $V_4 = W_3$. This equality means that the reaction is $R = 50\%$. If the work factor, Γ, equals 1.0, then the exit swirl is zero. As the work factor increases, the exit swirl increases. Since the reaction of 50% leads to a high total efficiency, this design is useful if the exit swirl is not counted as a loss as in the initial and intermediate stages.

Impulse Turbine

The impulse turbine is the simplest type of turbine. It consists of a group of nozzles followed by a row of blades. The gas is expanded in the nozzle, converting the high thermal energy into kinetic energy. This conversion can be represented by the following relationship:

$$V = \sqrt{2\Delta h} \qquad (4-44)$$

The high-velocity gas impinges on the blade, where a large portion of the kinetic energy of the moving gas stream is converted into turbine shaft work. Figure 4-47 shows a diagram of a single-stage impulse turbine. The static pressure decreases in the nozzle with a corresponding increase in the absolute velocity. The absolute velocity is then reduced in the rotor, but the static

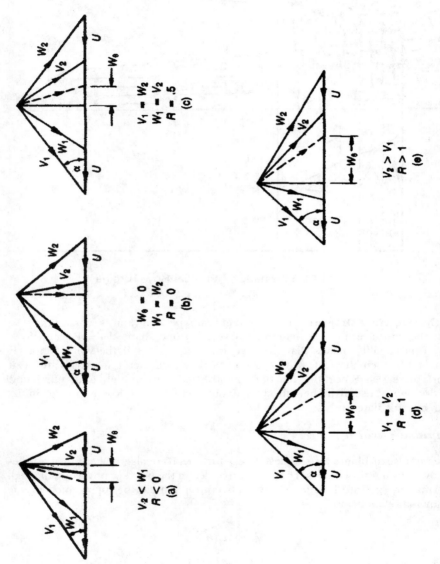

Figure 4-45. Velocity Triangles for Various Degrees of Reaction

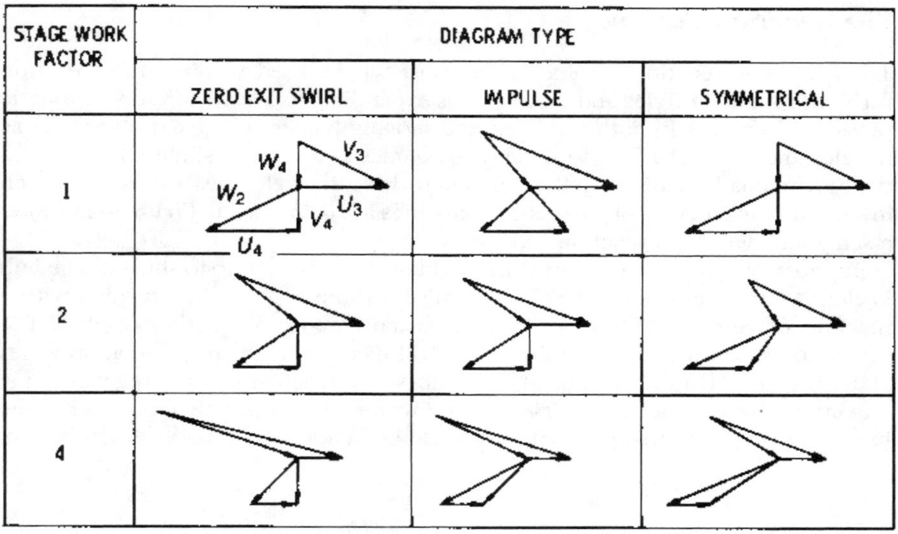

STAGE WORK FACTOR	DIAGRAM TYPE		
	ZERO EXIT SWIRL	IMPULSE	SYMMETRICAL
1	W_2 W_4 V_3 V_4 U_3 U_4		
2			
4			

Figure 4-46. Effect of Swirl and Reaction on the Work Factor

pressure and the relative velocity remain constant. To get the maximum energy transfer, the blades must rotate at about one-half the velocity of the gas jet velocity. By definition, the impulse turbine has a degree of reaction equal to zero. This degree of reaction means that the entire enthalpy drop is taken in the nozzle, and the exit velocity from the nozzle is very high. Since there is no change in enthalpy in the rotor, the relative velocity entering the rotor equals the relative velocity exiting from the rotor blade. For the maximum utilization factor, the absolute exit velocity must be axial.

The power developed by the flow in an impulse turbine is given by the Euler equation:

$$P = (\dot{m}_a + \dot{m}_f)(U_1 V_{\theta 1} - U_2 V_{\theta 2}) = (\dot{m}_a + \dot{m}_f)U(V_{\theta 1} - V_{\theta 2}) \qquad (4-45)$$

The relative velocity, W, remains unchanged in a pure impulse turbine, except for frictional and turbulence effect. This loss varies from about 20% for very high-velocity turbines (3000 fps) to about 8% for low-velocity turbines (500 fps). Since for maximum utilization, the blade speed ratio is equal to $U/V = (\cos \alpha)/2$, the energy transferred in an impulse turbine can be written as:

$$P = (\dot{m}_a + \dot{m}_f)U(2U_2 - 0) = 2(\dot{m}_a + \dot{m}_f)U^2 \qquad (4-46)$$

The Reaction Turbine

The axial-flow reaction turbine is the most widely used turbine. In a reaction turbine, both the nozzles and blades act as expanding nozzles. Therefore, the static pressure decreases in both the fixed and moving blades. The fixed blades act as nozzles and direct the flow to the moving blades at a velocity slightly higher than the moving blade velocity. In the reaction turbine, the velocities are usually much lower, and the entering blade relative velocities are nearly axial. Figure 4-48 shows a schematic view of a reaction turbine.

In most designs, the reaction of the turbine varies from hub to shroud. The hub is close to an impulse turbine. The impulse turbine is a reaction turbine with a reaction of zero ($r = 0$). The shroud is usually above a reaction of 50%. The utilization factor for a fixed nozzle angle will increase as the reaction approaches 100%. For $r = 1$, the utilization factor does not reach unity but reaches some maximum finite value. The 100% reaction turbine is not practical because of the high rotor speed necessary for a good utilization factor. For reaction less than zero,

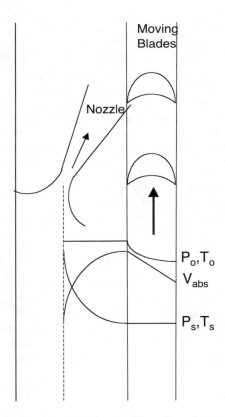

Figure 4-47. Schematic of an Impulse Turbine Showing the Variation of the Thermodynamic and Fluid Mechanic Properties

An Overview of Gas Turbines • 205

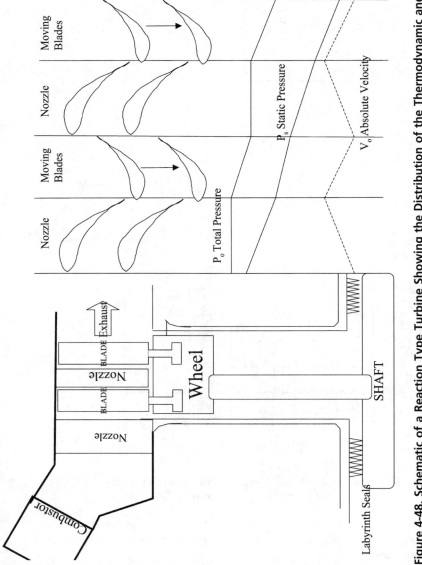

Figure 4-48. Schematic of a Reaction Type Turbine Showing the Distribution of the Thermodynamic and Fluid Mechanic Properties

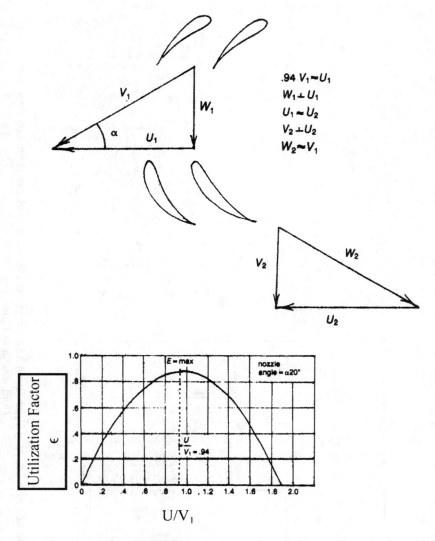

Figure 4-49. The Effect of Exit Velocity and Air Angle on the Utilization Factor for a 50% Reaction Axial Turbine Stage

the rotor has a diffusing action. Diffusing action in the rotor is undesirable, since it leads to flow losses.

The 50% reaction turbine has been used widely and has special significance. The velocity diagram of a 50% reaction is symmetrical and, for the maximum utilization factor, the exit velocity, V_4, must be axial. Figure 4-49 shows a velocity diagram of a 50% reaction turbine and the effect on the utilization factor. From the diagram $W_3 = V_4$, the angles of both the stationary and rotation blades are identical. Therefore, for maximum utilization,

$$\frac{U}{V_3} = \cos \alpha \qquad (4-47)$$

the 50% reaction turbine has the highest efficiency of all the various types of turbines. Equation 4-47 shows the effect on efficiency is relatively small for a wide range of blade speed ratios (0.6 to 1.3).

The power developed by the flow in a reaction turbine is also given by the general Euler equation. This equation can be modified for maximum utilization for a 50% reaction turbine with an axial exit; the Euler Equation reduces to:

$$P = (\dot{m}_a + \dot{m}_f)U(U-0) = (\dot{m}_a + \dot{m}_f)U^2 \qquad (4-48)$$

The work produced in an impulse turbine with a single stage running at the same blade speed is twice that of a reaction turbine. Hence, the cost of a reaction turbine for the same amount of work is much higher, since it requires more stages. It is a common practice to design multistage turbines with impulse stages in the first few stages to maximize the work and to follow it with 50% reaction turbines. The reaction turbine has a higher efficiency due to blade suction effects. This type of combination leads to an excellent compromise, since otherwise, an all-impulse turbine would have a very low efficiency, and an all-reaction turbine would have an excessive number of stages.

Turbine Blade Cooling Concepts

The turbine inlet temperatures of gas turbines have increased considerably over the past years and will continue to do so. This trend has been made possible by advancement in materials and technology, and the use of advanced turbine blade cooling techniques. The cooling air is bled from the compressor and is directed to the stator, the rotor, and other parts of the turbine rotor and casing to provide adequate cooling. The effect of the coolant on the aerodynamics depends on the type of cooling involved, the temperature of the coolant compared to the mainstream temperature, the location and direction of coolant injection, and the amount of coolant. In high-temperature gas turbines, cooling systems need to be designed for turbine blades, vanes, endwalls, shroud, and other components to meet metal temperature limits. The concepts underlying the following five basic air-cooling schemes are in Figure 4-50:

1. Convection cooling
2. Impingement cooling
3. Film cooling
4. Transpiration cooling
5. Water/Steam cooling.

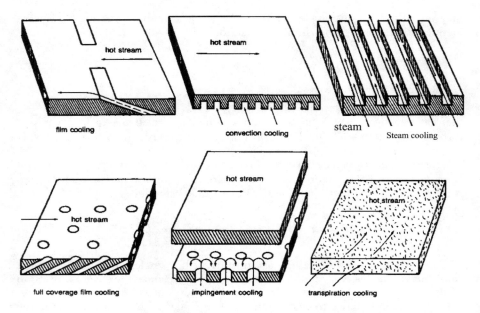

Figure 4-50. Types of Cooling Schemes Used in Gas Turbine for Cooling the Hot Section

Until the late 1960s, convection cooling was the primary means of cooling gas turbine blades; some film cooling was occasionally employed in critical regions. However, in the year 2000, steam cooling is being introduced in production frame type engines used in combined cycle applications. The new turbines have very high pressure ratios and this leads to compressor air leaving at very high temperatures, which affects their cooling capacity.

Convection Cooling. This form of cooling is achieved by designing the cooling air to flow inside the turbine blade or vane and remove heat through the walls. Usually, the airflow is radial, making multiple passes through a serpentine passage from the hub to the blade tip. Convection cooling is the most widely used cooling concept in present-day gas turbines.

Impingement Cooling. In this high-intensity form of convection cooling, the cooling air is blasted on the inner surface of the airfoil by high-velocity air jets, permitting an increased amount of heat to be transferred to the cooling air from the metal surface. This cooling method can be restricted to desired sections of the airfoil to maintain even temperatures over the entire surface. For instance, the leading edge of a blade needs to be cooled more than the mid-chord section or trailing edge, so the gas is impinged.

Film Cooling. This type of cooling is achieved by allowing the working air to form an insulating layer between the hot gas stream and the walls of the blade. This film of cooling air protects an airfoil in the same way combustor liners are protected from hot gases at very high temperatures.

Transpiration Cooling. Cooling by this method requires the coolant flow to pass through the porous wall of the blade material. The heat transfer is directly between the coolant and the hot gas. Transpiration cooling is effective at very high temperatures, since it covers the entire blade with coolant flow.

Steam Cooling. Steam is passed through a number of tubes embedded in the nozzles or blades. This method keeps blade metal temperatures below 1000°F.

Turbine Blade Cooling Design

The incorporation of blade cooling concepts into actual blade designs is very important. The most frequently used blade cooling designs are:

1. Convection and impingement cooling
2. Film and convection cooling
3. Steam cooling.

Convection and Impingement Cooling/Strut Insert Design. The strut insert design shown in Figure 4-51 has a mid-chord section, which is convection-cooled through horizontal fins, and a leading edge that is impingement cooled. The coolant is discharged through a split trailing edge.

The airflows up the central cavity formed by the strut insert and through holes at the leading edge of the insert to impingement cool the blade leading edge. The air then circulates through horizontal fins between the shell and strut and discharges through slots in the trailing edge. The temperature distribution for this design is shown also in Figure 4-51.

The stresses in the strut insert are higher than those in the shell, and the stresses on the pressure side of the shell are higher than those on the suction side. Considerably more creep strain takes place toward the trailing edge than the leading edge. The creep strain distribution at the hub section is unbalanced. This imbalance can be improved by a more uniform wall temperature distribution.

Film and Convection Cooling Design. This type of blade design with the temperature distribution for film and convection cooling design is shown in Figure 4-52. The mid-chord region is convection-cooled, and the leading edges are both convection- and film-cooled. The cooling air is injected through the blade base into two central and one leading edge cavity. The air then circulates up and down a series of vertical passages. At the leading edge, the air passes through a series of small holes in the wall of the adjacent vertical passages, and then impinges on the inside surface of the leading edge and passes through film cooling holes. The trailing edge is convection-cooled by air discharging through slots. From the cooling distribution diagram, the hottest section can be seen to be the trailing edge. The web, which is the most highly stressed blade part, is also the coolest part of the blade.

A similar cooling scheme with some modifications is used in some of the latest gas turbine designs. The firing temperature of GE FA units — 2350°F (1288°C) — is the highest in the power generation industry. To accommodate this increased firing temperature, the FA employs advanced cooling techniques developed by GE Aircraft Engines. The first- and second-stage blades, as well as all three-nozzle stages, are air-cooled. The first-stage blade is convectively cooled by means of an

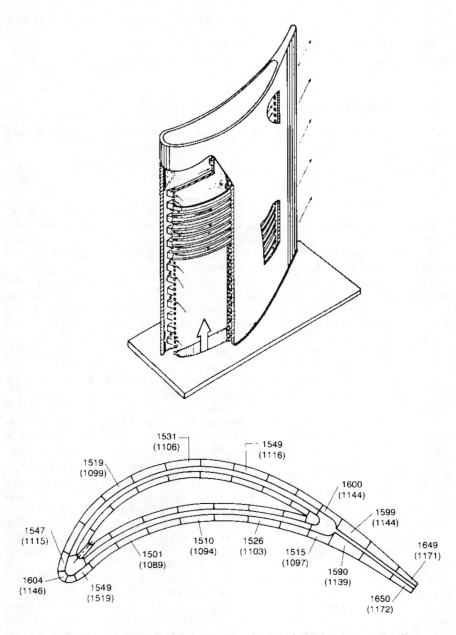

Figure 4-51. Cooling Based on Convection and Impingement Cooling. Temperature Distribution Based on Uncooled °F and (Cooled °F) Blades

An Overview of Gas Turbines • 211

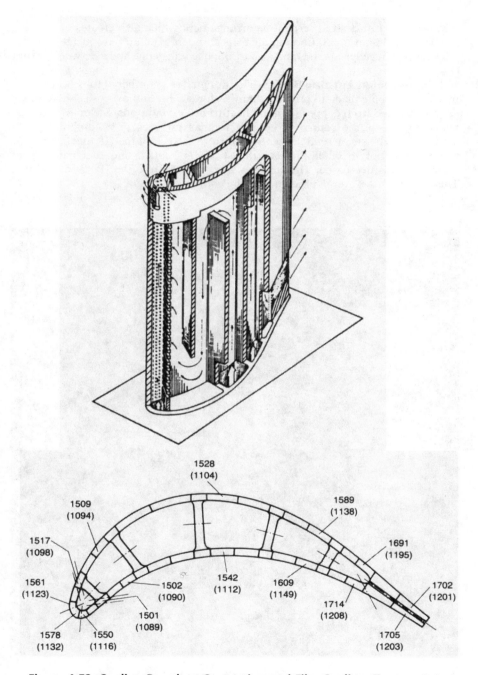

Figure 4-52. Cooling Based on Convection and Film Cooling. Temperature Distribution Based on Uncooled °F and (Cooled °F) Blades

advanced aircraft-derived serpentine arrangement which also creates turbulence as shown in Figure 4-53. Cooling air exits through axial airways located on the bucket's trailing edge and tip, and also through leading edge and sidewalls for film cooling.

Steam-Cooled Turbine Blades. This design has a number of tubes embedded inside the turbine blade to provide channels for steam. In most cases, these tubes are constructed from copper for good heat-transfer conditions. Steam injection is becoming the prime source of cooling for gas turbines in a combined cycle application. The steam, which is extracted from the exit of the HP turbine, is sent through the nozzle blades, where the steam is heated, and the blade metal temperature is decreased. The steam is then injected into the IP steam turbine. This increases the overall efficiency of the combined cycle.

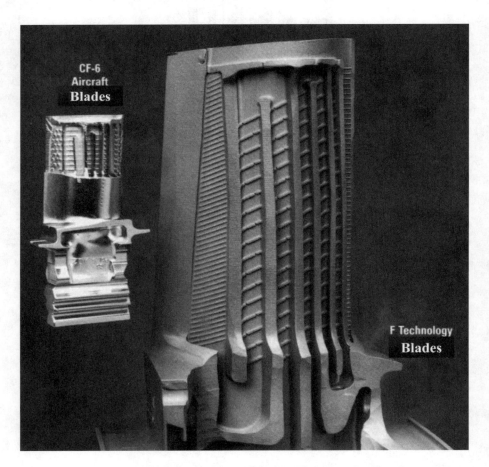

Figure 4-53. Internal of the Frame FA Blades, Showing Cooling Passages (Courtesy of GE Power Systems)

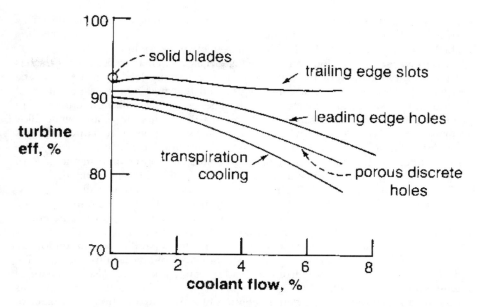

Figure 4-54. The Effect of Various Types of Cooling on Turbine Blade Efficiency

Steam cooling in combined cycle power plants holds great promise for the turbines of the future, in which turbine inlet temperatures of 3000°F (1649°C) are possible. This type of cooling should keep blade metal temperatures below 1000°F (538°C) so that there will be no hot-corrosion problems.

Cooled-Turbine Aerodynamics

The injection of coolant air in the turbine rotor or stator causes a slight decrease in turbine efficiency; however, the higher turbine inlet temperature usually makes up for the loss of the turbine component efficiency, giving an overall increase in cycle efficiency.

Total pressure surveys were made downstream of the stators in both the radial and circumferential directions to determine the effect of coolant on stator losses. The wake traces for the stator with discrete holes and the stator with trailing edge slots show that there is a considerable difference in total pressure loss patterns as a function of the type of cooling and the amount of cooling air supplied. As the coolant flow for the porous blades increases, the disturbance to the flow pattern and the wake thickness increases. Consequently, the losses increase. In a blade with trailing edge slots, the loss initially starts to increase with coolant flow as the wake thickens. However, as the coolant flow is increased, it tends to energize the wake and reduce losses. For a higher coolant flow, the coolant pressures must be higher, resulting in an energization of the flow.

By comparing the various cooling techniques as seen in Figure 4-54, it becomes obvious that a blade with trailing edge slots is thermodynamically the most

214 • COGENERATION AND COMBINED CYCLE POWER PLANTS

efficient. The porous stator blades decrease the stage efficiency considerably. This efficiency indicates losses in the turbine, but does not take into account cooling effectiveness. The porous blades are more effective for cooling.

Materials. The development of new materials, as well as cooling schemes, has seen the rapid growth of the turbine firing temperature leading to high turbine efficiencies. The stage 1 blade must withstand the most severe combination of temperature, stress, and environment; it is generally the limiting component in the machine. Figure 4-55 shows the trend of firing temperature and blade alloy capability. Since 1950, turbine bucket material temperature capability has advanced approximately 850°F (472°C), approximately 20°F/10°C/year. The importance of this increase can be appreciated by noting that an increase of 100°F (56°C) in turbine firing temperature can provide a corresponding increase of 8% to 13% in output, and 2% to 4% improvement in simple-cycle efficiency. Advances in alloys and processing, while expensive and time-consuming, provide significant incentives through increased power density and improved efficiency.

The composition of the new and conventional alloys discussed is shown in Table 4-3. The increases in blade alloy temperature capability accounted for the majority of the firing temperature increase until air-cooling was introduced, which decoupled firing temperature from the blade metal temperature. Also, as the metal temperatures approached the 1600°F (870°C) range, hot corrosion of blades became more life-limiting than strength until the introduction of protective coatings. During the 1980s, emphasis turned toward two major areas: improved materials technology, to achieve greater blade alloy capability without sacrificing alloy corrosion resistance; and advanced, highly sophisticated air-cooling technology to achieve the firing temperature capability required for the new generation of gas turbines. The use of steam cooling to further increase combined-cycle efficiencies in combustors was introduced in the mid to late 1990s. Steam

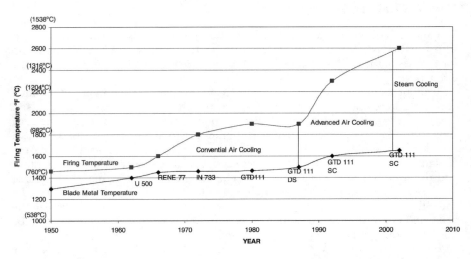

Figure 4-55. Firing Temperature Increase with Blade Material Improvement

Table 4-3. High Temperature Alloys (Courtesy of GE Power Systems)

Component	Cr	Ni	Co	Fe	W	Mo	Ti	Al	Cb	V	C	B	Ta
Turbine Blades													
U500	18.5	BAL	18.5	-	-	4	3	3	-	-	0.07	0.006	-
RENE 77 (U700)	15	BAL	17	-	-	5.3	3.35	4.25	-	-	0.07	0.02	-
IN738	16	BAL	8.3	0.2	2.6	1.75	3.4	3.4	0.9	-	0.10	0.001	1.75
GTD111	14	BAL	9.5	-	3.8	1.5	4.9	3.0	-	-	0.10	0.01	2.8
Turbine Nozzles													
X40	25	10	BAL	1	8	-	-	-	-	-	0.50	0.01	-
X45	25	10	BAL	1	8	-	-	-	-	-	0.25	0.01	-
FSX414	28	10	BAL	1	7	-	-	-	-	-	0.25	0.01	-
N155	21	20	20	BAL	2.5	3	-	-	-	-	0.20	-	-
GTD-222	22.5	BAL	19	-	2.0	2.3	1.2	0.8	-	0.10	0.008	1.00	-
Combustors													
SS309	23	13	-	BAL	-	-	-	-	-	-	0.10	-	-
HAST X	22	BAL	1.5	1.9	0.7	9	-	-	-	-	0.07	0.005	-
N-263	20	BAL	20	0.4	-	6	2.1	0.4	-	-	0.06	-	-
HA-188	22	22	BAL	1.5	14.0	-	-	-	-	-	0.05	0.01	-
Turbine Wheels													
ALLOY 718	19	BAL	-	18.5	-	3.0	0.9	0.5	5.1	-	0.03	-	-
ALLOY 706	16	BAL	-	37.0	-	-	1.8	-	2.9	-	0.03	-	-
Cr-Mo-V	1	0.5	-	BAL	-	1.25	-	-	-	-	0.30	-	-
A286	15	25	-	BAL	-	1.2	2	0.3	-	0.25	0.08	0.006	-
M152	12	2.5	-	BAL	-	1.7	-	-	-	0.3	0.12	-	-
Compressor Blades													
AISI 403	12	-	-	BAL	-	-	-	-	-	-	0.11	-	-
AISI 403 + Cb	12	-	-	BAL	-	-	-	-	0.2	-	0.15	-	-
GTD-450	15.5	6.3	-	BAL	-	0.8	-	-	-	-	0.03	-	-

cooling in blades and nozzles will be introduced in commercial operation in the year 2002.

In the 1980s, IN 738 blades were widely used. IN-738 was the acknowledged corrosion standard for the industry. New alloys, such as GTD-111, was developed and patented by GE in the mid-1970s. It possesses about a 35°F (20°C) improvement in rupture strength as compared to IN-738. GTD-111 is also superior to IN-738 in low-cycle fatigue strength.

The design of this alloy was unique in that it utilized phase stability and other predictive techniques to balance the levels of critical elements (Cr, Mo, Co, Al, W and Ta), thereby maintaining the hot corrosion resistance of IN-738 at higher strength levels without compromising phase stability. Most nozzle and blade castings are made by using the conventional equiaxed investment casting process. In this process, the molten metal is poured into a ceramic mold in a vacuum, to prevent the highly reactive elements in the super alloys from reacting with the oxygen and nitrogen in the air. With proper control of metal and mold thermal conditions, the molten metal solidifies from the surface to the center of the mold, creating an equiaxed structure.

Directional solidification (DS) is also being employed to produce advanced technology nozzles and blades. First used in aircraft engines more than 25 years ago, it was adapted for use in large airfoils in the early 1990s. By exercising careful control over temperature gradients, a planar solidification front is developed in the bade, and the part is solidified by moving this planar *front* longitudinally through the entire length of the part. The result is a blade with an oriented grain structure that runs parallel to the major axis of the part and contains no transverse grain boundaries, as in ordinary blades. The elimination of these transverse grain boundaries confers additional creep and rupture strength on the alloy, and the orientation of the grain structure provides a favorable modulus of elasticity in the longitudinal direction to enhance fatigue life. The use of directionally solidified blades results in a substantial increase in the creep life, or substantial increase in tolerable stress for a fixed life. This advantage is due to the elimination of transverse grain boundaries from the bucket, the traditional weak link in the microstructure. In addition to improved creep life, the directionally solidified blades possess more than ten times the strain control or thermal fatigue compared to equiaxed blades. The impact strength of the DS blades is also superior to that of equiaxed, showing an advantage of more than 33%.

In the late 1990s, single-crystal blades have been introduced in gas turbines. These blades offer additional, creep and fatigue benefits through the elimination of grain boundaries. In single-crystal material, all grain boundaries are eliminated from the material structure and a single crystal with controlled orientation is produced in an airfoil shape. By eliminating all grain boundaries and the associated grain boundary strengthening additives, a substantial increase in the melting point of the alloy can be achieved, thus providing a corresponding increase in high-temperature strength. The transverse creep and fatigue strength is increased, compared to equiaxed or DS structures. The advantage of single-crystal alloys compared to equiaxed and DS alloys in low-cycle fatigue (LCF) life is increased by about 10%.

Blade life comparison is provided in the form of the stress required for rupture as a function of a parameter that relates time and temperature (the Larson-Miller parameter). The Larson-Miller parameter is a function of blade metal temperature and the time the blade is exposed to those temperatures. Figure 4-56 shows the comparison of some of the alloys used in blade and nozzle application. This parameter is one of several important design parameters that must be satisfied to ensure proper performance of the alloy in a blade application, especially for long service life. Creep life, high- and low-cycle fatigue, thermal fatigue, tensile strength and ductility, impact strength, hot corrosion and oxidation resistance, producibility, coatability and physical properties must also be considered.

Coatings. Blade coatings are required to protect the blade from corrosion, oxidation and mechanical property degradation. As super alloys have become more complex, it has been increasingly difficult to obtain both the higher strength levels that are required and a satisfactory level of corrosion and oxidation resistance without the use of coatings. Thus, the trend toward higher firing temperatures increases the need for coatings. The function of all coatings is to provide a surface reservoir of elements that will form very protective and adherent oxide layers, thus protecting the underlying base material from oxidation and corrosion attack and degradation.

The main requirements of a coating are to protect blades against oxidation and/or corrosion problems. Other benefits of coatings include thermal fatigue

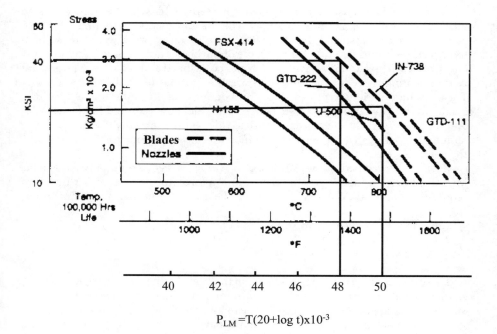

$$P_{LM} = T(20 + \log t) \times 10^{-3}$$

Figure 4-56. Larson Miller Parameter for Various Types of Blades

from cyclic operation, surface smoothness and erosion in compressor coatings, and heat flux loading when one is considering thermal barriers. A secondary consideration, but perhaps rather more relevant to thermal barriers, is their ability to tolerate damage from light impacts without spalling to an unacceptable extent because of the resulting rise in the local metal temperatures. Coatings also extend life, provide protection by enduring the operational conditions, and protect the blades by being sacrificial by allowing the coating to be restripped and recoated on the same base metal.

Life of coatings depends on composition, thickness, and the standard of evenness to which it has been deposited. Coatings help extend the life of bladings by protecting them against oxidation, corrosion, thermal fatigue, temperature and foreign object damage (FOD) damage. Oxidization is a prime consideration in "clean fuel" regime, while corrosion is due to higher metal temperatures, and emphasis in not so clean a fuel.

For a given combination of loadings, coating life is governed by:

1. Composition of the coating, which includes environmental properties and mechanical properties, such as thermal fatigue.
2. Coating thickness, where there is a greater protective reservoir if thicker; however, thicker coatings may have lower thermal fatigue resistance.
3. Standard of deposition, such as thickness uniformity or defined thickness variation and coating defects.

There are three basic types of coatings: thermal barrier coatings, diffusion coatings, and plasma sprayed coatings. The advancements in coating have also been essential in ensuring that the blade base metal is protected at these high temperatures. Coatings ensure that the life of the blades are extended, and in many cases, are used as sacrificial layer, which can be stripped and recoated. The general type of coatings is very little different from the coatings used 10 to 15 years ago. These include various types of diffusion coatings such as Aluminide Coatings, originally developed nearly 40 years ago. The thickness required is between 25- and 75- m thick, these coatings consisted of Ni/Co = about 30% Al. The new aluminide coatings with platinum (Pt) increase the oxidation resistance and also the corrosion resistance.

Coatings developed some 30 years to 35 years ago, commonly known as MCrAlY, have a wide range of composition tailored to the type of performance required, and are Ni/Co-based as shown in these three common types of coatings:

1. Ni, 18% Cr, 12% Al, 0.3% Y
2. Co, 29% Cr, 3% AI, 0.3% Y
3. Co, 25% Ni, 20% Cr, 8% Al, 0.3% Y.

These coatings are usually 75- to 500- m thick, and sometimes have other minor element additions used to improve environmental resistance such as Pt, Hf, Ta, and Zr. Carefully chosen these coatings can give very good performance.

The thermal barrier coatings have an insulation layer of 100- to 300- m thick, and are based on ZrO_2-Y_2O_3 and can reduce metal temperatures by 90°F to 270°F

(50°C to 150°C). This type of coating is used in combustion cans, transition pieces, nozzle guide vanes, and also blade platforms.

The interesting point to note is that some of the major manufacturers are switching away from corrosion protection biased coatings towards coatings, which are not only oxidation-resistant, but also oxidation-resistant at higher metal temperatures. Thermal barrier coatings are being used on the first few stages in all the advanced technology units. The use of internal coatings is getting popular due to the high temperature of the compressor discharge, which results in oxidation of the internal surfaces. Most of these coatings are aluminide type coatings. The choice is restricted due to access problems to slurry based, or gas phase/chemical vapor deposition. Care must be taken in production; otherwise, internal passages may be blocked. The use of pyrometer technology on some of the advanced turbines has located blades with internal passages blocked, causing that blade to operate at metal temperatures of 50°F to 100°F (28°C to 56°C) higher than the neighboring blades.

Instrumentation and Controls

The advanced gas turbines are all digitally controlled and incorporate on-line condition monitoring. The addition of new on-line monitoring requires new instrumentation. The use of pyrometers is to sense blade metal temperatures, which are the real concern, and not the exit gas temperature. The use of dynamic pressure transducers for detection of surge and other flow instabilities in the compressor and also in the combustion process, especially in the new low NO_x combustors, are growing. The use of accelerometers to detect high-frequency excitation of the blades is being used to prevent major failures in the new highly loaded gas turbines.

The use of pyrometers in control of the advanced gas turbines is being investigated. Presently, all turbines are controlled based on gasifier turbine exit temperatures, or power turbine exit temperatures. By using the blade metal temperatures of the first section of the turbine, the gas turbine is being controlled at its most important parameter, the temperature of the first-stage nozzles and blades. In this manner, the turbine is being operated at its real maximum capability.

The use of dynamic pressure transducers gives early warning of problems in the compressor. The very high pressure in most of the advanced gas turbines cause these compressors to have a very narrow operating range between surge and choke. Thus, these units are very susceptible to dirt and blade vane angles. The early warning provided by the use of dynamic pressure measurement at the compressor exit can save major problems encountered due to tip stall and surge phenomenon.

The use of dynamic pressure transducer in the combustor section, especially in the low NO_x combustors, ensures that each combustor can is burning evenly. This is achieved by controlling the flow in each combustor can till the spectrums obtained from each combustor can match. This technique has been used and found to be very effective and ensures smooth operation of the turbine.

Performance monitoring not only plays a major role in extending life, diagnosing problems, and increasing time between overhauls, but also can provide major savings on fuel consumption by ensuring that the turbine is being operated at its most efficient point. Performance monitoring requires an in-depth understanding of the equipment being measured. The development of algorithms for a complex train needs careful planning, understanding of the machinery and process characteristics. In most cases, help from the manufacturer of the machinery would be a great asset. For new equipment, this requirement can and should be part of the bid requirements. For plants with already installed equipment, a plant audit to determine the plant machinery status is the first step.

Chapter 5

AN OVERVIEW OF STEAM TURBINES

A steam turbine may be defined as a form of heat engine in which the energy of the steam is transformed into kinetic energy by means of expansion through nozzles and the kinetic energy of the resulting jet is in turn converted into force doing work on rings of blading mounted on a rotating part. The basic idea of steam turbines was conceived as early as 120 BC, yet it was in 1883 that the first practical steam turbine was developed by De Laval.

A typical steam turbine power plant is divided up into its heat sources, the boiler or steam generator, and the turbine cycle, which includes the turbine, generator, condenser pumps, and feedwater heaters. The steam turbine operates on the Rankine Cycle.

The Rankine Cycle

The Rankine Cycle employing water-steam as the working fluid is the most common thermodynamic cycle utilized in the production of electrical power. A schematic of a steam power plant is shown in Figure 5-1. Water enters the boiler feed water pump at point 1 and is pumped isentropically into the boiler. The compressed liquid at 2 is heated until it becomes saturated at 2A, after which it is evaporated to steam at 2B and then superheated to 3. The steam leaves the boiler at 3, expands isentropically in the ideal engine to 4, and passes to the condenser. Circulating water condenses the steam to a saturated liquid at 1, from which state the cycle repeats itself.

The thermodynamic diagrams corresponding to the steam power plant in Figure 5-1, showing the thermodynamic states, are shown in the pressure-volume (PV) diagram in Figure 5-2, and the temperature entropy (T-S) diagram in Figure 5-3.

The work done by the steam turbine W_{st} is given by

$$W_{st} = \dot{m}_{st}(h_3 - h_4) \qquad (5-1)$$

where M_s is the mass flow of the steam and h_3, h_4 are enthalpies at points 3 and 4.

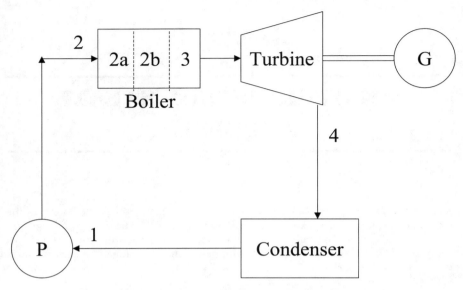

Figure 5-1. Schematic of a Steam Turbine Power Plant

The net work produced by the system, W_{net}, is the turbine work less the pump work, W_p, required to raise the water to the desired pressure and is expressed below:

$$W_{net} = W_{st} - W_p \qquad (5-2)$$

$$W_{net} = M_s(h_3 - h_4) - W_p \qquad (5-3)$$

The above analysis assumes an ideal isentropic expansion from points 3 to 4. In a steam turbine, the actual process is not isentropic and some loss does occur. The actual expansion is from points 3 to 4. The pump work is much smaller than the turbine work and can be neglected when estimating the overall performance and efficiency of steam plants.

The energy-input requirement to the system, Q_{in}, is given by

$$Q_{in} = \dot{m}_{st}(h_3 - h_B) \qquad (5-4)$$

The thermal efficiency of the system, η, is then given by:

$$\eta = \frac{W_{net}}{Q_{in}} \qquad (5-5)$$

$$\eta = \frac{\dot{m}_{st}(h_3 - h_4) - W_p}{\dot{m}_{st}(h_3 - h_B)} \qquad (5-6)$$

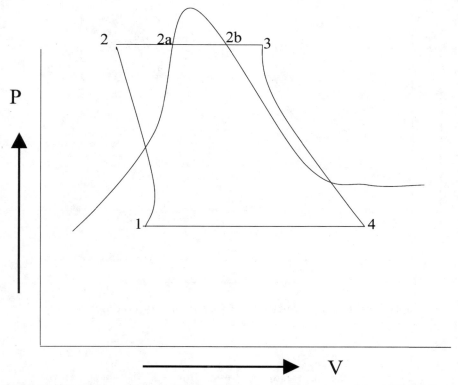

Figure 5-2. Pressure – Volume Diagram of a Typical Steam Turbine Power Plant

The overall efficiency of the system can be increased by pre-heating boiler feed water and the incoming combustion air by using hot exhaust from the boiler or gas turbine if available.

The Steam Regenerative-Reheat Cycle

It is evident from the Rankine Cycle shown in Figure 5-1 that considerable amount of heat is required to raise the temperature of the water from 2 to 2a. The Rankine Cycle has the disadvantage in that the fluid temperature at the pump discharge is much lower than the fluid temperature at the turbine inlet. One way of overcoming this disadvantage is to use the internal system heat rather than the external heat to minimize this difference in the temperatures. This concept is called the regenerative heating. In the gas turbines, the regenerative heating is accomplished by using the high-temperature exhaust gases. In the steam turbines, intermediate pressure steam rather than exhaust steam is used for heating feed water.

224 • COGENERATION AND COMBINED CYCLE POWER PLANTS

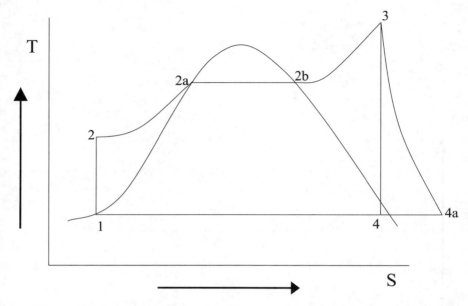

Figure 5-3. Temperature-Entropy Diagram of a Typical Steam Turbine Power Plant

As the high-pressure steam progresses through the steam turbine, the steam gets very wet at low pressures. This wet steam is detrimental for a turbine, it results in the reduction of efficiency and also in nozzle and blade erosion. The reheat cycle involves heating of the steam withdrawn after partial expansion. This idea, combined with regenerative heating for improved thermal efficiency, is commonly practiced among central power plants.

A simplified concept of the regenerative reheat steam cycle is depicted in Figures 5-4 and 5-5. The water enters the first pump at point 1 from where it enters the feedwater heater at point 2. In the feedwater heater, the pressurized condensate, is heated by part of the steam extracted from the high-pressure turbine at an intermediate pressure, point 6. The rest of the extracted steam is reheated in the reheater and enters the turbine at point 7. The heated water enters the second pump at point 3 from which it enters the boiler at point 4. The compressed liquid at 4 is heated until it becomes saturated at 4a, after which it is evaporated to steam at 4b, and then superheated to 5. The steam leaves the boiler at 5, expands isentropically in the ideal engine to 6, and in the real case to 6a where it is extracted for regeneration and reheat. The superheated steam now leaves the reheater at 7, expands isentropically in the ideal engine to 8 and in the real case to 8a where it passes to the condenser. Circulating water condenses the steam to a saturated liquid at 1, from which state the cycle repeats itself.

An Overview of Steam Turbines • 225

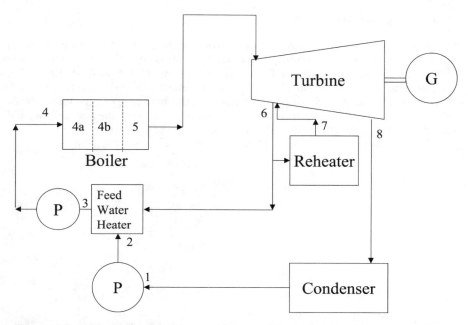

Figure 5-4. Schematic of a Regenerative – Reheat Steam Turbine Power Plant

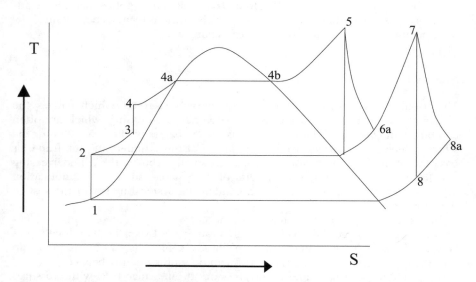

Figure 5-5. Temperature-Entropy Diagram of a Regenerative – Reheat Steam Turbine Power Plant

Heat Rate and Steam Rate

The heat rate is a modified reciprocal of the thermal efficiency and is in much wider use among steam-power and turbine engineers. The heat rate for a turbine is defined as the heat chargeable in Btu per kilowatt hour of horsepower-hour turbine output. Again, the basis upon which the turbine output is taken should be specified. Turbine heat rate should not be confused with the heat rate of the steam power plant known as the station heat rate. The station heat rate, like station thermal efficiency, takes into account all the losses from fuel to switchboard.

The heat rate for a straight condensing or non-condensing turbine is

$$HR = \frac{(h_3 - h_{f2})3415}{W_{net}} \quad \text{Btu per kw-hr} \tag{5-7}$$

or

$$HR = \frac{(h_3 - h_{f2})2544.4}{W_{net}} \tag{5-8}$$

where
h_1 = enthalpy of throttle steam
h_{f2} = enthalpy of liquid water at exhaust pressure
W_{net} = output on any specified basis, Btu per lb of steam at throttle

Steam rate is defined as the mass rate of steam flow in pounds per hour divided by the power or rate of work development of the turbine in kilo power-hour. The steam rate, therefore, is the steam supplied per kilo watt-hour or horsepower-hour unit of output. The heat rate may be obtained by multiplying the steam rate by the heat chargeable.

Steam Turbine

The usual turbine consists of four basic parts: the rotor, which carries the blades or buckets; the stator consisting of cylinder and casing, which are often combined and within which the rotor turns; the nozzles or flow passages for the steam, which are generally fixed to the inside of the cylinder; and the frame or base for supporting the stator and rotor. In small turbines, the cylinder casing and frame are often combined. Several other systems, such as the lubrication systems, steam piping systems and a condensing system, make up the rest of the turbine.

In most large plants, the steam turbine consists of three sections; a high-pressure turbine stage (HP) with pressures ranging between 1450 and 4500 psia (100 - 310 bar), with temperatures reaching as high as 1212°F (656°C), an intermediate-pressure turbine stage (IP) with pressures ranging between 300 and 600 psia (20.6 - 41.7 bar), and a low-pressure turbine stage (LP) with pressures ranging between 90 and 60 psia (6.1 - 3.1 bar). The steam exiting from the high-pressure stage is usually reheated to about the same temperature as the steam entering the HP stage before it enters the intermediate stage. The steam exiting the IP stage enters directly into the LP stage after it is mixed with steam coming from the LP superheater.

The steam from the LP turbine enters the condenser. The condenser is maintained at a vacuum ranging between 0.13 and 0.033 bar. The increase in back pressure in the condenser will reduce the power produced. Care must be taken to ensure that the steam leaving the LP stage blades does not carry a high content of liquid in the steam to avoid erosion of the LP blading.

Units which have been operated at 4350 psia (300 bar) with temperatures of 1112°F (600°C) for all three units with reheat have about an 8% thermal efficiency increase, as compared to a conventional unit operating at 1885 psia (130 bar) and 1000 (539°C) reheat system.

Steam turbines used in modern combined cycle power plants are simple machines. The important requirements for a modern combined cycle steam turbine are:

1. Ability to operate over a wide range of steam flows
2. High efficiency over a large operating range
3. Reheat possibilities
4. Fast startup
5. Short installation time
6. Floor-mounted installations.

The plants operate over a wide range of steam flows as the plants are now often cycled between base load and 50% of the base load in a 24-h period. Thus, this requires a high efficiency over a wide operating range. Combined cycle power plants operate at many pressure levels. It is not uncommon for manufacturing plants that use similar gas turbines to operate at two or three pressure levels. The combined cycle steam turbine has fewer or 0 bleed points compared to anywhere between 4 and 8 bleed points for feed water heaters.

Rapid startup is often very important since many of these plants are started up on a daily basis. Great care in design must be exercised due to rapid increase in temperatures. This does not allow for rotor wheels, which use a shrink fit on to the shaft. There have been cases during rapid start up the rotor wheel "walking" on the shaft due to the different growth rates between the shaft and the rotor wheel.

Reheat of the steam is being used in many steam turbines nowadays. Reheat improves the overall combined cycle efficiency.

Classification of Steam Turbines

Several classification schemes are available but most of them use the following chief characteristics:

1. Steam flow direction
    1.1. Axial flow turbines
    1.2. Radial flow turbines

228 • COGENERATION AND COMBINED CYCLE POWER PLANTS

    1.3. Tangential flow
    1.4. Mixed flow turbines

2. Steam passage between blades
    2.1. Impulse turbines
    2.2. Reaction or Parsons turbines
    2.3. Impulse and Reaction Combination

3. Arc of peripheral admission to the total circumference
    3.1. Steam Turbine Nozzles
    3.2. Full admission turbines
    3.3. Partial admission turbines

4. Turbines stages in series
    4.1. Single-stage or simple-impulse turbine
    4.2. Multi-stage turbine

5. Type of stage
    5.1. Pressure stage, Rateau-type impulse
    5.2. Velocity stage, Curtis-type impulse
    5.3. Pressure and velocity stage combination

6. General flow arrangement
    6.1. Single flow
    6.2. Double flow
    6.3. Compound flow
    6.4. Extraction flow

Steam Flow Directions. Steam enthalpy is converted into rotational energy as it passes through a turbine stage. A turbine stage consists of a stationary blade (or nozzle) and a rotating blade (or bucket). Stationary blades convert the potential energy of the steam (temperature and pressure) into kinetic energy (velocity) and direct the flow onto the rotating blades. The rotating blades convert the kinetic energy into impulse and reaction forces, caused by pressure drop, which result in the rotation of the turbine shaft or rotor.

Axial Flow. Axial flow signifies steam flow substantially parallel to the axis of rotation, among blades that are set radially. This is the only arrangement used in medium and large turbines and is also most commonly used in small turbines. It provides opportunity for almost any desired degree of expansion of the steam by increase in the length of the blades and the diameter at which they rotate, coupled with increase in the number of rows of blades.

Radial Flow. Radial flow is obtained when the steam enters at or near the shaft and flows substantially radially outward among blades, which are placed parallel to the axis of rotation. In one turbine of this type, the Ljongstrom turbine, as seen in Figure 5-6, successive rings of blades are attached alternately to two discs mounted on separate rotors, with these rotors turning in opposite directions on the same axis and driving separate generators.

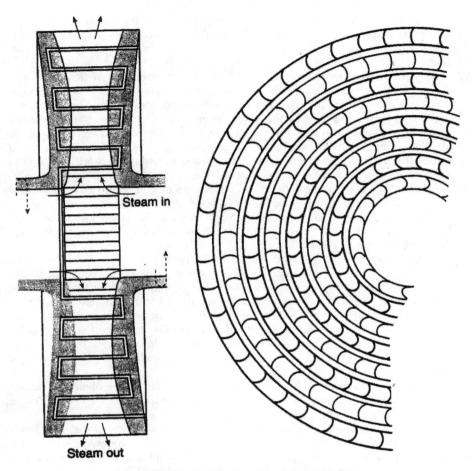

Figure 5-6. Radial Flow or Ljongstrom Turbine

Tangential Flow. Tangential flow is the term applied when the steam enters through a nozzle placed approximately tangent to the periphery and directed into semi-circular buckets milled obliquely into the edge of the wheel. Coupled with this is the action of a reversing chamber fitted closely to, but not touching, the periphery of the wheel, which has similar buckets milled into it. The latter receive the steam discharged from the wheel buckets and return it again a number of times to the wheel buckets in the proper direction to produce additional work, the steam following an approximately helical path. Any number of nozzles may be used, each with its reversing chamber, up to that number which will completely fill the periphery. This type of flow is also called helical.

Mixed Flow. Mixed flow is the term applied to the flow when it enters in the radial direction and leaves in the axial direction. These types of Radial inflow turbines have been widely used with gasses and in some cases with steam.

Steam Passage Between Blades. The most common type of flow for large power plants is axial flow. The flow is further divided into two categories, impulse and reaction. Impulse and reaction turbine designs are based on the relative pressure drop across the stage. There are two measures for pressure drop the pressure ratio and the percent reaction. The percent reaction is the percentage of isentropic enthalpy drop across the rotating element, as a percent of the total stage enthalpy drop. Some manufacturers utilize percent pressure drop across stage to define reaction.

Impulse Turbines may be defined as a system in which all steam expansion takes place in fixed nozzles and none occurs in passages among moving blades. In actual practice, there must be some pressure drop across the rotating blades to generate flow. A typical HP stage is usually an impulse stage but still has on the average about 5% reaction at full load.

Reaction or Parsons Turbine: In the turbines so far described, the steam expands only in fixed nozzles and flows through passages between blades arranged in rows, transferring its kinetic energy to these rows of blades and causing them to rotate against resistance. In the widely used reaction turbine proposed and first built by Sir Charles Parsons, the steam decreases in pressure and expands while it is passing through the moving blades as well as in its passage through the fixed nozzles. In a symmetric reaction design, equal pressure drop occurs across the stationary and rotating blades, amounting to a 50% reaction turbine.

The root and tip reactions are often different especially in long blades to counter effect the effect of centrifugal forces on the steam flow. The root is often more impulse and the tip nearer 50% reaction.

Impulse and Reaction Combination is the most common type of steam turbines found in the market. The first stages are usually Impulse type while the second stages are usually reaction types (50%). The impulse turbine produces about two times the power output as the 50% reaction-type turbine, the reaction turbine, on the other hand, is more efficient. This has been fully explained in Chapter 4 in the section on axial flow turbines. Thus, the combination of the first stages being impulse and the later stages being about 50% reaction produces a high-power and -efficiency turbine.

Arc of Peripheral Admission to the Total Circumference

Steam Turbine Nozzles: All steam turbines use nozzles of some form. It is in the nozzle that a change in enthalpy is converted into a change in kinetic energy. The two equations of some importance in steady state flow through nozzles are the energy equation and the continuity equation and these are shown below.

For a stationary nozzle
The energy equation:

$$h_1 + \frac{V_1^2}{2g_c J} = h_2 + \frac{V_2^2}{2g_c J} \qquad (5-9)$$

The continuity equation:

$$\dot{m} = A_1 V_1 \rho_1 = A_2 V_2 \rho_2 \qquad (5-10)$$

where
  A = Area
  V = Velocity
  ρ = Density
  J = Mech. equiv. of heat
  g_c = Gravitational constant
  h = Enthalpy
  $\dot{m}$ = Mass flow

The flow per unit area can be written as follows:

$$\frac{\dot{m}}{A} = \sqrt{\frac{\gamma}{R}} \frac{P}{\sqrt{T}} \frac{M}{\left(1 + \frac{\gamma-1}{2}M^2\right)^{\frac{\gamma+1}{2(\gamma-1)}}} \qquad (5-11)$$

where the Mach number (M) is defined as:

$$M = \frac{V}{a} \qquad (5-12)$$

it is important to note that the Mach no. is based on static temperature.

The acoustic velocity (a) in a gas is given by the following relationship

$$a^2 = \left.\frac{\partial P}{\partial \rho}\right|_{s=c} \qquad (5-13)$$

for an adiabatic process (s = entropy = constant) the acoustic speed can be written as follows:

$$a = \sqrt{\frac{\gamma g_c R T_s}{MW}} \qquad (5-14)$$

where T_s = static temperature.

In the steam turbine (Rankine Cycle), the expansion processes are adiabatic and isentropic processes. Thus, for an isentropic adiabatic process $\gamma = \frac{c_p}{c_v}$; where c_p and c_v are the specific heats of the gas at constant pressure and volume, respectively, and can be written as:

$$c_p - c_v = R \qquad (5-15)$$

where

$$c_p = \frac{\gamma R}{\gamma - 1} \quad \text{and} \quad c_v = \frac{R}{\gamma - 1} \qquad (5-16)$$

It is important to note that the pressure measured can be either total or static, however, only total temperature can be measured. The relationship between total and static conditions for pressure and temperature are as follows:

$$T_o = T_s + \frac{V^2}{2c_p} \quad (5-17)$$

where T_s = static temperature and V = gas stream velocity, and

$$P_o = P_s + \rho \frac{V^2}{2g_c} \quad (5-18)$$

Equations (5-17) and (5-18) can be written in terms of the Mach number as follows:

$$\frac{T_o}{T_s} = \left(1 + \frac{\gamma-1}{2} M^2\right) \quad (5-19)$$

and

$$\frac{P_o}{P_s} = \left[1 + \frac{\gamma-1}{2} M^2\right]^{\frac{\gamma}{\gamma-1}} \quad (5-20)$$

The point in the nozzle where the minimum area occurs and the Mach number = 1, is known as the throat. This reduces the above relationships as follows:

$$\frac{\dot{m}}{A} = \sqrt{\frac{\gamma}{R}} \frac{P}{\sqrt{T}} \frac{1}{\left(1 + \frac{\gamma-1}{2}\right)^{\frac{\gamma+1}{2(\gamma-1)}}} \quad (5-21)$$

$$\frac{T_o}{T_s} = \left(1 + \frac{\gamma-1}{2}\right) = 1.1640 \quad (5-22)$$

where $\gamma = 1.329$ for steam

$$\frac{P_o}{P_s} = \left[1 + \frac{\gamma-1}{2}\right]^{\frac{\gamma}{\gamma-1}} = 1.847 \quad (5-23)$$

To reach the Mach number above $M = 1.0$, a convergent-divergent nozzle would have to be used. The speed of sound (acoustic velocity) results from three-dimensional effects, accurate results occur from one-dimensional effect of small disturbances. Large disturbances such as shock waves propagate at a much higher velocity. These small disturbances, which are pressure waves, propagate in a gaseous medium in this case super heated steam.

Figure 5-7 shows the aero-thermal properties of the flow in a convergent divergent nozzle. The flow leaving the nozzle is supersonic. A convergent-divergent nozzle is designed to handle an expansion between certain expansion states. If the back pressure of the discharge region is less than the design pressure at the discharge boundary an underexpansion occurs. A free expansion occurs after the steam leaves the nozzle and since the steam is supersonic an expansion wave will occur. If the back pressure of the discharge region is higher than the design pressure at the discharge boundary an overexpansion occurs. A standing shock wave occurs in the divergent area of the nozzle. Across this shock, there is a sharp irreversible increase in pressure, an increase in entropy, and a decrease in velocity.

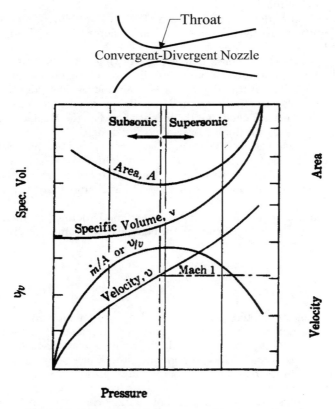

Figure 5-7. An Ideal Convergent Divergent Nozzle

The convergent-divergent nozzle may have any cross-sectional shape to fit the application. The elements of the surface of the divergent nozzle are generally straight for ease of manufacture for use in steam turbines. The nozzle is proportioned after the nozzle throat area is determined. The throat area is the smallest area in the nozzle and where the flow reaches a Mach number equal to one. The flare of the sides of the divergent section should be within good fluid dynamic limits so that no separation of the flow occurs, which is an inclusive angle between 12° and 15°.

Full Admission Turbines: One aspect of the classification of steam turbines as mentioned in the classification scheme, is according to the ratio of the arc of peripheral admission to the total circumference. If steam admission is over the full arc, we have a full admission turbine.

In any design of a steam turbine, two types of losses must be considered. They are the losses encountered by the Mach number, and the losses encountered by partial admission. These losses are shown in Figures 5-8 and 5-9. It can be seen that for high Mach numbers, the loss factor increases but it is more or less constant up to approximately Mach 1. The maximum limit for efficient design is around Mach number 1.15. Excessive Mach numbers result in severe shocks, choking of passages and high mach number losses.

Figure 5-9 shows the relation between efficiency and different configurations for partial entry of the flow. Usually, low admission ratios result in high partial admission losses.

Partial Admission Turbines: Usually, in the high-pressure end of a steam turbine, the nozzles are located only partially around the circumference (partial admission). At the intermediate stages, the increasing volume of steam necessitates that larger circumferential arcs be occupied by nozzles until a full admission situation is reached. It is the balance between the Mach number and the partial entry losses which finally determines the design.

Turbine Stages in Series. There are two major types of turbines: the impulse turbine and the reaction turbine. The steam volume increases whenever the pressure decreases, but the resulting velocity changes depend on the type of turbine. These velocity changes are distinguishing characteristics of the different types of turbines.

The degree of reaction in an axial-flow turbine is defined as the ratio of the change of enthalpy drop in the rotor to the change in total enthalpy drop across the stage:

$$R = \frac{H_{\text{rotor}}}{H_{\text{stage}}} \quad (5-24)$$

By definition, the impulse turbine has a degree of reaction equal to zero. This degree of reaction means that the entire enthalpy drop is taken in the nozzle, and the exit velocity from the nozzle is very high. In practice, there must be a pressure drop across the rotating blades to generate flow. Since there is no change in enthalpy in the rotor, the relative velocity entering the rotor equals the relative velocity exiting from the rotor blade. Most steam turbine high-pressure (HP) stages are typically impulse stages by design but average a reaction of 5% reaction at full load.

An Overview of Steam Turbines • 235

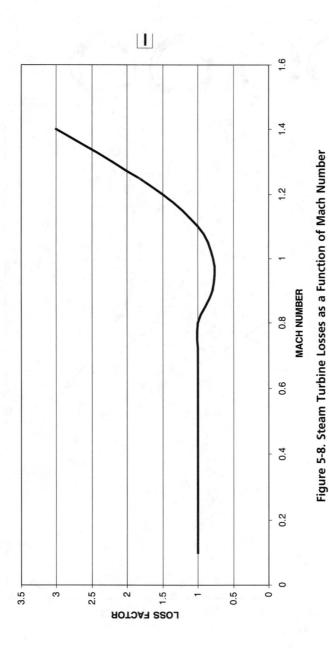

Figure 5-8. Steam Turbine Losses as a Function of Mach Number

236 • **COGENERATION AND COMBINED CYCLE POWER PLANTS**

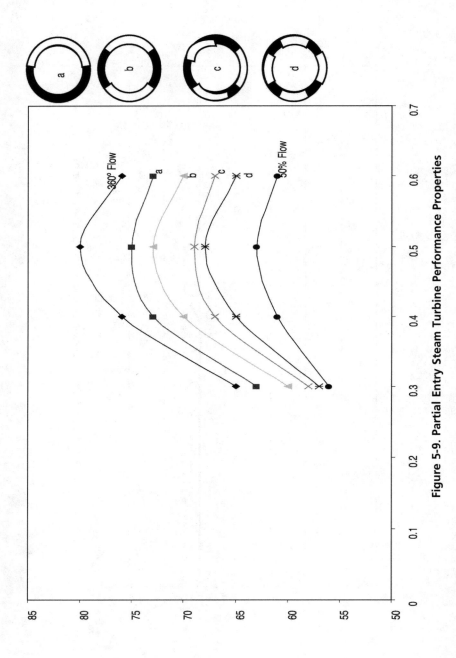

Figure 5-9. Partial Entry Steam Turbine Performance Properties

For a symmetric flow (50% reaction), the enthalpy drop in the rotor is equal to the drop in the stationary part of the turbine. This also leads to equal pressure drop across the stationary and rotating parts. Due to the difference in the turbine blade diameters at the tip and the root of the blade, the reaction percentages are different to counteract the centrifugal forces acting on the steam flow. If this was not done, too much flow would migrate to the blade tips. In intermediate pressure (IP) and low- pressure (LP) turbines, the basic design is reaction, however in these turbines the design is about 10% reaction at the root, progressing to about 60% to 70% reaction at the tips in the LP turbine.

Single-Stage or Simple-Impulse Turbine. The simple-impulse turbine is one of the simplest forms of a turbine. In this turbine type, the steam expands from its initial to its final pressure in one nozzle (or one set of nozzles, all working at the same pressure), resulting in a steam jet of high velocity which enters the blade passages and, by exerting a force on them due to being deflected in direction, turns the rotor. Energy of all forms that remain in the steam after it leaves the single row of blading is lost.

A single stage impulse turbine is shown in Figure 5-10. The variation of the absolute velocity, and total and static pressure, changes as the steam is accelerated through the nozzle and then impinges on the turbine wheel are shown in the figure. The total pressure and temperature remain unchanged in the nozzle, except for minor frictional losses.

Multi-Stage Impulse Type Turbine. In this type of turbine, there are as many steps as there are chambers, each being called a pressure stage. The resultant steam velocity in each stage is relatively small, allowing reasonably low blade velocities and preventing excessive loss by steam friction. The pressure drops in each stage and the steam volume increases; the steam velocity is high at exit from the nozzles and is low at exit from the blades. This arrangement is sometimes termed a rateau turbine, and the separate stages, rateau stages.

Velocity and Pressure Stage Combination. Turbines with combinations of velocity and pressure staging are commonly used and are of several types. The wheel in each pressure stage might have two (or even three) of rows of blades instead of one. The turbine has as many pressure stages as there are wheels, and each pressure stage has as many velocity stages as there are rows of blades on the wheel in that stage. This arrangement results in a small, short and cheap turbine, at more or less sacrifice of efficiency. Commercial turbines of this type are called Curtis turbines, after the original inventor, and the individual pressure stages, each with two or more velocity stages are often called Curtis stages.

Velocity-Stage, Curtis-Type Impulse Turbine. The Curtis-type turbine has one set of nozzles, with several rows of blades following it. In passing from the nozzle exit through one set of blades, the velocity of the steam is lowered by virtue of the work done on the blades but is still high. It then passes through a row of fixed guide blades, which change the direction of the steam until it flows approximately parallel to the original nozzle direction, discharging it into a second row of blading fixed to the same wheel. This second row again lowers the steam velocity by virtue of the work delivered to the wheel. A second set of guide blades and a third row of moving blades are sometimes used.

Figure 5-11 shows a Curtis-type stage, a velocity compounded turbine, in which there are several moving rows to absorb the kinetic energy coming from the

238 • COGENERATION AND COMBINED CYCLE POWER PLANTS

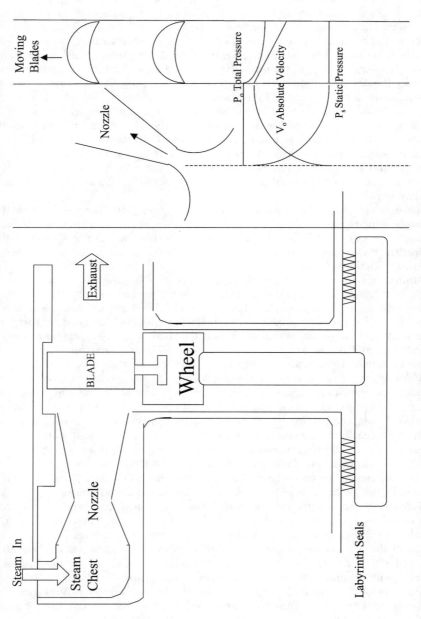

Figure 5-10. A Simple Impulse Steam Turbine

An Overview of Steam Turbines • 239

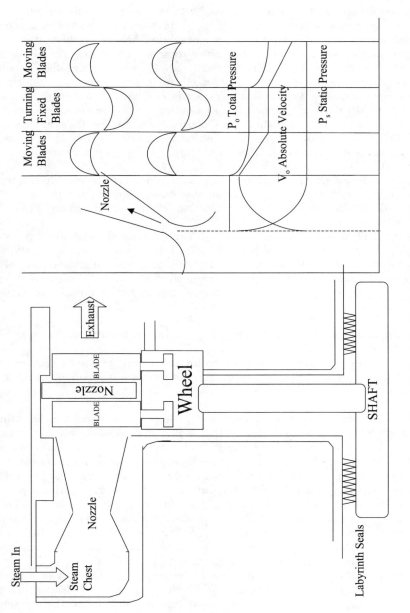

Figure 5-11. A Curtis Type Steam Turbine

nozzles. The absolute velocity, with the total and static pressure distribution is also shown in this figure.

Pressure-Stage, Rateau Type Impulse Turbine. Another impulse turbine is the pressure compound or Ratteau turbine. In this turbine, the work is broken down into various stages. Each stage consists of a nozzle and blade row where the kinetic energy of the jet is absorbed into the turbine rotor as useful work. The air, which leaves the moving blades, enters the next set of nozzles where the enthalpy decreases further, and the velocity is increased and then absorbed in an associated row of moving blades. The turbine has a series of chambers formed by parallel disc-shaped partitions called diaphragms has a simple-impulse turbine enclosed in it, all wheels being fastened to the same shaft. Each chamber receives the steam, in turn, through groups of nozzles placed on arcs, the last chamber in most cases discharging to the condenser. The pressure drop is divided into the various stages. Figure 5-12 shows a Ratteau turbine.

Reaction-Type Turbine. Figure 5-13 shows a reaction-type turbine. In this type, the blade passages are shaped so that another convergent nozzle is formed. The nozzle will give an additional force due to the acceleration of the jet and this is the basis of the reactive principal.

General Flow Arrangement. Although early machines were one section or cylinders (because of design and manufacturing limitations), most common power plant turbines consist of two or more sections designated high pressure (HP), intermediate pressure (IP) and low pressure (LP) in order to improve overall efficiency. The optimum number of stages depends on a number of factors including:

1. The amount of available energy from boiler conditions, allowable discharge pressures, and temperatures.
2. Stage percent reaction, Impulse rotor (near zero reaction) produces double the energy of a 50% reaction rotor, the rotor mean diameter, and rotor speed.

There are five basic types of shaft and casing arrangements: single casing, tandem-compound, cross-compound, double flow, and extraction steam turbines. Figure 5-14 shows schematics of various steam turbine arrangements.

Single-Flow Single Casing Turbines. In a single-flow turbine, the steam enters at one end, flows once through the blading in a direction approximately parallel to the axis, emerges at the other end, and enters the condenser. In turbines with a single casing, all sections are contained within one casing and the steam path flows from throttle to exhaust through that single casing. Figure 5-14a shows a simple path where steam enters a turbine and is exhausted to the atmosphere or a condenser it also shows an extraction for cogeneration purposes. This is the most common arrangement in small and moderately large turbines.

Compound-Flow or Tandem-Compound Turbine. Compound-flow or tandem-compound turbine is the term applied to a machine in which the steam passes in sequence through two or more separate units, expanding in each. The two units arranged in a tandem-compound design have both casings on a single shaft and driving the same electrical unit. In most cases, the HP exhaust is returned for reheating before entering the IP turbine. Most often, the high-pressure and the

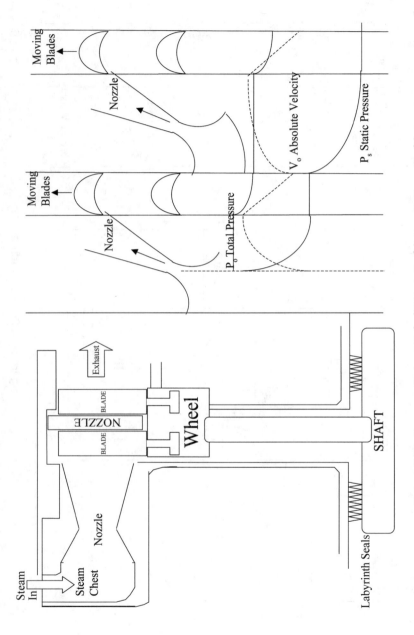

Figure 5-12. Pressure Compounded Steam Turbine — Ratteau Turbine

242 • COGENERATION AND COMBINED CYCLE POWER PLANTS

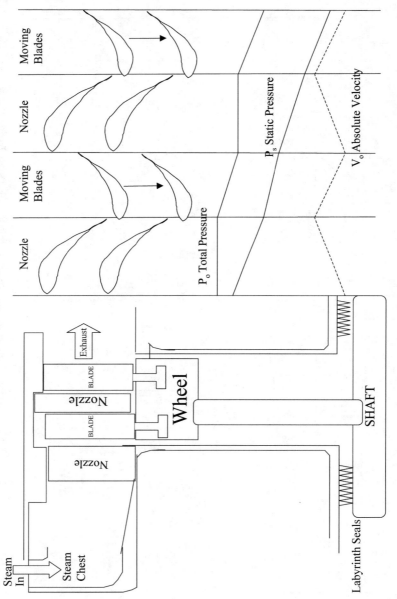

Figure 5-13. Reaction Turbine

An Overview of Steam Turbines • 243

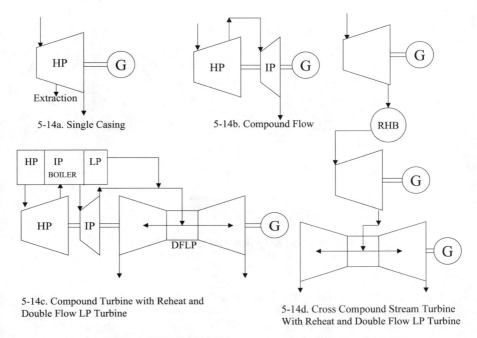

Figure 5-14. A Schematic of Arrangements of Steam Turbines

intermediate-pressure portions are in one casing and the low-pressure portion in another. The IP exhausts into crossover piping and on to one or more low-pressure turbines. In Figure 5-14b, two sections are used, an HP and an LP section. Figure 5-14c indicates a condition where split steam is used in a double flow, low-pressure (DFLP) section. Figure 5-15 shows a typical tandem-compound turbine with two casings — one HP and IP, and the second, a two-flow LP.

Cross-Compound Turbine. A cross-compound design typically has two or more casings, coupled in series on two shafts, with each shaft connected to a generator. In cross-compound arrangements, the rotors can rotate at different speeds but cannot operate independently as they are aerodynamically coupled. The cross-compound design is inherently more expensive than the tandem-compound design, but has a better heat rate, so that the choice between the two is one of economics. Figure 5-14d shows a turbine setup where there are three casings each with their own generator and a reheat between the HP and the IP section.

Double-Flow Turbines. LP turbines are typically characterized by the number of parallel paths available to the steam. The steam path through the LP turbines is split into parallel flows because of steam conditions and practical limitations on blade length. Typically, several LP flows in parallel are required to handle the large volume of flow rates. The steam enters at the center and divides, the two portions passing axially away from each other through separate sets of blading on the same rotor. This type of unit is completely balanced against end thrust and gives large area of flow through two sets of low-pressure blading.

244 • COGENERATION AND COMBINED CYCLE POWER PLANTS

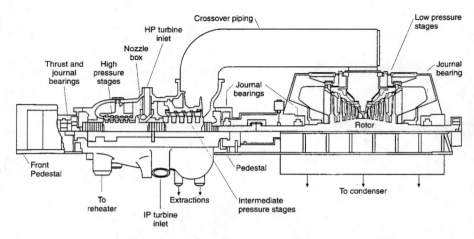

Figure 5-15. A Typical Compound Turbine

Extraction Flow Turbine. The extraction flow stream turbine is the term applied to a turbine where part of the flow is extracted for various reasons such as steam for the plant or for absorption-type chillers or for any other plant process. Figure 5-14a shows a schematic of an extraction-type turbine. These turbines maybe back pressure or condensing depending on the application. These types of turbine are used most commonly in a cogeneration application.

Steam Turbine Characteristics

Understanding the effect of the steam operating conditions on efficiency and load is very important in operating steam turbines at their optimum operating conditions. The two types of steam turbines are condensing steam turbines and the back-pressure steam turbines. Steam inlet temperature and pressure, and turbine exhaust pressure and vacuum, are the significant operating parameters of a steam turbine. The variations in these parameters effect steam consumption and efficiency. In a 100-MW steam turbine at a pressure of about 600 psia (41.4 bar) and 660°F (350°C), a 1% reduction in steam consumption can cost US$500,000/ year and about $900,000/year. This is based on a boiler efficiency of about 85% and LHV of fuel at 18,900 Btu/lb (10,500 Kcal/kg).

Turbine steam inlet pressure is a major parameter which affects turbine performance. To obtain the design efficiency, the steam inlet pressure should be maintained. Lowering steam inlet pressure reduces turbine efficiency and increases steam consumption. A 10% increase in steam pressure will reduce the steam consumption by about 1% in a condensing steam turbine, and will reduce the steam consumption by about 4% in a back-pressure steam turbine. The effect on efficiency for 10% increase in pressure for a condensing steam turbine is about 1.5% and 0.45% for a back-pressure steam turbine.

Turbine steam inlet temperature is another major parameter which affects turbine performance. Reducing steam inlet temperature reduces the enthalpy, which is a function of both the inlet temperature, and pressure. At higher steam inlet temperature, the heat extraction by the turbine will also be increased. An increase of about 100°F (55°C) will reduce the steam consumption by about 6.6% in a condensing steam turbine, and 8.8% in a back-pressure turbine. The effect on efficiency for a 100°F (55°C) will be an increase of 0.6% in efficiency for a condensing steam turbine, and 0.65% in efficiency for a back-pressure turbine. It should be noted that the overall efficiency in most cases for a condensing steam turbine (30% to 35%), is about twice that of a back-pressure turbine (18% to 20%).

In condensing or exhaust back-pressure steam turbines, the increase of this back pressure will reduce the efficiency and increase the steam consumption keeping all other operating parameters. In condensing steam turbines, the condenser vacuum temperature will also increase if the removal of heat from the condenser is reduced. Thus, in a water cooled condenser, if the temperature of the inlet water is increased, the power produced by the turbine is decreased because the back pressure will be increased.

In summary, the condensing steam turbines are more efficient and produce more power than back-pressure steam turbines. The condensing steam turbine is also more efficient. The cost of a condensing steam turbine is about $25/kW more than a back-pressure turbine.

Features and Structure of HP and IP Blades

High-Pressure (HP) and Intermediate-Pressure (IP) Rotating Blades. The blades in the HP turbine are small because of the low-volumetric flow. Rotating HP blades are usually straight, however, the use of leaned and bowed blades has recently introduced a three dimensional aspect to designs. Shrouds (also called covers or connecting bands) provide a sealing surface for radial steam seals and are used to minimize losses due to leakage. Shrouds also tie the blades together structurally and allow for some control over the damping and natural frequencies of the blades. The shrouds are typically attached either by peened tension or are integral with the blade airflow (or vane).

Blades are connected at the root to the rotor or disc by several configurations. Figure 5-16 shows the most common types of root attachments. The choice of type of attachment will depend upon a number of factors. A side-entry fir tree root design is used in the HP control stage for ease of replacement, if required, because of solid particle erosion. For longer blades in the control stage, however, a triple pin construction is sometimes used as the side-entry design has too many modes close to the nozzle wake frequency.

The blade roots may be of the "serrated" or "fir tree" configuration, inserted into individual axial slots in the disc or a similar serrated or T-shape, inserted into a continuous circumferential slot in the disc (this requires a special insertion gap), or may comprise one or several flat "fingers" fitting into circumferential slots in the disc and secured by axially inserted pins. Serrated or T-roots, furthermore, may be of male or female type.

A particular challenge in HP blading design is the first (control) stage where operation with partial arc admission leads to high dynamic stresses. Design factors such as choice of leading edge configuration and blade groupings are chosen to reduce vibratory stresses produced.

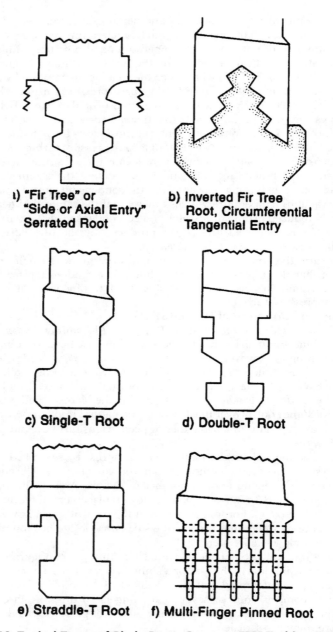

Figure 5-16. Typical Types of Blade Roots Courtesy EPRI Turbine Steam Path Damage. Theory and Practice

Blades in IP turbines are very similar in design to those in the HP; with somewhat more twist, plus bowing and leaning to account for greater radial variation in the flow, thus providing also flow radial equilibrium.

HP and IP Nozzles. In HP and IP turbines, stationary blades (or nozzles) can be classified into two general design categories: a wheel-and-diaphragm construction is used for impulse stages, the drum-rotor construction for reaction stages.

A diaphragm, used in impulse stages, consists of:

1. nozzles or stationary blades
2. a ring which locates them in the casing and
3. a web which extends down between the rotor wheels and supports the shaft packing.

Figure 5-17 indicates the typical construction. Diaphragms in the HP and IP are typically of welded construction. In the control (first) stage, nozzles are divided into segments, arranged in separate nozzle "chests" or "boxes" and each segment has an associated control valve or control valve group.

In reaction stages, stationary blades or nozzles are manufactured in a manner similar to that for rotating blades with a root attachment and, in some cases, a sealing shroud. The blades are fitted by the root attachment on a blade carrier, which is located in the outer casing.

Nozzles and diaphragms are typically exposed to pressure differentials, which bend them in the plane perpendicular to the turbine axis. These pressure

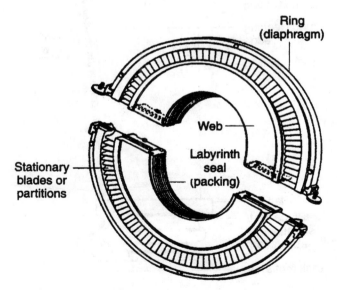

Figure 5-17. Construction of a Typical Steam Turbine Diaphragm (Courtesy EPRI Turbine Steam Path Damage. Theory and Practice)

differentials are highest in the HP, although the shorter blade length limits the bending stresses that develop.

Features and Structure of LP Blades. Figure 5-18 shows the nomenclature for a rotating LP blade. Rotating LP turbine blades may be "free standing," that is, not connected to each other in any way; they may be connected in "groups" or "packets" each comprising several blades, generally between two and eight, or all blades in the whole row may be "continuously" connected.

Free-standing blades have the following characteristics compared to grouped blades:

1. they have less inherent damping at the blade tips,
2. their resonances are more easily defined (i.e. no mechanical interactions with neighboring blades),

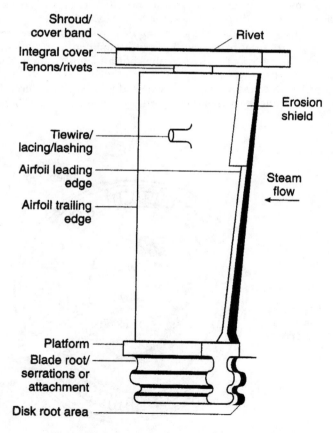

Figure 5-18. Nomenclature for a Typical LP Turbine Blade
(Courtesy EPRI Turbine Steam Path Damage. Theory and Practice)

3. they have more aerodynamic interactions, and
4. they are easier to install and disassemble, as there are no welds or rivets.

In connected blades, there are a number of design choices. The connections may consist of shrouds (or bands) over the tips of the blades, or of tie-wires (or lashing or lacing wires) located along the blade height. As with the HP blades, connections made at the blade tip are termed shrouding. Shrouds may be inserted over tenons protruding above the blade tips, and these tenons then riveted down to secure the shrouds, or they may consist of integrally forged stubs in a welded or brazed together assembly. Other types of riveted connections are also used. Tie-wires may consist of either integrally forged stubs welded or brazed together, or cylindrical "wires" or rods inserted through a hole (usually in a forged boss) in each blade foil.

In order to add mechanical damping, some wire or rod-type lashing wires are left loose in their holes, and there are also some shroud-type connections that merely abut each other and are not permanently attached. Continuous connections must make some provision to accommodate thermal expansion. Shrouds and lacing wires sometimes are introduced to decrease vibratory stresses, but can act as "dirt traps."

The airfoil of blades may be of constant section for short blades, and constant width, but twisted for longer ones. The longest blades for the last few rows of the LP are twisted to match the aerodynamics at different radii and improve aerodynamic efficiency. Most recently, designs have also begun to be leaned or bowed, thus introducing radial variation as well.

Nozzles or stationary blades in LP stages are typically arranged in diaphragms like those for HP and IP impulse stages. However, the construction may be simpler than those in the HP and IP, consisting, for example of only fixed blades, constrained by inner and outer (hub and rim) annular bands. Diaphragms in the LP are of cast or welded construction. In wet stages, diaphragms may be made with hollow blades vanes or other design features as a means of drawing off moisture that would otherwise lead to liquid droplet erosion.

Required Material Characteristics

Choosing the optimum blade material is an ongoing tradeoff between desirable material characteristics. The demands placed on HP and LP blades emphasize different material characteristics, as seen in Table 5-1. In addition, it is important that blading material be weldable, particularly last stage LP blading, as many designs require that cover blades, tie-wires and erosion shields be attached by thermal joining. Weldability is also important for blade repairs. It is important to note that up to about 1990 welding of blades and rotors was considered "impossible" by many major manufacturers.

- **Creep strength and creep-fatigue resistance.** Creep resistance is important in the first two or three rows of the HP and IP turbines to resist elongation and accumulation of strain at the higher operating temperatures, particularly at stress concentrations such as at the blade-to-rotor attachments.

Table 5-1. Important Blade Material Characteristics

Material Characteristics	HP and IP Turbine Blades	LP Turbine Blades
Creep Strength and Creep Fatigue Resistance	x	
Tensile Strength	x	x
Corrosion Resistance		x
Ductility and Impact Strength	x	x
Fatigue Strength	x	x
Corrosion Fatigue Resistance		x
Notch Sensitivity		x
Material Damping	x Partial Admission blades	x
Erosion Resistance	x solid particle erosion	x Liquid droplet erosion

- **Tensile strength.** Tensile strength is required to withstand steady centrifugal and steam bending loads.
- **Corrosion resistance.** Corrosion resistance is important to maintain blade life in the turbine environment.
- **Ductility and impact strength.** Ductility is required for three reasons:

  1. To allow for plastic deformation that can occur from blade rubs or foreign body impact.
  2. To allow for rivet formation where blades are attached to shrouds.
  3. To allow localized plastic flow to relieve stress peaks and concentrations that can occur at the blade root, and inter-blade connections such as lashing wires and shrouds.

- **Fatigue strength.** Fatigue strength is important to prevent failures from the vibratory stresses imposed by steam flow and system resonance. There are two types of fatigue failures high-cycle low stress, and high-stress and low-cycle failures.
- **Corrosion fatigue resistance.** In LP blades, even more important than simple fatigue strength is the resistance of the material to cyclic loads in aggressive or corrosive environments in order to avoid corrosion fatigue.
- **Notch Sensitivity.** Notch sensitivity is the effect of stress concentration on fatigue strength. Low notch sensitivity is a desirable characteristic; however, high tensile strength materials which are good for improved strength, also lead to high notch sensitivity.
- **Erosion Resistance.** The HP and IP stages suffer from erosion by solid particles, which are formed by oxides exfoliated from super heater and reheat tubing due to high-temperature operation. Liquid erosion occurs mostly in the LP stage where they are subjected to liquid droplet erosion.

To avoid this, the amount of water in the steam especially on the last LP stages must remain below 10%.
- **Blade Damping.** Damping is of primary concern in any blade design. There are three types of damping:

  1. Material damping is an inherent part of the blade material and is dependent on the pre-load stress, dynamic stress, temperature, and frequency.
  2. Mechanical or Coulomb friction damping. This phenomenon occurs from the relative motion between contacting parts.
  3. Aerodynamic damping occurs as a result of work done on the gas stream, or by the work done by the gas stream on the rotating airfoil.

Blade Materials. The most common material for HP and IP rotating and stationary blades and nozzles is 12Cr martensitic stainless steel. Three generic martensitic stainless steels are widely used for turbine blading, most commonly Type AISI 422 for HP blading and Types AISI 403 and AISI 410 in LP blading. There are numerous specific applications materials where turbine manufacturers have customized the generic grade by the addition or deletion of specific alloying elements, or by modification of the production or heat treating process. The final properties of these steels are strongly influenced by tempering temperature.

The austenitic stainless steels (AISI series 300) are used in some high-temperature applications. The austenitic stainless steels, which have a higher content chromium and tungsten then the martensitic stainless steel, have excellent mechanical properties at elevated temperatures and are typically readily weldable. There is a thermal expansion coefficient difference between matensitic and austenitic stainless steels so that care is required when designing attachment clearances for fitting austenitic blades into martensitic discs. Also there is a potential for stress corrosion cracking when 300 series stainless steels are used in wet steam conditions.

The rings and webs of HP and IP nozzle diaphragms are commonly manufactured from stainless steels, although if the working steam temperature does not exceeds 350°C (660°F), then welded diaphragms can be made from carbon steels.[6]

Early LP blades materials included cartridge brass (72% Cu, 28% Zn) nickel brass (50% Cu, 10% Ni, 40% Zn), and Monel (66% Ni, 31% Cu, 3% Fe). With the advent of larger turbines, most LP turbine blades have been manufactured from a 12% Cr stainless steel; typically, Types AISI 403, 410, or 410-Cb have been chosen depending on the strength required. Types 403 and 410 have better corrosion resistance than Type 422, an important characteristic for use in the wet stages of the LP turbine.

There are numerous, specifically customized versions of these generic materials, for example, Carpenter H-46 and Jethete M152. Jethete M152 has higher hardness and is thus more resistant to liquid droplet erosion in the LP than Types 403 and 410. So far, it has only been used in LP turbines, but could be used in the HP and IP if needed. European designations for 12% Cr blading alloys include: X20CrMoV121 and X20Cr13.

More recently, the precipitation hardened stainless steel designated 17-4 PH (AISI 630) was developed by one manufacturer for the last blades of the LP turbine in the largest 3600 rpm machines. It has a nominal composition, that is 17% Cr and 4% Ni. The hardening temperature can control a wide range of mechanical properties. Alloy 17-4 PH is somewhat difficult to weld and requires post-weld heat treatment.

Titanium alloys, chiefly Ti-6A1-4V (6% aluminum and 4% vanadium), have been used for turbine blades since the early 1960s. The use of titanium in the last few rows of the LP offers a number of advantages over other materials.

The advantages to titanium include:

- Titanium has about half the density of 12Cr steels which allows for longer last stage blades without an increase in centrifugal stresses in the blades and thus an increase in annular area and improved turbine efficiency. The capability of LP turbines to produce power is limited by the long last row of blading and the strength of the rotor to support the blade. The practical limitation for blades constructed of 12% Cr martensitic steel was reached with 840 mm (33.5 in.) blades operating in 3600 rpm machines and 1200 mm (48 in.) blades operating in 3000 rpm machines. In contrast, titanium offers an opportunity to go to 1000 mm (40 in.) and 1350 mm (54 in.) blades for 3600 and 3000 rpm machines, respectively. This represents a marked increase in power and makes possible a new generation of LP steam turbines.
- Titanium has particularly favorable mechanical characteristics in applications involving high stresses at low temperatures. Because titanium has half the density and about half the elastic modulus of steel, the frequencies and mode shapes of titanium blades are very similar to those made of steel. Note, however, that the elastic modulus is dependent on the particular titanium composition.
- Titanium has greater corrosion resistance and, as a result, may have better performance in dry/wet transition phase regions on the LP.
- Titanium also has excellent resistance to impact and water droplet erosion damage and, in many applications, can be used without erosion shields.

The drawbacks to titanium include:

- Higher cost than steel, even though Titanium's lower density means that more blades can be manufactured for a given mass of material which somewhat offsets the higher cost per pound of the material.
- More difficult to machine.
- More difficult to weld. Titanium requires a high state of cleanliness and an inert welding atmosphere.
- Poor resistance to sliding wear, which can allow fretting corrosion in some conditions, although fretting has not been found to be as much of a problem as was once anticipated.
- Lower internal material damping than stainless steel.
- A major disadvantage of Titanium for blades is that recent high cycle fatigue studies have shown an endurance limit in air and in steam smaller

than for 12% Cr stainless steels. Shot peening has been used to restore the fatigue life lost after machining, production, or repair processes.

Duplex stainless steels are those stainless steels that contain very high levels of chromium and about equal amounts of ferrite and austenite. They have been evaluated for use for LP blading, primarily in Europe. There are a variety of types of duplex stainless steels with ferrite contents ranging from about 45 to 75. The duplex stainless steels have excellent corrosion fatigue characteristics. Their primary drawback for blading may be somewhat lower yield strength than Type 403/410 (in general, although the specific characteristics for the Ferralium alloy indicate that good yield and tensile strength can be achieved) and they do show some long time service embrittlement at temperatures about 300°C (570°F).

Surface Treatments. Coatings of surface hardening are frequently used to improve the surface characteristics of turbine blades. Such protective schemes are most commonly used to improve erosion resistance in susceptible locations. LP blade coatings and surface treatments are used for improved performance against environmentally assisted mechanisms and liquid droplet erosion.

Mechanical Efficiency. The mechanical efficiency of a turbine is the ratio of the brake output to the internal output. The mechanical efficiency is an index of the external losses.

Turbine Efficiency

The engine efficiency is the ratio of the real output of the turbine to the ideal output. The engine efficiency is, primarily, of interest to the designer as a means of comparing the real turbine with the ideal. Just as in the case of thermal efficiencies, the output basis upon which the engine efficiency is determined must be designated.

Advantages and Disadvantages of Steam Turbines

A steam power plant offers several advantages as a prime mover for the production of electrical power. One big advantage is that the process involves two-phase fluid and the working fluid being pumped from low pressure to high pressure is liquid. The work required to bring the water to the desired pressure (Point B) is much less than the work required to raise the pressure of gas by compressing it. In fact, the work required to pump the water is frequently neglected when determining the output and efficiency of large steam power plants.

A second advantage of steam power plant is its ability to burn a wide variety of fuels. Since the working fluid does not come in direct contact with the combustion products, any fuel ranging from natural gas to heavy residual fuel and solid fuels as coal or refuse can be burned in the boiler.

The disadvantage of the steam plant is the large amount of equipment required. Since the water is heated indirectly in the furnace and the heat exchangers must be very large to raise the water to the desired temperature. The boiler requires a large supporting structure and foundation. In order to obtain high efficiencies with a Rankine Cycle the steam must be condensed at the turbine exhaust. The steam condenser requires a large surface area to cool the exhaust steam. The condenser also requires cooling water with its accompanying cooling towers. A multi-megawatt steam power plant requires approximately 30 to 36 months for complete installation. The current estimate cost for a steam power plant is approximately $800/kW.

While the use of a boiler offers the advantage of multiple fuel capabilities, the use of some of these fuels presents problems. If heavy fuels are burned, the liquids must be heated prior to introduction into the burners. The use of solid fuels creates even more problems and the handling equipment can be a sizeable portion of the overall plant equipment. Fuels such as Bunker C oil and coal cause corrosion and fouling problems in the heat exchanger due to fuel impurities.

Chapter 6

AN OVERVIEW OF PUMPS

A pump is a device used to deliver fluids from one location to another. Over the years, many designs have evolved to meet the many different applications for pumps. The basic requirements to define the applications are suction and delivery pressure, and the flow rate.

When selecting pumps for any service, it is necessary to know the liquid to be handled, the total dynamic head, the suction and discharge heads, and, in most cases, the temperature, viscosity, vapor pressure, and specific gravity. In the power industry, the task of pump selection is frequently complicated by the pressure of the liquid, and liquid corrosion characteristics requiring special materials of construction.

Range of Operation

Because of the wide variety of pump types and the number of factors, which determine the selection of any one type for a specific installation, the designer must first eliminate all but those types of reasonable possibility. Since range of operation is always an important consideration, Figure 6-1 should be of assistance. The boundaries shown for each pump type are at best approximate, as unusual applications for which the best selection contradicts the chart will arise. In most cases, however, Figure 6-1 will prove useful in limiting consideration to two or three types of pumps. The positive displacement pumps are used for high pressure and low flow. Gear-type pumps are a very classic example of a positive displacement-type pump. The centrifugal pumps cover a very wide range and are the most commonly used pumps. The axial flow pumps are low-pressure and high-flow-type pumps.

Suction limitations of a pump occurs when the pressure in a liquid drops below the vapor pressure corresponding to its temperature, thus causing the liquid to vaporize. When this occurs in an operating pump, the vapor bubbles will be carried into the point of higher pressure in the pump. These bubbles will suddenly implode, causing high pressure points. This imploding of the vapor bubbles is called cavitation. Cavitation in pumps should be avoided as it is accompanied by metal removal, vibration, reduced flow, loss in efficiency, and noise. When the absolute suction pressure is low, cavitation can also occur in the pump inlet, causing damage in the pump suction and the leading edge of the impeller blades.

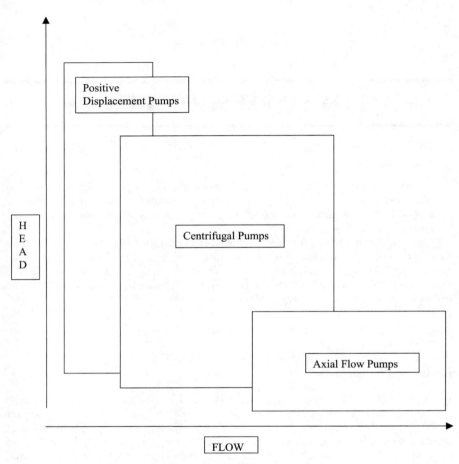

Figure 6-1. Categories of Pumps

To avoid the cavitation in a pump, the pump should be designed such that the net suction head available is higher than the design net suction head of the pump for the desired capacity. The net positive suction head (NPSH) is defined as the total head of the pump less the vapor pressure. The available net positive suction head can be computed as follows:

$$(\text{NPSH})_A = h_{s1} - h_{fs} - p_v \qquad (6-1)$$

where
 h_{s1} = static suction head, which is the vertical distance measured from the free surface of the liquid line to the pump center line, plus the absolute pressure at the liquid source

h_{fs} = suction frictional head, which is the pressure required to overcome the resistance in the pipe and fittings

p_v = vapor pressure at the corresponding temperature of the liquid

The NPSH for an existing installation can be determined by the following relationship:

$$(\text{NPSH})_A = P_{atm} + h_{gs} - p_v + h_{vs} \qquad (6-2)$$

where

P_{atm} = Atmospheric ambient pressure
h_{gs} = Gauge reading at the suction
h_{vs} = Velocity head at the suction ($V^2/2g_c$)

The work expended in pumping the fluid in a pump to force it into a vessel at a higher pressure requires it to raise the fluid to the pressure plus overcome any frictional losses in the piping. This is the total head (ft·lb$_f$/lb$_m$; N·m/kg) developed in the pump required to force the fluid into the pressure vessel. The total power expended to drive the pump is as follows:

$$\text{Power (HP)} = \dot{m}H \times 5050 = QsH/3960 \qquad (6-3)$$

where

$\dot{m}$ = mass flow rate, lb$_m$/h or kg/h
H = total dynamic head, ft·lb$_f$/lb$_m$ or N·m/kg
s = specific heat of the fluid
Q = capacity, gal/min

The performance data consists of pump flow rate and head. Figure 6-2 shows a typical pump performance curve that includes, in addition to pump head and flow, the break horsepower required, and pump efficiency. If detailed manufacturer-specified performance curves are not available for a different size of the pump or operating condition, a best estimate of the off-design performance of pumps can be obtained through similarity relationship or the affinity laws. These are:

1. Capacity (Q) is proportional to impeller rotational speed (N).
2. Head (h) varies as square of the impeller rotational speed.
3. Power varies as the cube of the impeller rotational speed.

The above relationships can be expressed mathematically as shown in Table 6-1.

System Curves. In addition to the pump design, the operational performance of a pump depends upon factors such as the downstream load characteristics, pipe friction, and valve performance. Typically, head and flow follow the following relationship:

$$\frac{Q_2^2}{Q_1^2} = \frac{H_2}{H_1} \qquad (6-4)$$

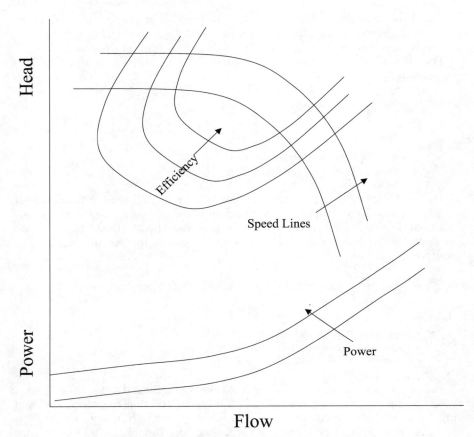

Figure 6-2. A Typical Performance Map for a Pump

where the subscript 1 refers to the design condition and 2 to the actual conditions. The above equation indicates that head will change as a square of the flow rate.

Table 6-1. Pump Laws

Pump Characteristics	Constant Impeller Diameter	Constant Impeller Speed
Capacity	$\dfrac{Q_1}{Q_2} = \dfrac{N_1}{N_2}$	$\dfrac{Q_1}{Q_2} = \dfrac{D_1}{D_2}$
Head	$\dfrac{H_1}{H_2} = \dfrac{N_1^2}{N_2^2}$	$\dfrac{H_1}{H_2} = \dfrac{D_1^2}{D_2^2}$
Power	$\dfrac{HP_1}{HP_2} = \dfrac{N_1^3}{N_2^3}$	$\dfrac{HP_1}{HP_2} = \dfrac{D_1^3}{D_2^3}$

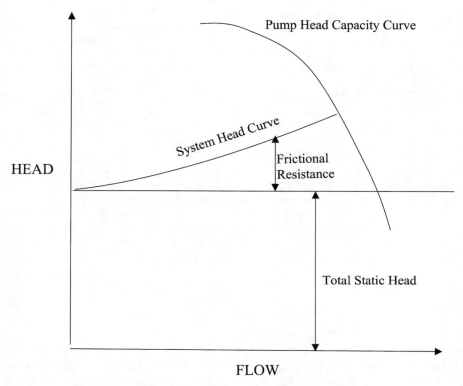

Figure 6-3. Steady State Response of a Pump System

The pump has to overcome, in most cases, the pressure difference between the two points, including the frictional resistance of the piping and valves in the system. Figure 6-3 shows the pump performance in a piping system. The frictional resistance is a function of the square of the flow velocity.

Pump Selection

One of the parameters that is extremely useful in selecting a pump for a particular application is specific speed, N_s. Specific speed of a pump can be evaluated based on its design speed, flow, and head:

$$N_s = \frac{NQ^{0.5}}{H^{0.75}} \qquad (6-5)$$

where
 N = rpm
 Q = flow rate, gpm
 H = head, ft·lb$_f$/lb$_m$

260 • COGENERATION AND COMBINED CYCLE POWER PLANTS

Table 6-2. Specific Speed of Different Pumps

Pump Type	Specific Speed Range
Process Pumps and FeedPumps	Below 2,000
Turbine Pumps	2,000 – 5,000
Mixed Flow Pumps	4,000 – 10,000
Axial Flow Pumps	9,000 – 15,000

Specific speed is a parameter that defines the speed at which impellers of geometrically similar design have to be run to discharge 1 gal/min against a 1-ft (0.3m) head. In general, pumps with a low specific speed have a low capacity, while those with high specific speed have a high capacity. Specific speeds of different types of pumps are shown in Table 6-2 for comparison.

Another parameter that helps in evaluating the pump suction limitations, such as cavitation, is suction specific speed (N_{sc}):

$$N_{sc} = \frac{NQ^{0.5}}{(\text{NPSH})^{0.75}} \qquad (6-6)$$

Typically, for single-suction pumps, suction-specific speed above 11,000 is considered excellent, below 7000 is poor and in the 7000-9000 range is of an average design. Similarly, for double-suction pumps, suction-specific speed above 14,000 is considered excellent, below 7000 is poor, and 9000-11,000 is average.

A typical pump-selection diagram, such as that shown in Figure 6-4, calculates the specific speed for a given flow, head, and speed requirements. Based on the calculated specific speed, the optimal pump design is indicated. The figure clearly indicates that, as the specific-speed increases, the ratio of the impeller outer diameter, D_2, to inlet or eye diameter, D_1, decreases, tending to become unity for pumps of axial-flow type.

Typically, axial-flow pumps are of high-flow and low-head type and have a high specific speed. On the other hand, purely radial pumps are of high-head and low flow-rate capability and have a low specific speed. Obviously, a pump with a moderate flow and head has an average specific speed.

Pump Materials

In the power industry, the selection of pump materials of construction is dictated by considerations of corrosion, erosion, personnel safety, and liquid contamination. The experience of pump manufacturers is often valuable in selecting materials.

An Overview of Pumps • 261

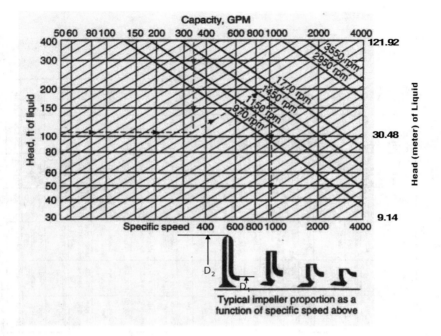

Figure 6-4. A Typical Pump Selection Diagram

Metal pumps are the most widely used. Although they may be obtained in iron, bronze, and iron with bronze fittings, an increasing number of pumps of ductile-iron, steel, and nickel alloys are also being used. Pumps are also available in glass, glass-lined iron, carbon, rubber, rubber-lined metal, ceramics, and a variety of plastics, such units usually being employed for special purposes.

Table 6-3 is the API 610 materials table for various pump parts.

Reference and general notes for Table 6-3:

1. Austenitic stainless steels include ISO Types 683-13-10/19 (AISI Standard Types 302, 303, 304, 316, 321, and 347). If a particular type is desired, the purchaser will so state.
2. For vertically suspended pumps with shafts exposed to liquid and running in bushings, the shaft shall be 12% chrome, except for Cl. S-9, A7, A-8, and D-1. Cantilever (Type VS5) pumps may utilize AISI 4140 where the service liquid will allow.
3. Unless otherwise specified, the need for hard-facing and the specific hard-facing material for each application shall be determined by the vendor and described in the proposal. Alternatives to hard-facing may include opening running clearances (2.6.4) or the use of non-galling materials, such as Nitronic 60 and Waukesha 88, depending on the corrosiveness of the pumped liquid.

Table 6-3. Materials for Various Pump Parts (Courtesy API 610)

Part	Full Compliance Material?	I-1 CI	I-2 BRZ	S-1 CI STL	S-3 STL	S-4 STL	S-5 STL 12% CHR	S-6 STL	S-8 STL	S-9 STL	C-6 12% CHR	A-7 AUS	A-8 316AUS	D-1 DUPLEX
Pressure Casing	Yes	Cast iron	Cast iron	Carbon steel	Ni-RESIST	Carbon steel	Carbon steel	Carbon steel	316 AUS	Carbon steel	MONEL	12% CHR	AUS (1&2)	316 AUS (1&2)	DUPLEX
Inner case parts (bowls, diffusers, diaphragms)	No	Cast iron	Bronze	Cast iron	Ni-resist	Cast iron	Carbon steel	12% CHR	316 AUS	Monel	12% CHR	AUS	316 AUS	Duplex
Impeller	Yes	Cast iron	Bronze	Cast iron	Ni-resist	Cast iron	Carbon steel	12% CHR	316 AUS	Monel	12% CHR	AUS	316 AUS	Duplex
Case wear rings	No	Cast iron	Bronze	Cast iron	Ni-resist	Cast iron	12% CHR	12% CHR Hardened	Hard Faced 316 AUS (2)	Monel	12% CHR hardened	Hard Faced AUS (3)	Hard Faced 316 AUS (3)	Duplex (3)
Impeller wear rings	No	Cast iron	Bronze	Cast iron	Ni-resist	Cast iron	12% CHR	12% CHR Hardened	Hard faced 316 AUS (3)	Monel	12% CHR hardened	Hard Faced AUS (3)	Hard Faced 316 AUS (3)	Duplex (3)
Shaft (2)	Yes	Carbon steel	Carbon steel	Carbon steel	Carbon steel	Carbon steel	AISI 4140	AISI 4140 (4)	316 AUS	K-Monel	12% CHR	AUS	316 AUS	Duplex
Shaft sleeves, packed pumps	No	12% CHR hardened	Hard bronze	12% CHR hardened	12% CHR hardened or hard faced	12% CHR hardened or hard faced	12% CHR hardened or hard faced	12% CHR hardened or hard faced	Hard Face 316 AUS (3)	K-Monel, hardened	12% CHR hardened or hard faced	Hard Faced AUS (3)	Hard Faced 316 AUS (3)	Duplex (3)
Shaft sleeves, mechanical seals	No	AUS or 12% CHR	AUS or 12% CHR	AUS or 12% CHR	AUS or 12% CHR	AUS or 12% CHR	AUS or 12% CHR	AUS or 12% CHR	AUS or 12% CHR	K-Monel, hardened	AUS or 12% CHR	AUS	316 AUS	Duplex
Throat bushings	No	Cast iron	Bronze	Cast iron	Ni-resist	Cast iron	12% CHR	12% CHR	316 AUS	Monel	12% CHR hardened	AUS	316 AUS	Duplex
Interstage sleeves	No	Cast iron	Bronze	Cast iron	Ni-resist	Cast iron	12% CHR hardened	12% CHR hardened	Hard Faced 316 AUS (3)	K-Monel, hardened	12% CHR hardened	Hard Faced AUS (3)	Hard Faced 316 AUS (3)	Duplex (3)
Interstage bushings	No	Cast iron	Bronze	Cast iron	Ni-resist	Cast iron	12% CHR hardened	12% CHR hardened	Hard Faced 316 AUS (3)	K-Monel, hardened	12% CHR hardened	Hard Faced AUS (3)	Hard Faced 316 AUS (3)	Duplex (3)
Seal gland	Yes	316 AUS (5)	316 AUS (5)	316 AUS (5)	316 AUS (5)	316 AUS (5)	316 AUS (5)	316 AUS (5)	316 AUS (5)	Monel	316 AUS (5)	316 AUS (5)	316 AUS (5)	Duplex (5)
Case and gland studs	Yes	Carbon steel	Carbon steel	AISI 4140 steel	AISI 4140 steel	AISI 4140 steel	AISI 4140 steel	AISI 4140 steel	AISI 4140 steel	K-Monel, hardened (8)	AISI 4140 steel	AISI 4140 steel	AISI 4140 steel	Duplex (8)
Case gasket	No	AUS, spiral wound (6)	AUS, spiral wound (6)	AUS, spiral wound (6)	AUS, spiral wound (6)	AUS, spiral wound (6)	AUS, spiral wound (6)	AUS, spiral wound (6)	316 AUS, spiral wound (6)	Monel, spiral wound, PTFE filled (6)	AUS, spiral wound (6)	AUS, spiral wound (6)	316 AUS spiral wound (6)	316 AUS SS spiral wound (6)
Discharge head / suction can	Yes	Carbon steel	Carbon steel	Carbon steel	Carbon steel	Carbon steel	Carbon steel	Carbon steel	Carbon steel	Carbon steel	AUS	AUS	316 AUS	Duplex
Column / bowl shaft bushings	No	Nitrile (7)	Bronze	Filled carbon	Nitrile (7)	Filled carbon	Filled carbon	Filled carbon	Filled carbon	Filled carbon	Filled carbon	Filled carbon	Filled carbon	Filled carbon
Wetted fasteners (bolts)	Yes	Carbon steel	Carbon steel	Carbon steel	Carbon steel	Carbon steel	316 AUS	316 AUS	316 AUS	K-Monel	316 AUS	316 AUS	316 AUS	Duplex

[a] The abbreviation above the diagonal line indicates the case material; the abbreviation below the diagonal line indicates trim material
Abbreviations are as follows: BRZ = bronze, STL - steel, 12% CHR = 12% chrome, AUS = austenitic stainless steel, CI = cast iron, 316 AUS = Type 316 austenitic stainless steel

4. For Cl. S-6, the shaft shall be 12% chrome if the temperature exceeds 175°C (350°F) or if used for boiler feed service.
5. The gland shall be furnished with a non-sparking floating throttle bushing of a material such as carbon graphite or glass-filled PTFE. Unless otherwise specified, the throttle bushing shall be premium carbon graphite.
6. If pumps with axially split casings are furnished, a sheet gasket suitable for the service is acceptable. Spiral wound gaskets should contain a filler material suitable for the service.
7. Alternate materials may be substituted for liquid temperatures greater than 45°C (110°F) or for other special services.
8. Unless otherwise specified, AISI 4140 steel may be used for non-wetted case and gland studs.

Types of Pumps

Process Pumps

This term is usually applied to single-stage pedestal-mounted units with single-suction overhung impellers and with a single packing box. These pumps are ruggedly designed for ease in dismantling and accessibility, with mechanical seals or packing arrangements, and are built especially to handle corrosive or otherwise difficult-to-handle liquids.

Most pump manufacturers now build to national standards the various types of horizontal and vertical pumps. American National Standards Institute (ANSI) Standards B73.1 — 1977 and B73.2 — 1975, and API 610 apply to the horizontal, and vertical in-line pumps, respectively.

The horizontal pumps are available for capacities of up to 4000 gal/min (900 m^3/h); the vertical in-line pumps, for capacities of up to 1400 gal/min (320 m^3/h). Both horizontal and vertical in-line pumps are available for heads up to 400 ft (130 m). The intent of each ANSI specification is that pumps from all vendors for a given nominal capacity and total dynamic head at a given rotative speed shall be dimensionally interchangeable with respect to mounting, size, location of suction and discharge nozzles, input shaft, base plate, and foundation bolts.

The vertical in-line pumps, although relatively new additions, are finding considerable use in power plants around the world, especially for cooling water systems. An inspection of the two designs will make clear the relative advantages and disadvantages of each.

Sump Pumps

These pumps are used for services requiring heads (pressures) higher than can be generated by a single impeller. All impellers are in series, the liquid passing from one impeller to the next and finally to the pump discharge. The total head then is the summation of the heads of the individual impellers. In combined cycle power plants, multistage pumps are required for various services such as deep-well pumps, high-pressure water-supply pumps, boiler-feed pumps, fire pumps.

Multistage pumps may be of the volute type, with single- or double-suction impellers, or of the diffuser type. They may have horizontally split casings or, for extremely high pressures, 3000 to 6000 psi (200 to 400 bar), vertically split barrel-type exterior casings with inner casings containing diffusers, and interstage passages.

Axial-Flow Pumps

These pumps are essentially very high-capacity, low-head units. Normally, they are designed for flows in excess of 2000 gal/min (450 m^3/h) against heads of 50 ft (17 m) or less. They are used to great advantage in closed-loop circulation systems in which the pump casing becomes merely an elbow in the line.

Turbine Pumps

The term "turbine pump" is applied to units with mixed-flow (part axial and part centrifugal) impellers. Such units are available in capacities from 100 gal/min (20 m^3/h) upward for heads up to about 100 ft (34 m) per stage. Turbine pumps are usually vertical.

A common form of turbine pump is the vertical pump, which has the pump element mounted at the bottom of a column that serves as the discharge pipe. Such units are immersed in the liquid to be pumped and are commonly used for wells, condenser circulating water, and large-volume drainage. Another form of the pump has a shell surrounding the pumping element, which is connected to the intake pipe. In this form, the pump is used on condensate service in power plants.

Regenerative Pumps

Also referred to as turbine pumps because of the shape of the impeller, regenerative pumps employ a combination of mechanical impulse and centrifugal force to produce heads of several hundred feet (meters) at low volumes, usually less than 100 gal/min (20 m^3/h). The impeller, which rotates at high speed with small clearances, has many short radial passages milled on each side at the periphery. Similar channels are milled in the mating surfaces of the casing. Upon entering, the liquid is directed into the impeller passages and proceeds in a spiral pattern around the periphery, passing alternately from the impeller to the casing and receiving successive impulses as it does so.

These pumps are particularly useful when low volumes of low-viscosity liquids must be handled at higher pressures than are normally available with centrifugal pumps. Close clearances limit their use to clean liquids. For very high heads, multistage units are available.

Gear Pumps

These types of pumps are usually part of the lubrication system for the turbines in most combined cycle power plants. These pumps are positive displacement types. When two or more impellers are used in a rotary-pump casing, the impellers will take the form of toothed-gear wheels of helical gears, or of lobed cams. In each case, these impellers rotate with extremely small clearance between them and between the surfaces of the impeller and the casing. The pumped liquid flows into the spaces between the impeller teeth as these cavities pass the suction opening. The liquid is then carried around the casing to the discharge opening, where it is forced out of the impeller teeth.

These positive displacement pumps are available in two general classes, interior bearing and exterior bearing. The interior-bearing type is used for handling liquids of a lubricating nature, and the exterior-bearing type is used with non-lubricating liquids. The interior-bearing pump is lubricated by the liquid being pumped; the exterior-bearing type is oil-lubricated from an exterior source.

The use of spur gears in gear pumps will produce discharge pulsations having a frequency equivalent to the number of teeth on both gears multiplied by the speed of rotation. This is known as the gear mesh frequency. The amplitude of these disturbances is a function of tooth design. The pulsations can be reduced markedly by the use of rotors with helical teeth. This, in turn, introduces end thrust, which can be eliminated by the use of double-helical or herringbone teeth.

Screw Pumps

A modification of the helical gear pump is the screw pump. Both gear and screw pumps are positive displacement pumps. In the two-rotor version, the liquid is fed to either the center or the ends, depending upon the direction of rotation, and progresses axially in the cavities formed by the meshing threads or teeth. In three-rotor versions, the center rotor is the driving member, while the other two are driven.

Screw pumps, because of multiple dams that reduce slip, are well adapted for producing higher pressure rises around 1000 psi (69 bar), especially when handling viscous liquids such as heavy oils. These pumps are used as fuel feed in gas turbines. The all-metal pumps are generally subject to the same limitations on handling abrasive solids as conventional gear pumps. In addition, the wide bearing spans usually demand that the liquid have considerable lubricity to prevent metal-to-metal contact.

Centrifugal Pumps

The centrifugal pump is the type most widely used in the Power industry for transferring liquids and steam for general services of water supply, boiler feed, condenser circulation, condensate return, and cooling water. These pumps are available through a vast range of sizes, in capacities from 2 gal/min to 10^5 gal/min (0.5 to 2×10^4 m³/h), and for discharge heads (pressures) from a few psi to approximately 7000 psi (483 bar). The size and type best suited to a particular application can be determined only by an engineering study of the problem.

The primary advantages of a centrifugal pump are simplicity, low first cost, uniform (non-pulsating) flow, small floor space, low maintenance expense, quiet operation, and adaptability for use with a motor or a turbine drive.

A centrifugal pump, in its simplest form, consists of an impeller rotating within a casing. The impeller consists of a number of blades, either open or shrouded, mounted on a shaft that projects outside the casing. Its axis of rotation may be either horizontal or vertical, to suit the work to be done. Closed-type, or shrouded, impeller are generally the most efficient. Open- or semi-open-type impellers are used for viscous liquids or for liquids containing solid materials and on many small pumps for general service. Impellers may be of the single-suction or the double-suction type. A single-suction type impeller has the liquid entering from one side; a double-suction impeller if the liquid enters from both sides.

A centrifugal pump in its simplest form is shown in Figure 6-5. Power from an outside source is applied to the shaft A, rotating the impeller B within the stationary casing C. The blades of the impeller in revolving produce a reduction in pressure at the entrance or eye of the impeller. This causes liquid to flow into the impeller from the suction pipe D. This liquid is forced outward along the blades at increasing tangential velocity. The velocity head it has acquired when it leaves the blade tips is changed to pressure head as the liquid passes into the volute chamber and, hence, out the discharge E.

Most of these pumps can be classified into the following major categories such as overhung, between bearings and vertical pumps. Figure 6-6 shows such a classification. Figure 6-7 shows the different configurations of over hung pumps. Figure 6-8 shows the various type of between bearing pumps, and Figure 6-9 shows the various types of vertical pumps.

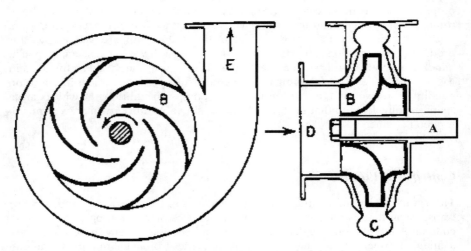

Figure 6-5. A Typical Centrifugal Pump Showing the Volute and the Impeller

An Overview of Pumps • 267

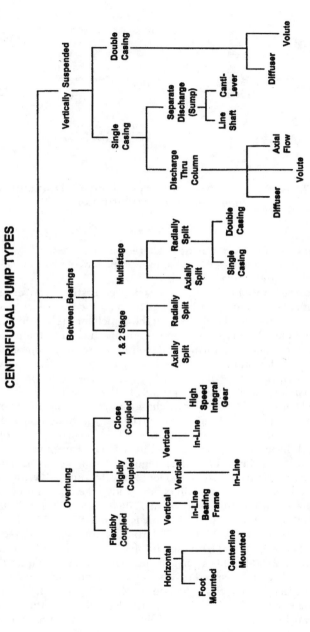

Figure 6-6. Classification of Centrifugal Pump Types. Chapter 3 [1] (Courtesy of the American Petroleum Institute)

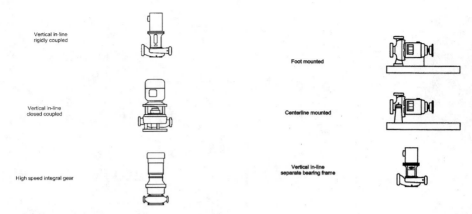

Figure 6-7. Overhung Centrifugal Pumps. Chapter 3 [1] (Courtesy of the American Petroleum Institute)

Centrifugal Compressor Casings. There are three general types of casings, but each consists of a chamber in which the impeller rotates, provided with inlet and exit for the liquid being pumped.

Circular casing is the simplest form of casing, consisting of an annular chamber around the impeller; no attempt is made to overcome the losses that arise from eddies and shock when the liquid leaving the impeller at relatively high velocities enters this chamber. This type of casing is not much used in the power industry.

Volute casings take the form of a spiral, increasing uniformly in cross-sectional area as the outlet is approached. The volute efficiently converts the velocity energy imparted to the liquid by the impeller into pressure energy.

Diffuser casing is used in turbine pumps. In this type, guide vanes or diffusers are interposed between the impeller discharge and the casing chamber. Losses are kept to a minimum in a well-designed pump of this type, and improved efficiency is obtained over a wider range of capacities. This construction is often used in multistage high-head pumps such as the boiler feed water pumps.

Centrifugal-Pump Characteristics. Figure 6-10 shows a typical characteristic curve of a centrifugal pump used as a boiler feed water pump. It is important to note that at any fixed speed the pump will operate along the curve known as the speed line. For instance, on the curve shown, at 770 gal/min (175 m^3/h), the pump will generate a 2100-ft (700-m) head. If the flow is decreased to 352 gal/min (80 m^3/h), the head will be increased to 2280 ft (750 m). It is not possible to reduce the flow capacity to 352 gal/min (80 m^3/h) at the 2100-ft (700-m) head unless the discharge is throttled so that a 2100-ft (700-m) head is actually generated within the pump.

On pumps with variable-speed drives, such as steam turbines, it is possible to have multiple speed lines such as the one shown in Figure 6-2. The head depends upon the velocity of the fluid, which, in turn, depends upon the capability of the impeller to transfer energy to the fluid. This is a function of the fluid viscosity and the impeller design. It is important to remember that the head produced will be the same for any liquid of the same viscosity. The pressure rise, however, will vary in

Axially split, 1 and 2 stage

Radially split, 1 and 2 stage

Axially split, multistage

Radially split, multistage:
Single casing
Double casing

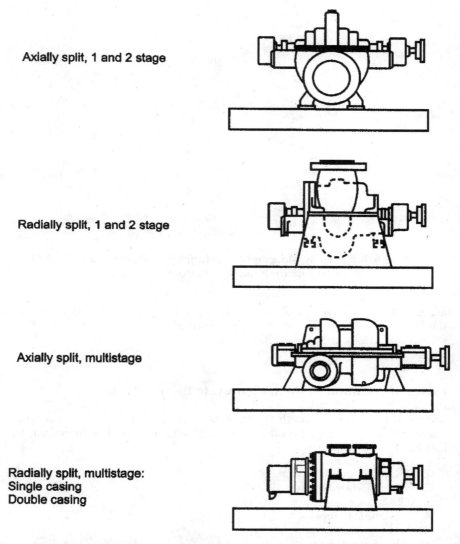

Figure 6-8. Between Bearing Type Centrifugal Pumps. Chapter 3 [1]
(Courtesy of the American Petroleum Institute)

proportion to the specific gravity. For quick pump selection, manufactures often give the most essential performance details for a whole range of pump sizes. The performance data consists of a usually contains pump flow rate and head. Figure 6-10 shows a more detailed pump performance curve that includes, in addition to pump head and flow, the break horsepower required, the NPSH required, and pump efficiency. If detailed manufacturer-specified performance curves are not

270 • COGENERATION AND COMBINED CYCLE POWER PLANTS

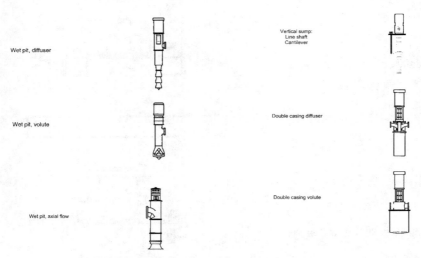

Figure 6-9. Vertically Mounted Centrifugal Type Pumps. Chapter 3 [1] (Courtesy of the American Petroleum Institute)

available for a different size of the pump or operating condition, a best estimate of the off-design performance of pumps can be obtained through similarity relationship or the affinity laws as described earlier.

Pump Application in Combined Cycle Power Plants

Most pumps used to pump the steam or the condensate are centrifugal pumps. There are six types of major pumps in a combined cycle power plant system. They consist of the following:

1. IP-LP Circulating Pumps
2. HP Feed Water Pumps
3. HP Circulating Pumps
4. Condenser Pumps
5. Cooling water Pumps
6. Lubrication Pumps
7. Fuel Pumps (Liquid Fuel Applications)

The IP-LP Circulating Pump

The IP-LP circulating pump circulates the water between the deaerator and the LP evaporator. There is always one spare pump in most systems as they are very critical to the operation of the plant. In plants with two or more HRSG units, there could be one pump for each unit and one pump spare.

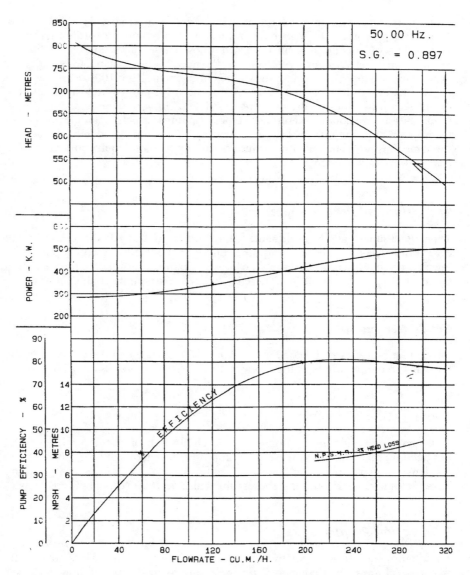

Figure 6-10. A Typical Characteristic Performance Map of a Boiler Feed Water Pump

HP Feed Water Pumps

The HP boiler feed water pump (HP-BFP) is a multistage centrifugal pump. A pump of this kind pumps the heated water between the deaerator and the HP economizer. These pumps are large in size and can be as large as 1500 kW in size. These pumps operate at high temperature and corrosion problems as well as cavitation problems often occur in these pumps. These pumps are usually driven by single-speed electric motors. However, the need for large plants to be operated over a large range of flow due to the cycling of the plant from 100% design load to about 40% design load has made variable-speed drive pumps important, and the use of steam turbines as drives being considered in various applications. Field modifications are conducted on these types of pumps, which include trimming of the impellers, thus modifying the performance of the pump by reducing the head at the same flow rate. The total importance of this application requires that there be a spare pump in the system.

The HP Circulating Pump

The HP circulating pump circulates the fluid from the HP drum to the HP evaporator. These pumps behave quite similarly to the IP-LP circulating pumps. In most cases, they are much larger in the flow they pump. It is not uncommon to find two of these pumps per HRSG. There is always a spare pump in this application.

Condenser Pumps

This pump type drives the condensate from the condenser to the dearetor in the HRSG. The temperatures in these pumps are low. These pumps are also spared.

Cooling Water Pumps

These pumps provide the cooling water, which flows through the condenser. It is not uncommon for these pumps to be pumping brackish or salt water through the condensers. In many applications, especially when these pumps are pumping water from lakes or deep wells, these pumps also serve as vertical pumps.

Lubrication Pumps

These pumps are usually positive displacement type of pumps. They are usually gear-type pumps. Most codes require these pumps to be spared, and the spare pumps have a separate source of power since these turbines would be very badly damaged if the lubrication were to stop. A third pump is also required — usually this is a DC drive pump and does not have the flow capacity of the two main pumps. Instead of a DC pump, some plants have a special lubrication tank, which is placed at a height above the turbine; thus, in case of a lubrication pump failure, the tank would, due to gravity, provide the lubrication for a safe coast down of the system.

Lubrication systems, aside from providing lubrication for the turbines, also provide cooling. Thus, it is very important that the pumps should be operated for a period of at least 20 minutes after the turbines have been shut down since some of the highest temperatures are reached in that time span after shutdown. To prevent sagging, many large turbines will also, after shutdown, be operated on turning gear and will need lubrication for that.

Fuel Pumps

These pumps are, in most cases, positive displacement pumps since they require a very high head, but the flow is relatively small. The fuel must be injected into the combustor at about 60 psia (4.0 bar), above the pressure of the gas turbine compressor discharge air. The fuel, in some cases, is very viscous and has to be heated. Details of the fuel requirements are given in the chapter on fuels in the book.

Chapter 7

HEAT RECOVERY STEAM GENERATORS

The heat recovery steam generator (HRSG) is a critically important subsystem of a combined cycle or cogeneration power plant. In most of these plants the HRSG uses the exhaust gas from the gas turbine as the energy source for the production of high-pressure and-temperature steam. The main difference in these plants is that in a combined cycle power plant, the steam generated in the HRSG is used solely in the production of power while in a cogeneration plant, the steam can be used for process as well as power production. The combined cycle power plant uses the steam in a large condensing steam turbine that produces about 40% of the power generated at design conditions, while in a cogeneration application it is not uncommon to bleed the steam from an extraction steam turbine for process purposes. In the cogeneration mode, if a steam turbine is used, these extraction type steam turbines are usually smaller and may be of a back-pressure type.

The combined cycle power plant, in most cases, consists of the combination of the Brayton and Rankine Cycles, one of the most efficient cycles in operation for practical power generation systems. The Brayton Cycle is the gas turbine cycle and the Rankine Cycle is the steam turbine cycle. In most combined cycle applications, the gas turbine is the topping cycle and the steam turbine is the bottoming cycle. Thermal efficiencies of the combined cycles can reach as high as 60%. In the typical combination, the gas turbine produces about 60% of the power and the steam turbine about 40%. Individual unit thermal efficiencies of the gas turbine and the steam turbine are between 30% and 40%. The steam turbine utilizes the energy in the exhaust gas of the gas turbine as its input energy. The energy transferred to the HRSG by the gas turbine is usually equivalent to about the rated output of the gas turbine at design conditions. At off-design conditions, the inlet guide vanes (IGV) are used to regulate the air so as to maintain a high temperature to the HRSG.

In the traditional combined cycle plant, air enters the gas turbine where it is initially compressed and then enters the combustor where it undergoes a very high rapid increase in temperature at constant pressure. The high temperature and high pressure air then enters the expander section where it is expanded to nearly atmospheric conditions. This expansion creates a large amount of energy, which is used to drive the compressor used in compressing the air, plus the generator where power is produced. The compressor in the gas turbine uses about 50% to 60% of the power generated by the expander.

The air upon leaving the gas turbine, is essentially at atmospheric pressure conditions, and at a temperature between 950°F and 1200°F (510°C to 650°C). The free oxygen content of the gas ranges from 14% to 16% by volume if steam/water is not injected into the turbine, and from 12% to 14% if steam is injected.

The gas turbine exhaust gases enter the HRSG, where the energy is transferred to the water to produce steam. There are many different configurations of the HRSG units. Most HRSG units are divided into the same amount of sections as the steam turbine. In most cases, each section of the HRSG has a pre-heater, an economizer and feed-water, and then a Superheater. The steam entering the steam turbine is superheated.

In most large plants the steam turbine consists of three sections: a high-pressure turbine stage (HP) with pressures between 1500 and 1700 psia (100.7-114.2 Bar), an intermediate-pressure turbine stage (IP) with pressures between 350 and 550 psia (23.51-36.9 Bar), and a low-pressure turbine stage (LP) with pressures between 90 and 60 psia (6-4 Bar). The steam exiting from the high-pressure stage is usually reheated to about the same temperature as the steam entering the HP stage before it enters the intermediate stage. The steam exiting the IP stage enters directly into the LP stage after it is mixed with steam coming from the LP superheater.

The steam from the LP turbine enters the condenser. The condenser is maintained at a vacuum of between 2 and 0.5 psia (0.13-0.033 Bar). The increase in back pressure in the condenser will reduce the power produced. Care is taken to ensure that the steam leaving the LP stage blades does not have a high content of liquid in the steam to avoid erosion of the LP blading.

In the past, the HRSG was viewed as a separate "add-on" item. This view is being changed with the realization that good performance, both thermodynamically and in terms of reliability, grows out of designing the heat recovery system as an integral part of the overall system. The concept that the combined cycle power plant is always base loaded has also been altered and with that, some of the design philosophy behind the HRSG.

HRSG Units could be classified into three major categories:

1. The Horizontal Type HRSG
2. The Vertical Type HRSG
3. Once Through Steam Generator

The Horizontal Type HRSG

The Horizontal Type HRSG is usually a modular type construction consisting of a superheater, reheater, evaporator, economizer and feedwater heater sections. All tubes of the heat transfer coils are welded into upper and lower headers forming groups of platen assemblies which may consist of two or three rows of tubes (parallel to the gas flow) designed for triangular (staggered) pitch or square (in-line) pitch.

All heat transfer sections are usually of an all-welded construction to enhance the availability and reliability. In addition, all heat transfer sections are top supported to minimize expansion of external piping, and all heat transfer sections are fully drainable.

Heat Recovery Steam Generators • 277

Evaporator modules are connected to steam drums via downcomer pipes, riser pipes and distribution manifolds utilizing all-welded construction.

Steam drums are usually equipped with two-stage purification systems consisting of primary vortex steam separators and secondary chevron dryers to meet stringent steam purity requirements.

The inlet duct and module side casing panels are usually internally insulated and covered by liner panels consisting of appropriate material and thickness as determined by exhaust gas temperatures and velocities.

Figure 7-1 shows a typical drum type HRSG. Figure 7-2 is a cross-sectional drawing showing the rough dimensions of a typical Drum type HRSG for a large Gas Turbine of about 70-100 MW. Figure 7-3 shows a top elevation view of the horizontal type HRSG Unit. Most HRSG units are divided into the same amount of sections as the steam turbine, which usually consist of an High Pressure Section, an Intermediate Section (often optional), and a Low Pressure Section, which is connected to a condensing unit. In most cases each section of the HRSG has a Pre-heater or Economizer, an Evaporator, and then a one or two stage Superheaters. The steam entering the steam turbine is superheated.

The Vertical Type HRSG

The Vertical type HRSG has the gas exiting the gas turbine flow upward as it crosses the various heat exchanger sections. It also consists of a preheater, an economizer, and evaporator with a drum and a superheater. Figure 7-4 is a cross section of the schematic of a typical vertical HRSG. It has a smaller footprint than the Horizontal Type. The water/steam mixture flowing through the evaporator is constantly controlled by a pump. The tubes of the heat exchangers are guided and supported horizontally by tube plates which form the back bones of the modules. The modules in turn are top supported from a structural frame.

Figure 7-1. A Typical HRSG for a Large Gas Turbine (Courtesy Voght-NEN Inc.)

278 • **COGENERATION AND COMBINED CYCLE POWER PLANTS**

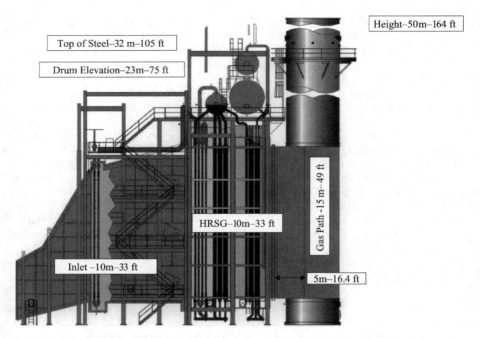

Figure 7-2. Drum Type HRSG Elevation View (Courtesy of Aalborg Industries)

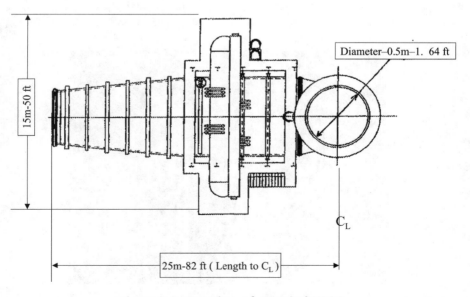

Figure 7-3. Top View of a Typical HRSG

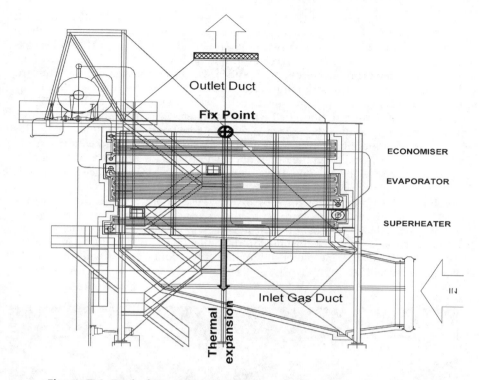

Figure 7-4. Typical Cross Section Schematic of a Vertical Type HRSG (Courtesy CMI International)

A stack is located on top of the structure and an inlet duct connects the modules to the gas turbine outlet. Interconnecting piping, valves and instruments complete the HRSG. The HRSG is externally insulated.

Vertical Type HRSGs are designed to handle heavy cycling (frequent shutting OFF and starting ON and load modulating) thus extending its life and permitting to run the power plant with a lot of flexibility. Vertical HRSGs are usually very easy to inspect and maintain because all components are accessible. Since all the internal parts are on a horizontal plan, they are very convenient to access through side doors. No scaffolding is usually necessary in these types of HRSGs. The {warm casing} design of the Vertical HRSG with external insulation provides additional resistance to corrosion, and is safer for the operators (no hot spots), and by allowing for extra flexibility the life of the HRSG is extended.

Once Through Steam Generator

The Once Through Steam Generators (OTSG) unlike other HRSGs do not have a defined economizer, evaporator, or superheater sections; nor do they have steam drums, as can be seen Figure 7-5 which is a photograph of the HRSGs of 4 GE LM 6000 aeroderivative gas turbines. The Once Through Steam Generators have a much smaller foot print as shown in Figure 7-6. This type of an HRSG consists basically of one tube; water enters at one end and steam leaves at the other end, eliminating the drum and circulation pumps. The heat input from the gas turbine determines the water to steam interface. This interface is free to move, depending also on the flow rates and pressures of the Feedwater, in the tube bank.

The HRSGs are also equipped with a complete Selective Catalyst Reduction (SCR) system to reduce the NOx emissions to a few part per million (PPM). This includes an ammonia skid, an ammonia injection grid and the catalyst. A gas side silencer is installed in the outlet duct. The HRSG also receives a CO catalyst system to reduce the CO emissions to a few PPM and also educe the Volatile Organic Compound (VOC) emissions.

Noise is a critical factor for most plants. HRSGs are designed to reduce the noise coming from the gas turbine by the use of a large silencer, external insulation, and reduced internal obstacles. A stack is located on top of the structure where a gas side silencer is installed in the outlet duct. An inlet duct connects the HRSG to the gas turbine outlet. Interconnecting piping, valves and instruments complete the HRSG.

Figure 7-5. A Photograph of a Once Through Steam Generators for LM 6000. Note the Compactness and Simplicity of Such a System (Courtesy of Innovative Steam Technologies Ltd.)

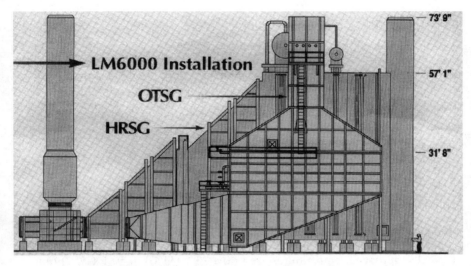

Figure 7-6. Comparison of a Drum-Type HRSG to a Once Through Steam Generator (Courtesy of Innovative Steam Technologies)

The gas from the Gas Turbine has a lot of excess oxygen, thus all these types of HRSGs can be further equipped with supplementary firing. This is done usually by the addition of a duct burner located in the inlet duct. The natural gas fired burner provides additional heat to the HRSG to handle the summer peak demand.

Depending on the steam requirements, HRSGs can take three forms:

1. Unfired
2. Supplementary-fired
3. Exhaust-fired

In unfired HRSGs, the energy from the exhaust is used as such, while in supplementary-fired and exhaust-fired HRSGs, additional fuel is inputted to the exhaust gas to increase steam production.

Design Considerations

In the design of an HRSG, the following are some of the major points that should be considered.

Multi-Pressure Steam Generators

These are becoming increasingly popular. In a single pressure boiler there is a limit to heat recovery because the exhaust gas temperature cannot be reduced below the steam saturation temperature. This problem is avoided by the use of multi-pressure levels.

Pinch Point

This is defined as the difference between the exhaust gas temperature leaving the evaporator section and the saturation temperature of the steam. Ideally, the lower the pinch point, the more heat recovered, but this calls for more surface area and, consequently, increases the back pressure and cost. Also, excessively low pinch points can mean inadequate steam production if the exhaust gas is low in energy (low mass flow or low exhaust gas temperature). General guidelines call for a pinch point of 40°F (22°C) to 60°F (33°C). The final choice is obviously based on economic considerations.

Approach Temperature

This is defined as the difference between the saturation temperatures of the steam and the inlet water. Lowering the approach temperature can result in increased steam production, but at increased cost. Conservatively high approach temperatures ensure that no steam generation takes place in the economizer. Typically, approach temperatures are in the 20°F (11°C) to 50°F (27°C) range. A typical temperature profile showing the pinch point and approach temperature is also shown in Figure 7-7. During the initial design stages, these two factors may be selected. In unfired units, the pinch point and approach point lie in the

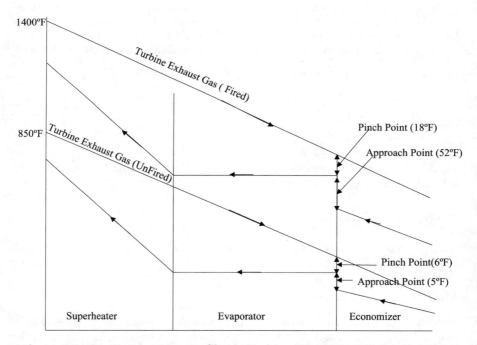

Figure 7-7. Gas Temperature Profiles in Turbine Waste-Heat Boiler Showing Pinch and Approach Temperatures

range of 50°F to 30°F. Low values of pinch and approach mean that the log-mean temperature differential in the evaporator and economizer are reduced, thereby increasing the surface area requirements, and adding to the gas pressure drop and cost of the HRSG. However, the exit gas temperature for the HRSG can be lower, leading to more steam production. An optimum combination of boiler surface (reflecting the initial cost), gas pressure drop and steam production (reflecting the operating cost/savings) can be determined using economic evaluations.

Once the temperature profile is selected in the unfired mode, the HRSG design is nearly fixed. Its performance at other operating conditions, namely different ambient temperatures and firing modes, has to be analyzed using a computer, as many iterative loops are involved in the calculation procedure.

Casing of the HRSG

The casings of HRSGs can be divided into two main design configurations:

1. Cold casing with inner insulation. This is usually used for natural circulation HRSGs. The casing does not impose any limits on start-up times due to thermal expansion of the casing. The casing withstands the exhaust gas pressure because the inner insulation keeps it at a low temperature. Many of today's vertical HRSGs also use cold casing design.
2. Hot casing with external insulation. This type of casing is mostly used in HRSGs which have forced circulation and are usually associated with vertical HRSGs. When fuels with high sulfur content are fired, a hot casing at the cold end of the HRSG can limit corrosion.

The new gas turbines have much higher exhaust temperatures than their predecessors. This makes the hot casing use high alloy materials in the hot end of the HRSG.

Finned Tubing

Heat transfer in the HRSG is mainly by convection. Finned tubes are employed to increase the heat transfer surface. Heat transfer on the waterside is much higher than on the exhaust gas side. Thus, the fins are used on the exhaust gas side to improve the heat transfer.

Gas turbines firing on natural gas have a fin density of 5 to 7 fins per inch (200 to 280 fins per meter). If the gas turbine uses heavier fuels than the fin, density is reduced to 3% to 4% per inch (120 to 160 fins per meter). This is due to the deposits that are usually caused by the use of heavy fuels.

Off-Design Performance

This is an important consideration for waste heat recovery boilers. Gas turbine performance is affected by load, ambient conditions and gas turbine health (fouling, etc.). This can affect the exhaust gas temperature and the airflow rate. Adequate considerations must be given to how steam flows (low pressure and high pressure) and superheat temperatures vary with changes in the gas turbine operation.

Evaporators

These usually utilize a fin-tube design. Spirally finned tubes of 1.25 to 2 in. outer diameter (O.D.) with 3 to 6 fins per inch are common. In the case of unfired designs, carbon steel construction can be used and boilers can run dry. As heavier fuels are used, a smaller number of fins per inch should be utilized to avoid fouling problems.

Forced Circulation System

Using forced circulation in a waste heat recovery system allows the use of smaller tube sizes with inherent increased heat transfer coefficients. Flow stability considerations must be addressed. The recirculating pump is a critical component from a reliability standpoint and standby (redundant) pumps must be considered. In any event, great care must go into preparing specifications for this pump.

Back Pressure Considerations (Gas Side)

These are important as excessively high back-pressures create performance drops in gas turbines. Very low pressure drops would require a very large heat exchanger and more expense. Typical pressure drops are 8 to 10 in. of water.

Supplementary Firing of Heat Recovery Systems

There are several reasons for supplementary firing a waste heat recovery unit. Probably the most common reason is to enable the system to track demand (i.e., produce more seam where the load swings upwards than the unfired unit can produce). This may enable the gas turbine to be sized to meet the base load demand, with supplemental firing taking care of high load swings.
    Raising the inlet temperature at the waste heat boiler allows a significant reduction in the heat transfer area and, consequently, the cost. Typically, as the gas turbine exhaust has ample oxygen, duct burners can be conveniently used.
    The presence of oxygen in the exhaust gas allows additional fuel to be added, either through duct burners, as in a supplementary-fired HRSG, or using conventional register burners in an exhaust-fired HRSG. The fuel input, exhaust temperature and gas analysis may be obtained using computer programs developed for this purpose. Quick estimates may be obtained, however, from the following relationship:

$$Q = 58.4 \, WO \qquad (7-1)$$

where
- Q = fuel input to the burner in Btu/hr on lower heating value basis
- W = exhaust gas flow in lb/hr and
- O = percent oxygen by volume consumed by the fuel

The above relationship is valid for gaseous and liquid fuels.

For example, by adding 30 mm Btu/hr to 150,000 lb/hr of gas, we note that the percent oxygen consumed is only 3.5. If the initial oxygen content were 16%, 12.5% is still left in the gas, which is significant.

The supplementary-fired HRSG typically raises the gas temperature to 1650°F (900°C) maximum. The construction is similar to that of the unfired HRSG except for the design of the firing duct, tube/fin geometry, and material and size of steam flow-related components such as the drum, piping and valves. If the steam production in the unfired HRSG is taken as 1, that of the supplementary-fired HRSG can be 2 to 2.5.

An advantage of supplemental firing is the increase in heat recovery capability (recovery ratio). A rule of thumb is that a 50% increase in heat input to the system increases the output by 94%, with the recovery ratio increasing by 59%. Figure 7-8 shows the relationship between the heat recovery ratio and the output of a supplemental-fired HRSG.

The performance of a 200×10^6 Btu/hr furnace when supplementally fired to 300×10^6 Btu/hr, increases from 118 to 229 mm Btu/hr. This is an increase of 94% for an input increase of 50%. The recovery ratio is increased from 59% to 76%. Figure 7-9 is a schematic of the effects of supplemental firing in a combined cycle power plant.

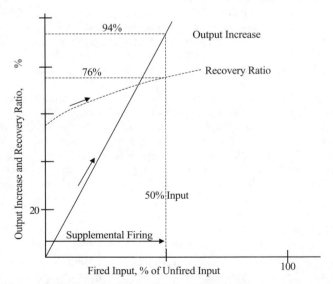

Figure 7-8. Heat Recovery Ratio and Output for a Supplementary-Fired HRSG

286 • COGENERATION AND COMBINED CYCLE POWER PLANTS

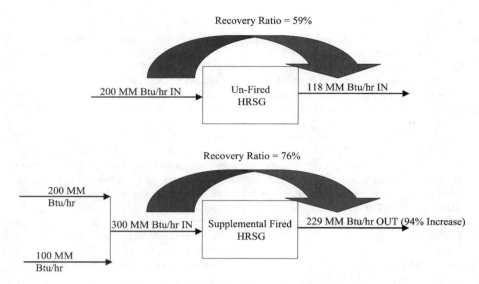

Figure 7-9. Effects of Supplemental Firing

The following are some important design guidelines to ensure a successful design of supplemental-fired HRSG:

1. Special alloys may be needed in the superheater and evaporator to withstand the elevated temperatures.
2. The inlet duct must be of sufficient length to ensure complete combustion and avoid direct flame contact on the heat transfer surfaces.
3. If natural circulation is utilized, an adequate number of risers and feeders must be provided as the heat flux at entry is increased.
4. Insulation thickness on the duct section must be increased.

The exhaust-fired HRSG resembles the conventional membrane wall fossil-fired boilers. Present technology permits firing of gaseous and liquid fuels only in duct burner, while solid, liquid and gaseous fuels can be fired in the exhaust-fired HRSG. The register burners treat the gas turbine as merely a source of hot combustion air. This application is being used in all re-powering of existing steam turbine plants to combined cycle power plants. If the steam production in the unfired HRSG is taken as 1, that of the exhaust-fired HRSG can be in the range of 4 to 5.

Design Features

An HRSG can be of natural or forced circulation design. An HRSG without supplementary firing is essentially a convective heat exchanger. The HRSG must be built for high reliability and availability. Provisions must be made for quick load changes and short start-up time. The most important development in the recent HRSG design is the move from single pressure to dual and triple pressure steam production. This maximizes the efficiency of the combined cycle. The dual pressure increases the cycle efficiency by over 4% from the single pressure and another 1% increase over the dual pressure for triple pressure units. The two types of HRSGs are forced circulation type, and natural circulation type.

The most common type of an HRSG in a large combined cycle power is a forced circulation type. These types of HRSGs are vertical, the exhaust gas flow is vertical with horizontal tube bundles suspended in the steel structure. The steel structure of the HRSG supports the drums. In a forced circulation HRSG, the steam/water mixture is circulated through evaporator tubes using a pump. These pumps increase the parasitic load and thus detract from the cycle efficiency. In this type of HRSG the heat transfer tubes are horizontal, suspended from uncooled tube supports located in the hot gas path. Some vertical HRSGs are designed with evaporators, which operate without the use of circulation pumps.

The circulation ratio in the forced circulation design is usually selected in the range of 3 to 6, while in the natural circulation design it can lie in the range of 10 to 25, depending on the steam pressure and circulation system resistance. Though the floor space occupied by the forced circulation HRSG is smaller, the main concern is the possibility of circulation pump failure. In today's plant, there is always a standby pump available at each circulation pump.

Figure 7-10 shows a forced circulation HRSG; this HRSG is in three stages: an HP stage, an IP stage and an LP stage. This feeds a similar three-stage steam turbine.

Extended surfaces are invariably used in HRSG as the size can be smaller and gas pressure drop can be within limits. Fins can be solid or serrated. Fin density ranges from 3 to 6 depending on fuel type used in the gas turbine and in HRSG burners. If the fuel is a clean gas, fin density can be as high as 6 and can be serrated or solid. Solid fins are preferred for oil fuels. With distillates, the fin density is limited from 4 to 5, while with heavier fuel oils it is limited to 3. Fin height ranges from 0.5 to 0.75 in., while the thickness ranges from 0.05 to 0.120 in.

A low fin height or a high thickness improves fin efficiency, and results in higher heat transfer coefficients and lower fin base and tip temperatures. Generally, higher fin density results in a higher gas pressure drop.

The main advantage of the vertical HRSGs are:

1. Smaller foot print arising from the vertical design
2. Less vulnerability to thermal cycling problems
3. Smaller HRSG volume because of the smaller diameter tubes due to the forced circulation pumps
4. Less sensitivity to steam blockage in economizers during startup

In the natural circulation design, there is no need for the pumps. This is the main advantage especially for the high-pressure designs above 1450 psia (100 Bar). The natural circulation HRSGs are horizontal type because circulation is established by the density differential between the down comer and riser circuits

288 • COGENERATION AND COMBINED CYCLE POWER PLANTS

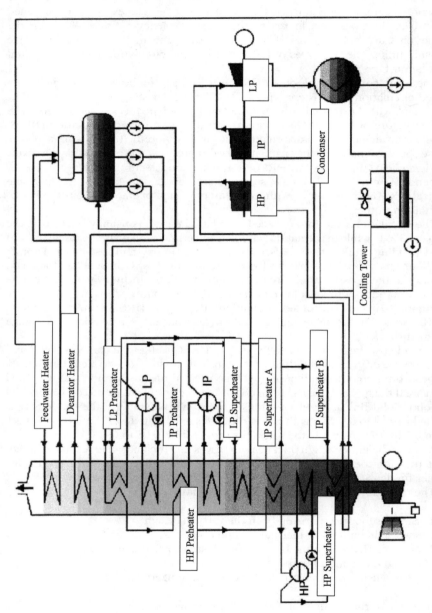

Figure 7-10. A Typical Large Combined Cycle Power Plant HRSG

and their hydraulic resistances. In this type of HRSG, the exhaust flow from the gas turbine is horizontal and the heat transfer tubes are vertical and essentially self-supporting. The steel structure is usually more compact than on the vertical HRSGs. The natural circulation boilers can be of single or two-pass design, depending on the space availability and layout criteria.

The main advantages of horizontal-type HRSG units are:

1. No pumps, thus less parasitic load
2. Vertical tubes do not allow dryout due to vigorous circulation

In North America, the use of horizontal-type, natural circulation HRSGs are used widely. Space requirements and start-up times are now nearly identical; water volumes in evaporators have been reduced with the use of smaller diameter tubes. Steam blockages are being better handled in modern natural circulation HRSGs.

The same pinch points can be achieved in high and intermediate pressure evaporator. The major difference occurs in large HRSGs with a tight low-pressure pinch-point. The new plants with the dual and triple pressure levels are using mostly vertical HRSGs.

Since the fin geometry affects the gas pressure drop (and every inch of additional gas pressure drop decreases the turbine power output by nearly 0.25%), surface area requirements and fin base and tip temperatures, the optimum configuration should be determined by evaluating various geometries.

Tubes may be arranged in-line or staggered. With clean gases both arrangements are used, while with dirtier fuels in-line arrangement with provision for soot blower lanes is preferred. Staggered arrangement results in higher gas pressure drop, but less surface area requirements. The total gas pressure drop in a typical HRSG from the turbine exhaust flange to the boiler stack can vary from 6″ H_2O (1.5 kPa) to 12″ H_2O (3.0 kPa).

In fired HRSGs with a superheater, the steam temperature can vary widely between unfired and fired modes, unless care is taken to locate the superheater beyond a screen section, in parallel with the main evaporator.

The screen section prevents direct radiation on the superheater tubes. Also, it helps reduce the gas temperature entering the superheater so that steam temperature fluctuations are minimal. The need for a screen section increases especially if the boiler is designed for a high firing temperature and also has to operate in unfired mode with a good steam temperature.

Another important aspect to consider in the design of a fired HRSG is the material selection of tubes and fins. Direct radiation from the cavities and non-luminous radiation have to be evaluated while determining fin base and tip temperatures. With finned tubes, the beam length for radiation my be small, but if the gas contains a large volume of water vapor, as in steam injected cycles (steam can be in the range of 10% to 12% by weight of air), or where steam/water is injected into the turbine for NO_x control (1% to 3% by weight), non-luminous radiation is significant.

The heat flux inside the tubes also becomes important due to the high ratio of tube external area to inside area in finned tubes, and can run up the heat flux significantly. Non-uniformity in gas temperature profiles has to be considered in evaluating the heat flux and HRSG performance. Departures from nucleate boiling (DNB) studies are performed if the steam pressure is high, after establishing the circulation in critical tube rows. Generally, a vertical tube can withstand a higher heat flux compared to an inclined or horizontal tube and hence, for the same heat flux a natural circulation design with vertical tubes offers more safety than other tube configurations. Critical heat flux, which is responsible for DNB, is a function of several variables such as pressure, mass flow through the tubes, steam quality, tube sizes and inlet sub-cooling. For a natural circulation HRSG, it can be in the range of 100 to 250,000 Btu/ft^2 hr.

Finned superheaters are widely used in HRSG designs. Since the steam side coefficient is several times lower than the boiling heat transfer coefficient, the fin base and tip temperatures can run high if a large fin density is used. Fin densities lie in the range of 2 to 3 per inch. If gas temperatures are high, a combination of bare and finned tubes may be used.

Once Through Steam Generators

The once through steam generators (OTSG), unlike other HRSGs, does not have a defined economizer, evaporator or superheater sections. Figure 7-11 is the schematic of an OTSG system, and a drum-type HRSG. The OTSG is basically one tube, water enters at one end and steam leaves at the other end, eliminating the drum and circulation pumps. The location of the water to steam interface is free to move, depending on the total heat input from the gas turbine, and flow rates and pressures of the feedwater, in the tube bank.

Unlike other HRSGs, the once-through units have no steam drums. Figure 7-12 is a schematic of a two-stage OTSG system. This is a major improvement of the cyclic capabilities of the combined cycle plant since the thick-walled drums are thermally sluggish and are also vulnerable to low-cycle fatigue. These systems also reduce water consumption, because there is no need for a blowdown system. This leads to smaller water treatment plants, giving large savings in operation and maintenance cost as well as first cost. In some units, the low-pressure system can be a conventional, natural circulation design with an LP drum on top of the boiler, while the high-pressure systems a once through type.

The simplicity in erection and operation means that the plant can be erected sooner and a less complicated system. This type of HRSG is controlled through a feedback loop, which adjusts the feedwater regulating valve, which is controlled by the steam temperature. A feed-forward signal senses changes in the gas turbine loading and pre-adjusts the feedwater regulator valve to the calculated new flows.

The small footprint enables these types of HRSG units to be installed in a confined area. Due to the ability of the OTSG to "run dry" without harming the tubing, the gas turbine can be operated, without the steam turbine, at full load conditions for long periods of time depending on the alloy of the tubes. This gets rid of the diverter valves, and bypass stacks, which are major problems in all installations due to the leakage in the diverter valve after a short operation period.

Heat Recovery Steam Generators • 291

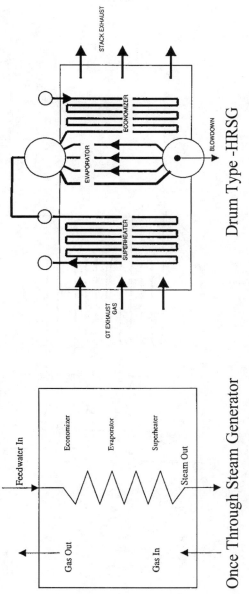

Figure 7-11. Schematic Comparison Between a Once Through Steam Generator and a Drum-Type HRSG

292 • COGENERATION AND COMBINED CYCLE POWER PLANTS

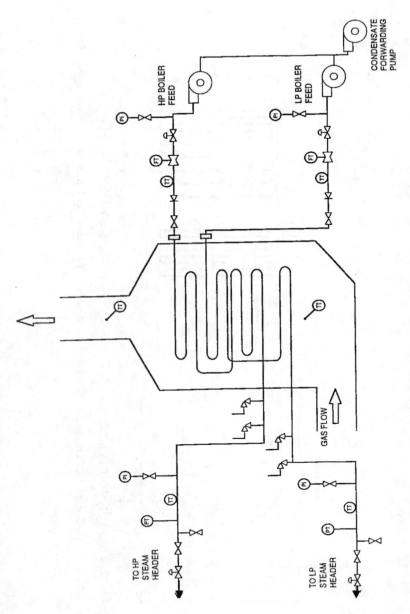

Figure 7-12. Schematic of an OSTG Two-Loop System

Materials used in these HRSGs are high nickel Incoloy 800 and 825 for the tubes, as seen in Figure 7-13. These alloys retain substantial amount of their strength and corrosion at the high temperatures. The tube fins are carbon steel for good heat transfer, and will oxide during extended dry operating periods. A major manufacturer of these types of HRSG units claims that the system can be operated dry with negligible effects on fin thickness.

OTSGs are expensive to manufacture due to the higher quality of the materials, but this is usually recovered in the installation and construction costs, which are greatly reduced by not having the bypass damper valve, or separate bypass stack. In most cases, the overall cost compares favorably with the drum-type HRSGs.

The OTSGs need better water treatment, as the units have no tolerance for solids. This makes many types of low-cost phosphate treatments used in drum boilers unacceptable. Flow acceleration corrosion, which reduces the pipe wall thickness, which then leads to leaks or ruptures in carbon steel piping, is a problem. The high velocity of the water or wet steam reacts with the protective coatings, usually a magnetite (Fe_3O_4) on carbon steel piping. This process reduces or erodes the oxide layer, allowing rapid erosion of the base metal. Most new once through boilers use oxygenated treatment of the water. The older all-volatile treatment is being replaced by this new oxygenated treatment.

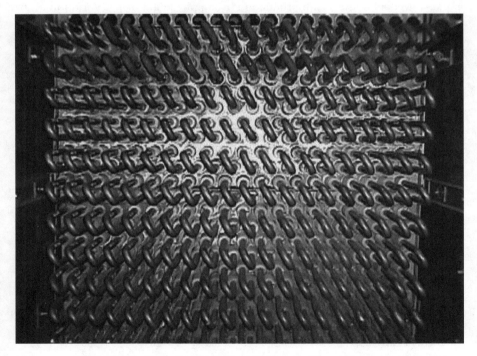

Figure 7-13. Coils Inside a Once Through Steam Generator
(Courtesy of Innovative Steam Technologies)

HRSG Operational Characteristics

A combined cycle power plant operates over a wide range of ambient temperatures and power settings. Many combined cycle power plants operate over a wide range of settings in a single day. Performance of the HRSG at off-design conditions is vital as problems such as steaming or underperformance can be expected if consultants or plant operators do not provide the HRSG designer with information on all possible modes of operation. Gas pressure drop, fuel input and steam output affect the operating costs of the system and in the long run their contribution can be significant. Evaluating designs based only on initial costs is not prudent, **life cycle cost** evaluation is the only way to evaluate a good HRSG design.

The exhaust from gas turbines in the simple cycle or regenerative mode represents a large amount of waste heat. Mass flow rates can vary depending on the size of the gas turbine from 4 lb/sec (1.8 kg/sec) for a 180-kW unit, to a tremendous 1430 lb/sec (650 kg/sec) for a large 250 MW gas turbine. In most cases, exhaust temperatures will between 800°F (427°C) and 1200°F (650°C). Gas turbines utilizing recuperators may have exhaust gas temperatures of around 950°F (510°C). As gas turbines operate with large amount of excess air, about 18% oxygen is available in the exhaust and this allows supplemental firing.

Temperatures after supplemental firing can be as high as 1600°F and it is often possible to double the steam production by supplemental firing. Variation in the exhaust gas specific heat for combustion at constant pressure (c_p) is linear with temperature, varying from 0.259 Btu/lb/R (1084 J/kg K) at 800°F (427°C) to about 0.265 Btu/lb/R (1109 J/kg K) at 900°F (482°C).

When computing exhaust heat usable for a cogeneration project, capability variations in the gas turbine performance must be taken into account. These include ambient temperature variations, altitude effects, etc. The effect of back-pressure on the gas turbine capability is also important. Back-pressure from heat recovery equipment should not exceed 10″ H_2O (2.5 kPa).

In Table 7-1, the exhaust gas flow, temperature, and the energy in the exhaust gas flow vary significantly with ambient conditions. The combination of high gas flow coupled with low inlet temperature to an HRSG results in a problem in the economizer, namely steaming.

The effect of hot ambient temperatures on gas turbine performance can be significant from a cogeneration perspective, as both electrical output and steam generation may drop off as seen in Table 7-1. Typically, there is about a 1% drop in

Table 7-1. Combined Cycle Power Plant Parameters as a Function of Ambient Temperature at Full Load

Ambient Temp, °C	−5	5	15	25	32
Total Load, MW	388	368	353	336	322
Exhaust Flow, kg/sec	672	639	615.75	602.9	603.7
Exhaust Gas Temp, °C	576.8	584.6	593	602.9	609.4
Energy In Exhaust Gas, MW	380	368	361	356	327
Gas Turbine Load, MW	254	238	225	212.5	205
Steam Turbine Load, MW	134	130	128	123.5	117

power output for every 2°F or 1°C rise in temperature. This drop has led to a number of schemes, such as evaporative cooling and chilling of the inlet air, to overcome the problem.

Table 7-2 shows the variation in the flow, temperature, and energy in the exhaust gases with the variation in the loading at a constant ambient temperature. The control at these off-design conditions is by the movement of the IGV; in this manner, the exhaust gas temperature is maintained near constant.

At off-design operations due to the lower heat availability in the gas turbine exhaust gas, the inlet temperature is maintained while the steam turbine inlet flow and pressure are reduced. This helps the effectiveness in the HRSG to remain constant, since the gas and steam temperature is constant. By holding the temperature on both the gas and steam side constant, the thermal conductance is kept constant. This helps maintain the heat transfer and therefore the high effectiveness of the HRSG at these off-design conditions.

Steaming of the economizer occurs when a boiler designed for high gas inlet temperatures is operated in low inlet gas temperatures, with the gas flow remaining nearly the same or more.

A fired HRSG is more likely to steam in the economizer at cold ambient unfired conditions, resulting in water hammer, chugging and possible vibrations. The reason for this is simple. Unlike a conventional fossil fuel-fired boiler where a decrease in load results in reduced gas flow, single-shaft gas turbines exhaust nearly the same gas flow at a lower temperature when the load decreases. Also, with a decrease in ambient temperature, the gas flow increases while the exhaust temperature drops. The energy transferred in the evaporator decreases as the gas inlet temperature and log-mean temperature reduces, while the energy-transferring ability of the economizer remains unchanged. This is because the overall coefficient is unchanged and log-mean temperature is not affected significantly.

With the reduced duty in the evaporator, less steam is produced and with a larger economizer to evaporator duty ratio, the water enthalpy rise in the economizer is higher, leading to steaming conditions or close to steaming.

There are a few ways economizer steaming can be avoided:

1. By proper design, which avoids steaming under the worst operating condition for the HRSG: the lowest ambient unfired operation. This may result in less steam production at normal operating conditions.

Table 7-2. Combined Cycle Power Plant Parameters as a Function of Load at Constant Ambient Temperature (32°C)

Percent Total Load, %	100	75	50
Total Load, MW	322	242	161
Overall Thermal Efficiency, %	51	48.28	43.51
Exhaust Flow, kg/sec	604	491	377
Exhaust Gas Temp, °C	610	621	649
Energy In Exhaust Gas, MW	327	274	224
Gas Turbine Load, MW	205	143	81
Steam Turbine Load, MW	117	99	80
Steam Turbine as Percent of Plant Load, %	36	41	50

2. Exhaust gas can be bypassed at low ambient conditions either at the HRSG inlet using a diverter or at the economizer using bypass dampers.
3. The waterside of the economizer can also be bypassed using valves.

These measures reduce the duty of the HRSG by decreasing the heat transfer coefficient or surface area. However, if the period of operation in the worst case is long, these measures are necessary.

HRSG Effectiveness

Using ASME power test code procedures, one can evaluate the efficiency of the HRSG. Uncertainties in the measurement of gas temperature and flow in particular have to be considered while proposing guarantees of performance. Leakage through diverters is also important.

Water Chemistry

Water treatment criteria are not as stringent for HRSG units, which operate at low pressures on the order of 100 psig (689 kPa) to 500 psig (3447 kPa), as in large utility combined cycle power plants, operating at high pressures of 1500 psig (10342 kPa) and steam temperatures of 1200°F (650°C). ASME/ABMA criteria are used as a minimum rate. Steam purity is very important when superheated steam at high temperature is injected into the gas turbine to generate more electrical output. Process steam is also generated by the HRSG at saturated conditions at low pressures, 180 psig (1241 kPa) to 220 psig (1516 kPa). Depending on the optimum ratio between process and injection steam, the flow of superheated and saturated steam can be varied by adding a duct burner to the system.

The superheater is located ahead of the burner. The drum is sized to accommodate wide load swings. Also, internal and external steam separators are incorporated to attain steam purity of less and 0.05 ppm solids. Superheater tubes are alonized to minimize the oxidation and possibility of exfoliation of particles into the turbine, as the tubes can run dry in some modes.

Vibration and Noise

One of the problems associated with flow of gases over tube banks is the possibility of tube vibration and noise. Whenever gas flows over tube banks, vortices are formed and shed behind the wake of the tubes at a frequency determined by the Strouhal number, which is a function of the tube pitch and geometry given as:

$$N_{ST} = \frac{I \varpi_v}{V_T} \qquad (7-2)$$

where

I = Characteristic length
ϖ_v = Vibration Frequency
V_T = Transitional Speed

When the vortex shedding frequency, which is obtained after determining the gas velocity, coincides with the tube natural frequency, tubes can vibrate leading to failure or pull-out from the joints. Depending on the boiler width and sonic velocity at the operating conditions, acoustic frequency may be determined. Noise problems arise if the acoustic frequency coincides with the vortex shedding frequency.

Baffles are used to correct noise problems, while tube geometry, length or method of attachment can be varied to prevent vibration problems. Noise and vibration of boiler tubes is geometry- and size-related and experience of similar-sized units can help in the prediction and correction of problems.

Filter Housing, Duct Work and Insulation

Figure 7-14 shows schematic of a combined cycle system showing a gas turbine, and HRSG system with inlet cooling and filter house, a bypass diverter damper system, and exhaust stacks.

Figure 7-14. A Schematic of a Gas Turbine – HRSG System (Courtesy of Braden Manufacturing LLC)

Filter Housing and Duct Work

The air filter housing is one of the most important components in the combined cycle power plant. The air entering the gas turbine must be as clean as possible. To achieve this goal, a very high efficiency filter system must be designed. The filter housing comprises a three- or four-stage filtration system. The entrance to the filter housing is located from about 10 ft (3 m) to 50 ft (15 m) above the ground level, to ensure that the air is least contaminated and at a slightly lower temperature. The filter system must also be designed so that a minimum drop in pressure takes place across the filter system. As a rule of thumb, 4' H_2O (7.47 mm Hg) pressure drop at the inlet of the turbine would result in a 1.6% drop in power and an increase of 0.65 % in the heat rate of a turbine.

A high-efficiency filtration system would consist of all or most of the following components:

Louvers and Vanes, Rain Screens: These are typically used in the first stages along with coalesce filters to remove water droplets, trash screens, and Inertial filters to ensure that no large pieces enter the high-efficiency filters and cause major damage. Figure 7-15 shows a typical filter house under construction.

Figure 7-15. Filter House with Rain-Shields or Weather Hoods
(Courtesy of Braden Manufacturing LLC)

Coalescers: These are constructed by the use of wire mesh, which acts to agglomerate the mist in the inlet air, thus removing the moisture from the air. Figure 7-16 is a typical coalescing filter system.

Inertial Filters: The objective here is to make the air change direction rapidly, causing separation of dust particles. These filters are permanently fixed and require minimal maintenance. Inertial filters typically operate at face velocities of 20 ft/sec.

Prefilters: These are medium-efficiency filters made of cotton fabric or spun fiberglass. They are relatively inexpensive and serve as "protection" for high-efficiency filters.

High-Efficiency Filters: These filters remove smaller particles of dirt. They are typically barrier or bag type filters, as seen in Figure 7-17.

Self-Cleaning, or the "Huff and Puff"-type Filters: These consist of a bank of high-efficiency media filters. Air is drawn through the media at a low velocity. At a pre-determined pressure drop, about 2.5" H_2O (4.67 mm Hg), a reverse blast of air is used to remove dust buildup. Figure 7-18 shows rows full of these filter cartridges in a typical air filter housing.

Air tightness is a must for any gas turbine inlet system because even the most efficient filtration system will be useless if unfiltered airflow enters the compressor. Some common causes of leakage are: bypass door leakage; poor gaskets and seals at flanged points; and modifications made on the inlet ducting.

Corrosion in carbon steel inlet ducts has also been a source of problems. At times the corrosion can be severe enough to cause a loss of integrity. Because of this, several users are now using 316L stainless steel for the filter houses and inlet ducts, which is a cost-effective thing to do in the long term. The effectiveness of a filtration system is impacted by its design, installation and maintenance.

Diverters, Silencers and Burners

Diverters, silencers and burners are vital auxiliaries for an HRSG, and their selection must be made with care.

Diverters

Diverters may be used to regulate the gas flow to the HRSG in case the steam output has to be varied in an unfired mode, as seen in Figure 7-19. The diverter shown is a triple-blade diverter. They may also be used to divert the gas either to the bypass stack or to the HRSG. Several diverter types are available, and the choice is based on system needs. Leakage across diverters (also called dampers) is an important source of energy loss. The leakage flow may be estimated using the following relationships:

$$W = 2484 A (100 - E)(h/T)^{0.5} \qquad (7-3)$$

where
$\quad W =$ Leakage flow
$\quad A =$ Area of damper, ft^2
$\quad E =$ Sealing efficiency of damper on area basis, %

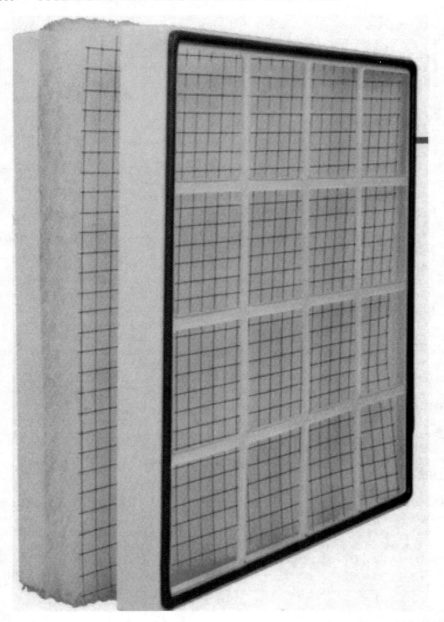

Figure 7-16. A Typical Coalescing Filter (Courtesy of Braden Manufacturing LLC)

H = Differential head between the gas and atmosphere, in. WC
T = Gas temperature, °F

Figure 7-17. Final Filters Designed for the Rigors of Gas Turbine Operations
(Courtesy of Braden Manufactirung LLC)

A Typical Case

- Gas Pressure at the damper = 10″ H_2O,
- Gas Temperature = 1000°F
- Area of damper cross-section = 10 ft^2
- Gas Flow = 14,000 lb/hr
- Sealing Efficiency = 99%

Figure 7-18. Rows of Self-Cleaning "Huff or Puff"-Type Filters Mounted in an Inlet Filtration System (Courtesy of Braden Manufacturing LLC)

Figure 7-19. A Photograph of a Triple Blade Diverter Damper
(Courtesy of Braden Manufacturing LLC)

Then,

$$W = 2484\,(10)\,(100 - 99)\,(10/1460)^{0.5} = 2055 \text{ lb/hr} \quad (7-4)$$

On a percent flow basis, the leakage = 2055/140,000 = 1.5%, which is significant.

Insulation

An HRSG is generally insulated on the inside of the casing using mineral and ceramic fiber depending on gas temperature inside the duct. One of the advantages is reduction in axial expansion of the duct. Liners are then placed over the insulation to protect it from the gas. Liners may be made of cortex or stainless steel, depending on gas temperatures. Refractory is not preferred due to the fact that frequent startups and shutdowns can result in their early failure.

One of the critical components in the HRSG is the duct work between the burner and the HRSG. Above 1600°F (871°C) to 1700°F (927°C), the selection of the liner material and its attachment to the insulation has to be done with great care. Due to the prolonged high temperatures, the liner undergoes expansion in all directions and sometimes curls up, exposing portions of the ceramic fiber to the gas flow, which can destroy them over a period of time. This leads to overheating of the casing, the support rods and eventually damage to the entire duct. Hence, provision must be made for movement of liners, especially at the corners of ducts by the use of expansion joints, such as that shown in Figure 7-20. Expansion joints are often the weakest link in the duct work. These expansion joints must be fully engineered to withstand the stresses of sudden thermal growth created by the gas turbine startup.

Duct Burners

Duct burners are located in the duct work and the gas flows over the burner elements. Sizing of elements, and the need for augmenting air is determined by the burner supplier. The airflow pattern ahead of and beyond the burner is important for proper mixing and uniform temperature profiles. The recommendations of the burner supplier must be taken into account while doing the layout of the duct work. Improper choice of firing duct length can cause flames to touch the HRSG tubes. Gas pressure drops in duct burners are not high and range from 0.25" H_2O (62 Pa) to 0.5" H_2O (124 Pa). However, if register burners are used in a wind box, the arrangement is different and burner pressure drops could be in the range of 2" H_2O (500 Pa) to 3" H_2O (750 Pa).

Silencers

Silencers are used to limit the noise levels of the exhaust gas when it flows through the bypass stack. Turbine exhaust gas characteristics must be known to select the proper silencer. Some consultants specify stringent noise criteria, which may not be warranted, especially if the percentage of time in bypass mode is very small and the plant is isolated from a populated region.

HRSG Reliability and Durability

HRSGs have become more efficient with the dual-and triple-pressure HRSGs. The flexibility required from the new operational parameters of the combined cycle power plants places considerably more stress on the HRSGs.

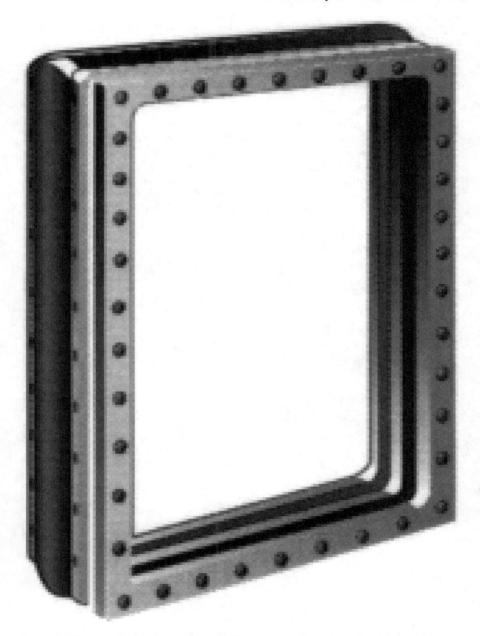

Figure 7-20. A Typical Expansion Joint (Courtesy of Braden Manufacturing LLC)

306 • COGENERATION AND COMBINED CYCLE POWER PLANTS

The principal operating problems which reduce the durability and reliability in an HRSG are:

1. Low cycle thermal fatigue. This is particularly true in the HP section of the HRSG, like the H-P superheaters, H-P steam drums, evaporator circuits, and low temperature economizers.
2. Rapid heat-up rates.
3. Corrosion-related problems, which include flow-accelerated corrosion, cold end gas side corrosion, and pitting from oxygenated Feedwater.
4. Failure of casing and expansion joints, and other thermal- and mechanical-related problems.
5. Low-temperature corrosion. All surface temperatures of materials that come into contact with the turbine exhaust gas must be above the dew point of sulfuric acid. If the exhaust gas is free of sulfur, then the temperature must be above the dew point.
6. Acid gas corrosion in the exhaust stack.
7. Condensate in the low headers.
8. Stratification of air flow during pre-start purges.

Chapter 8

CONDENSERS AND COOLING TOWERS

Condensers

Condensers are used in steam turbines to condensate the steam exiting from the steam turbine to water. These condensers are run at a very low vacuum of about 3 in. Hg. (mercury) or about 1.5 psia. The steam entering the condenser usually has a quality of about 92% to 96%, i.e., 8% to 4% liquid in the steam. The amount of heat removed from the condenser by the water or air depends on the condenser load Q_c, which is given by the following relationship:

$$Q_c = \dot{m}_s(h_s - h_c) \qquad (8-1)$$

where
 $\dot{m}_s$ = mass flow of steam entering the condenser
 h_s = enthalpy of steam leaving the steam turbine (this enthalpy is usually very difficult to determine since the quality of the steam is not a measurable quantity)
 h_c = enthalpy of the liquid leaving the condenser

The above equation is based on the assumption that no drains are entering the condenser.

To remove the heat from the steam, the rate of the air flow in the case of an air-cooled condenser, or the rate of the water in case of a water-cooled condenser is computed by the heat balance, equating the heat given up by the steam in the condenser, equal to that absorbed by the air or the cooling water:

$$\dot{m}_s(h_s - h_c) = \dot{m}_{air}(h_{2a} - h_{1a}) = \dot{m}_w c_{pw}(T_2 - T_1) \qquad (8-2)$$

where
 $\dot{m}_{air}$ = mass of cooling air
 h_{2a} = enthalpy of cooling air leaving the condenser
 h_{1a} = enthalpy of cooling air entering the condenser
 $\dot{m}_w$ = mass of cooling water

c_{pw} = specific heat at constant pressure of the cooling water, if the cooling water contains salt (such as sea water) the value of c_{pw} must be that of the sea water

T_2 = temperature of the cooling water leaving the condenser

T_1 = temperature of the cooling water entering the condenser

The overall thermal transmittance in a condenser is the heat transferred per unit of time, unit of surface area, degree of temperature difference. With surface condensers this quantity is the fundamental measure of the condenser's performance and can be written as follows:

$$U = \frac{\dot{m}_w c_{pw}}{A} \text{Ln}\left(\frac{\theta_1}{\theta_2}\right) \qquad (8-3)$$

where

θ_1 = difference in temperature of the steam and the entering cooling water temperature

θ_2 = difference in temperature of the steam and the exiting cooling water temperature

Thus, the logarithmic mean temperature difference between steam and water can be computed as follows:

$$\theta_m = \frac{T_2 - T_1}{\ln\left(\frac{\theta_1}{\theta_2}\right)} \qquad (8-4)$$

Condenser tubes foul in service. The ratio of the heat transfer characteristics of the design tubes to the tubes in service is designated as the cleanliness factor. The cleanliness factor is a unique value, since it applies to one specific operating condition, of circulating water temperature velocity of flow through the tubes, and the same external steam temperature and flow:

$$CF = \frac{U_u}{U_n} \qquad (8-5)$$

where

U_u = overall thermal transmittance of the used tubes

U_n = overall thermal transmittance of the design tubes

The above relationship is for an individual tube; the entire condenser cleanliness factor is given by the following relationship:

$$CF_{ov} = \frac{(\sum U_u)/N_u}{(\sum U_n)/N_n} \qquad (8-6)$$

where

N_u = number of tubes in operation

N_n = number of tubes as per design

Once the cleanliness factor data is obtained, the thermal resistance of the dirt, scale, or slime can be computed. The resistance of the tube metal to the flow of heat and the fluid film resistance on the two surfaces are approximately equal, for the design and used tubes. Therefore, their total resistance $1/U_u$ and $1/U_n$ differ by the resistance of the foreign matter, which caused the fouling of the tube in use. Thus, the fouling resistance can be expressed as follows:

$$R_f = \frac{1}{U_u} - \frac{1}{U_n} \qquad (8-7)$$

Types of Condensers

There are two types of condensers widely used in combined cycle power plants. They are the water-cooled condenser and the air-cooled condenser.

Water-Cooled Condenser. The water-cooled condenser is the most common type of condenser. The condenser consists of a bundle of tubes through which the cooling water flows and steam flows on the outside of the bundle, as shown in Figure 8-1. The cooling water removes the heat from the steam and condenses the steam to a liquid. The cooling water can be pumped from the sea or a lake and is then usually put in a holding pond before it is reintroduced into the sea or lake.

Air-Cooled Condenser. The air-cooled condenser is also used in many places where cooling water is hard to obtain. The turbine exhaust steam flows through a main steam duct to the air-cooled condenser. The condenser condenses the steam inside finned tubes, which are cooled by ambient air. The cooling air is provided by axial fans, which are driven by electric motors via speed-reducing gearboxes. The system of a direct air-cooled steam condenser with fan drive is shown schematically in Figure 8-2.

The air-cooled condenser consists of 18 modules, which are divided, in six streets of three modules. Each module contains six tube bundles, which are arranged in an "A" frame type (three bundles on each side) and which is served by one axial fan. On top of each module, a steam distribution manifold is connected to feed the tube bundles with steam. At the bottom outlet of the finned tubes, the condensate is collected in condensate manifolds and flows by gravity via a loop seal to the main condensate tank.

The first stage consists of approximately 96% of the tubes and is used to condense the steam; the second stage, also called air-cooling stage, which is only some 4%, is used to remove the air and non-condensables from the first stage.

The tube bundle is provided with four separate outlet headers which are connected to the condensate return line. To compensate for the pressure differences in the four condensate headers, a loop seal is introduced.

Each tube row has its own second stage. This second stage operates in reflux mode (i.e., steam + air flowing upwards and condensate flowing downwards) to cool the steam + air mixture. The second stage outlets are connected to the air take-off line which leads to the vacuum unit. Cold air blown by an axial fan strikes the tube rows and is warmed up.

Fan Units. The condensing heat from the steam is dissipated to the cooling air, which is provided by large axial fans (18 in total). Each fan (model ELF) of 9.9-m

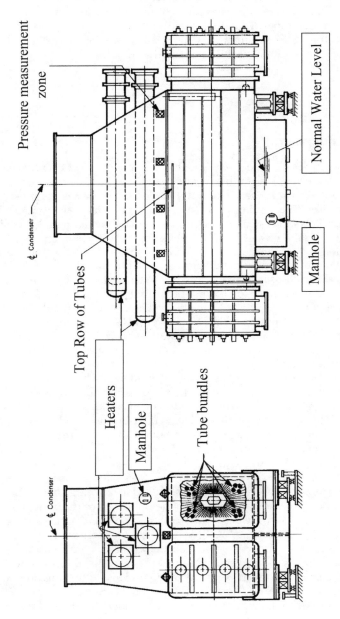

Figure 8-1. Typical Steam – Water Condenser

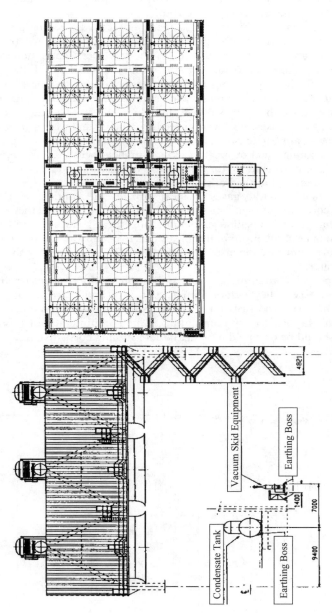

Figure 8-2. A Typical Air-Cooled Condenser

312 • COGENERATION AND COMBINED CYCLE POWER PLANTS

diameter has five glass fiber reinforced polyester (GRP) blades, which are mounted on a steel impeller hub by U-bolts. The fan blade pitch is manually adjustable during standstill. The fan speed is 100 rpm.

Each fan is driven by a 180-kW single-speed motor via a speed-reducing gearbox with parallel shafts. The motor shaft and gearbox are connected by a flexible coupling. The fan is directly mounted on the low-speed shaft underneath the gearbox. Each fan unit is provided with a vibration cut-out switch (trip setting at one turn from flush), which stops the fan in case of excessive vibration.

Treatment of Condensate. In most cases, condensate does not require treatment prior to reuse. Makeup water is added directly to the condensate to form boiler feedwater. In some cases, however, especially where steam is used in industrial processes, the steam condensate is contaminated by corrosion products or by the in-leakage of cooling water or substances used in the process. Hence, steps must be taken to reduce corrosion or to remove the undesirable substances before the condensate is recycled to the boiler as feedwater.

The presence of acidic gases in steam makes the condensate acidic with consequent corrosion of metal surfaces. In such cases, feeding to the boiler water chemicals that produce alkaline gases in the steam can reduce the corrosion rate. The addition of neutralizing and filming amines to boiler water or to condensate to minimize corrosion by condensate and feedwater is carried out by the control of pH of the water.

Many types of contaminants can be introduced to condensate by various industrial processes. They include liquids, such as oil and hydrocarbons, as well as all sorts of dissolved and suspended materials. Each installation must be studied for potential sources of contamination. The recommendations of a water consultant should be obtained to assist in determining corrective treatment.

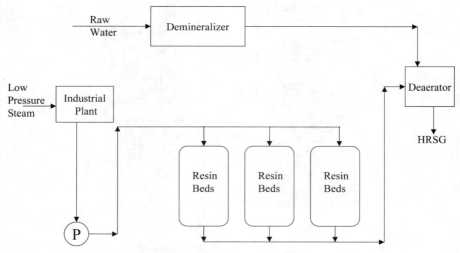

Figure 8-3. Condensate Purification System

Figure 8-3 shows condensate purification system used in a paper-mill boiler cycle. The resin beds not only remove dissolved impurities by ion exchange but also serve as filters to remove suspended solids. It is necessary to backwash and regenerate these resin beds periodically. Several types of condensate purification systems are available from various vendors. Some of these are capable of operation at temperatures as high as 275°F (135°C).

Condensate-Polishing Systems

Demineralizer systems, installed for the purpose of purifying condensate, are known as condensate-polishing systems. A condensate-polishing system is a requisite to maintain the purity required for satisfactory operation of once-through boilers. High-pressure drum-type boilers (over 2000 psi) can and do operate satisfactorily without condensate polishing. However, many utilities recognize the benefits of condensate polishing in high-pressure plants, including:

1. Improved turbine capability and efficiency
2. Shorter unit start-up time
3. Protection from the effects of condenser leakage
4. Longer intervals between acid cleanings.

Two types of condensate-polishing systems are available, both capable of removing suspended material, such as corrosion products, as well as ionized solids.

Deep-bed demineralizers operate at flow rates of 40 to 60 gpm/ft^2 of bed cross-section. This type requires external regeneration facilities. The deep-bed system has a higher initial cost with possible lower operating costs, especially during initial unit startup. Its greater capacity for removing ionized solids permits continued operation with small amounts of condenser leakage. The deep-bed system is usually operated with the cation resin in the hydrogen form, but the ammonium form can also be used. The cost of regenerating the resin in the ammonium form is greater than for the hydrogen form, but the time period between regenerations is much longer.

The cartridge-tubular type, such as the Powdex system, uses smaller amounts of disposable resins, eliminating the need for regeneration. The Powdex system uses cation resin in the ammonium form. Because of the many considerations involved, an evaluation of alternate types should be made before a system is selected for any given installation.

Condenser Fouling

Assuming that proper boiler and feedwater chemistry controls have been implemented in the unit, perhaps the most important source of impurities in the cycle is condenser leakage.

The 3-ppb limit for each in the highest purity units was derived based on solubility limits. Operators need to ensure that sufficient instrumentation is provided on the unit to monitor sodium and chloride (in the case of condenser leakage) in main or reheat steam. For drum boilers, an earlier indication can be determined by monitoring boiler water chloride or feedwater cation conductivity at the economizer inlet. The allowable level in the boiler is a function of the operating pressure of the unit and the chemical treatment type. Actions to confirm a suspected minor condenser leak will include:

1. Review plant chemistry control logs, on-line cycle chemistry records, or instrumentation alarms such as sodium analyzers, for evidence of the presence of chloride over an extended period, typically, a minimum of 2 years. In particular, the pH and cation conductivity at the economizer inlet should be reviewed.
2. Determine if, and when, the impurity levels were excessive for chloride or for sulfate. This action is important because if levels were outside the control limits, but could have been detected, then corrective measures such as procedural changes, training or equipment repairs are needed.

A large condenser leak especially ingress of seawater would be a major contamination event that will lead to a boiler water pH depression. Therefore, actions to confirm would include the examination of the records of pH levels, other instrumentation alarms, plant control logs, and other available information. This action is required to determine whether appropriate corrective measures have been used. For example, the actions required for an indication of drum boiler water pH < 8.0 during full load operation are to shut the unit down, confirm the source of the pH depression, and to examine the need to chemical clean.

Cooling Towers

A cooling tower is a specialized heat exchanger in which two fluids (air and water) are brought into direct contact with each other to affect the transfer of heat. In the "spray-filled" tower shown in Figure 8-4, this is accomplished by spraying a flowing mass of water into a rain-like pattern, through which an upward moving mass flow of cool air is induced by the action of a fan.

The heat gained by the air, neglecting negligible amount of sensible heat exchange that may occur through the walls (casing) of the tower, is equal to:

$$Q_{air} = \dot{m}_{air}(H_2 - H_1) \qquad (8-8)$$

where
$\quad \dot{m}_{air}$ = mass flow of dry air through the tower, lb/min
$\quad H_1$ = enthalpy (total heat content) of entering air, Btu/lb of dry air
$\quad H_2$ = enthalpy of leaving air, Btu/lb of dry air

Wet-bulb and dry-bulb temperatures values are easily measured. The temperature read with a dry thermometer is the dry-bulb air temperature. The bulb is surrounded by wet gauze; if the liquid is evaporated as air is blown over it, it will cause the temperature to drop in a process similar to the adiabatic saturation

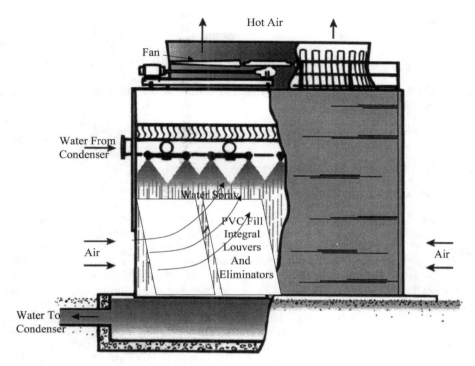

Figure 8-4. Schematic of Flow in a Cooling Tower

process. This lower reading when properly taken is called the wet-bulb temperature. The calculation of the enthalpy could then be done by using the psychometric chart.

The heat loss by the water is given as:

$$Q_{\text{water}} = \dot{m}_{\text{water}}(T_1 - T_2) \qquad (8-9)$$

where

$\dot{m}_{\text{water}}$ = mass flow of water entering the tower, lb/min
T_1 = hot water temperature entering the tower, °F
T_2 = cold water temperature leaving the tower, °F

Due to the evaporation that takes place within the tower, the mass flow of water leaving the tower is less than that entering it, and a proper heat balance must account for this slight difference. Since the rate of evaporation must equal the rate of change in the humidity ratio (absolute humidity) of the air stream, the rate of heat loss represented by this change in humidity ratio can be expressed as:

$$Q_{\text{vapor}} = (\omega_1 - \omega_2)(T_2 - 32) \qquad (8-10)$$

where

ω_1 = humidity ratio of entering air, lb vapor/lb dry air
ω_2 = humidity ratio of leaving air, lb vapor/lb dry air
$(T_2 - 32)$ = an expression of water enthalpy at the cold water temperature, Btu/lb (the enthalpy of water is zero at 32°F)

Including this loss of heat through evaporation, the total heat balance between air and water is:

$$Q_{air} = Q_{water} + Q_{vapor} \qquad (8-11)$$

Design of Cooling Towers

Optimum operation of a process usually occurs within a relatively narrow band of flow rates and cold water temperatures, which establishes two of the parameters required to size a cooling tower — namely, flow (gpm) and cold water temperature. The heat load developed by the process establishes a third parameter — hot water temperature coming to the tower. The heat load of the air is given by the change in enthalpy that, as mentioned earlier, can be represented by the wet and dry bulb temperatures. Thus, the four parameters that will affect the size of a tower are: (1) heat load Q_{water}; (2) the range, which is the differential in temperature between the hot temperature of the water and the cold temperature of the water; (3) the approach, which is the differential in temperature between the wet bulb temperature of the air and the cold water temperature; (4) wet-bulb temperature. If three of these parameters are held constant, changing the fourth will affect the tower size as follows:

1. Tower size varies directly and linearly with heat load.
2. Tower size varies inversely with range. Two primary factors account for this. Firstly, increasing the range also increases the driving force between the incoming hot water temperature and the entering wet-bulb temperature. Second, increasing the range (at a constant heat load) requires that the water flow rate be decreased, which reduces the static pressure opposing the flow of air.
3. Tower size varies inversely with approach. A longer approach requires a smaller tower. Conversely, a smaller approach requires an increasingly larger tower and, at 5°F (2.8°C) approach, the effect upon tower size begins to become asymptotic. For that reason, it is not customary in the cooling tower industry to guarantee any approach of less than 5°F (2.8°C).
4. Tower size varies inversely with wet-bulb temperature. When heat load, range, and approach values are fixed, reducing the design wet-bulb temperature increases the size of the tower. This is because most of the heat transfer in a cooling tower occurs by virtue of evaporation (which extracts approximately 1000 Btu for every pound of water evaporated), and air's ability to absorb moisture reduces with temperature.

At a given rate of air moving through a cooling tower, the extent of heat transfer that can occur depends upon the amount of water surface exposed to that air. In the tower depicted in Figure 8-4, total exposure consists of the cumulative surface areas of a multitude of random-sized droplets, the size of which depends largely upon the pressure at which the water is sprayed. Higher pressure will produce a finer spray and greater total surface area exposure. However, droplets contact each other readily in the overlapping spray patterns and, of course, coalesce into larger droplets, which reduces the net surface area exposure. Consequently, predicting the thermal performance of a spray-filled tower is difficult at best, and is highly dependent upon good nozzle design as well as a constant water pressure. To achieve maximum heat transfer, various types of "fills" are utilized to increase water surface area exposure and enhance thermal performance. Fill sheets include both louvers and drift eliminators as seen in Figure 8-5. The louvers keep the water on the fill sheets and also assure proper heat transfer throughout the wide variation of airflow. Drift eliminators prevent the costly nuisance of drift spotting on objects in the surrounding environment. Figure 8-6 is a photograph of a typical fill being installed.

The film fill design goal is to maximize thermal performance (a combination of heat and mass transfer), typically expressed as KAV/L (a dimensionless grouping of mass transfer coefficient K, interfacial surface area A, packed volume V, and water loading L), while minimizing pressure drop. By accomplishing this, an increase in the fill's thermal performance (ability to provide colder water or cooling greater volumes of water) using lower power consumption will be achieved.

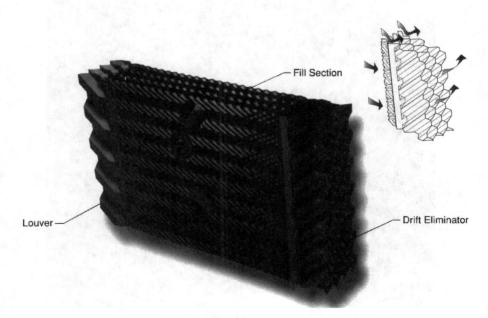

Figure 8-5. Fill Used in Cooling Towers

Figure 8-6. A Fill Being Installed in a Cooling Tower

Most film fill technology is fabricated from PVC plastic, meeting the Cooling Tower Institute (CTI) standard. Benefits of using PVC are that it is durable, self-extinguishing, inexpensive, has a long surface life and is able to create a uniform water film (wetting). Wetting is a condition where the surface allows water to form a film rather than droplets or beads.

There are two broad categories of film designs: cross-corrugated vertical flow and vertical flow. Cross-corrugated fills can be used in both counter-flow and cross-flow towers. They work by splitting water and air paths in opposing angles through the pack. Because of the cost and high water loading per ft^2 (m^2), the cross-corrugated configuration is not used in cross-flow towers with over a 16-ft (5 m) vertical packed height.

Vertical Fills. Vertical fills are only used in counter-flow cooling towers. Water enters the pack from the top and flows vertically, and within the first 2 in. (51 mm) to 4 in. (102 mm) makes an angular transition to other vertical channels. These patterns of angular transition occur continuously through the vertical pack depth.

Fills with vertical offset flow are also only used in counter-flow towers. They keep the water and air path oriented vertically, with no offsets, through the pack. Each of these designs offers variations in sheet shape and texture. Design considerations include variations in the microstructure of the small surface system embossed inside the sheet, corrugation angle, flow path, and sheet spacing (flute height). Each design variation affects fill thermal performance.

When equipped with a suitable microstructure, cross-corrugated fill gives high performance, its thermal performance is up to three times greater than earlier splash fills. This high-performance fill is typically selected for industrial applications where good water quality makeup is used. High-performance film fills are regularly used in counter-flow towers offering the benefits of this type of tower design.

Thermal improvement in cross-corrugated designs is achieved with a corresponding reduction in water film velocity through the pack. Water film velocity reduction is due to the non-vertical water path and the increase in filmed area. Lower water velocity is a key factor in increasing fouling potential. In lower water velocity areas, small obstructions can create a deposit cell combining gelatinous biological activity with airborne material, scale and suspended solids from the makeup water, and possibly process contamination. This cell can grow rapidly into an overall fouling condition. Light deposits on film fill should not have a significant effect on tower performance. More severe fouling inhibits uniform water film formation and obstructs airflow distribution, thereby adversely affecting cooling tower performance.

Mixed Fills. Many cooling towers have achieved optimal thermal performance by combining fill designs. The most commonly used is substituting a 12-in. (305 mm) layer of cross-corrugated or vertical offset fill above a vertical flow pack. Tower fouling rarely occurs in the top layer due to the washing effect of the spray system. Using mixed media fill pack, a combination of cross-corrugated or vertical offset fills with vertical flow fills will provide overall thermal performance improvement without appreciably increasing fouling potential. The benefit can be seen when a top layer of high-performance, cross-corrugated or vertical offset is used with a vertical flow pack. The net performance increase can be approximately

10% compared with the same depth of only a vertical flow fill pack. Using a cross-corrugated film fill pack as the top layer will also maximize water distribution uniformity.

Mixed fill possibilities also can combine a more open, large flute (30 to 50 mm) vertical fill design with a more dense (19 mm) design when unusually poor water quality or very difficult contaminants are present. This approach allows customizing the fill pack to the specific application and provides for maximum thermal performance and anti-fouling potential.

Splash Fills. Splash fills are the oldest fill media used in cooling towers. Most splash fills are currently used in cross-flow towers. Splash fill designs include wood lath, profile shapes, tubular vertical grids, and horizontal grid panels. Each of these fills provides anti-fouling characteristics. There are splash fill packs such as tubular, vertical grid, and horizontal grid panels that are used in counter-flow towers. Thermal performance of splash fills varies significantly, but all of the designs provide good protection against fill fouling. Also, pressure washing can be applied to most splash fills. However, their thermal performance is rated significantly below film fills. This lower thermal performance of splash fills will require a larger tower footprint along with higher operating energy.

The preponderance of splash fills are hung in the tower. It is typically considered more costly and difficult to install these hanging fills. The hanging system usually consists of PVC-coated steel wires. These hanging systems are not as durable as a bottom-supported system. Also, hanging systems can make maintenance more difficult since they restrict access inside the tower, even when the splash fill is removed.

Turbosplash Fill. The turbosplash bottom-supported fill is a new concept that can be used in either a counter-flow or cross-flow tower, and is a new design that provides up to twice the thermal performance compared to the most commonly used horizontal grid panels on 8-in. (200 mm) vertical centers in counter-flow applications. The system offers high thermal performance and is easy to install. Thermal performance of the new splash fill design is lower than film fills, but is much higher than other splash fills.

The higher thermal performance of the turbosplash fill not only allows a smaller tower to do the same amount of cooling as other large splash fill towers, it also allows ongoing savings by providing significantly lower pumping heads.

There are a number of installations that have mixed splash with film fill. Process engineers and tower designers would consider mixing splash and film fill media for the following reasons:

- Improving thermal performance. Higher thermal performance can be achieved by using a top layer of film with splash fill in counter-flow towers. There are designs that mix bays of splash fill with bays of film fill in cross-flow towers for improved thermal performance.
- High temperature applications. In counter-flow towers using PVC or higher temperature PVC film fill, the recommended maximum operating water temperatures are 125°F (52°C) to 135°F (57°C), respectively. For higher water temperatures, using an alternate fill in the top layers is recommended. By incorporating turbosplash fill in the top layers over PVC film fill, hot water temperatures of up to 180°F

(82°C) can be accommodated. The turbosplash pack will reduce the water temperature to the range that will not affect the PVC fill. An alternate to turbosplash is polypropylene film fill, viable up to 160°F (71°C).
- For expediting fill repair. Unexpected film fill damage can occur at the most inopportune time. It seems to occur when the process needs require maximum cold water. This damage is typically confined to the top fill layers. To expedite repair, maintenance personnel can use stacked horizontal grids or turbosplash packs to replace the film fill. Also, the turbosplash packs accordion fold to one-seventh of the installed film volume, so a supply can be easily stored on-site.

Fill is only about 25% to 30% of the cost of a cooling tower, but it should be regarded as its 'heart' with careful attention being given to its selection. By considering thermal performance of new fill products and matching it to the application, the end user can save on capital, yearly maintenance and operating costs.

Chemical Water Treatment

Regardless of the type of fill installed — cellular film, anti-fouling vertical high-performance, wood splash bar slats, or any of the extruded PVC splash bars — all can foul if chemical treatment is not utilized. Thus, good water treatment is very important. Good water treatment will help in maintaining clean heat transfer surfaces in all the equipment. Corrosion can damage equipment and lead to early replacement or retubing of condensers. Corrosion in the condenser piping can cause leaks and lead to costly replacement of system piping.

Water treatment plant upset, such as breakdown of makeup water treatment systems (ingress of demineralizer regeneration acid or caustic into boiler) or condensate polisher regeneration, leads to changing boiler water pH. Ingress of sulfuric acid or sodium hydroxide, which are used to regenerate ion exchange resin of condensate polishers or of the makeup system, can be sources of contamination. Problems with water treatment plant may also be initiated by mechanical failure, such as valves, and should normally be detected quickly by conductivity monitors or alarms in the feedwater train; however, operator error or poor maintenance can cause the alarms to be unreliable. Actions to confirm this source of impurity ingress include the following: evaluate results from, and reliability of, monitoring systems and alarms, particularly of cation conductivity measurements; determine if, and when, the impurity levels were excessive for chloride, sulfate or hydroxide. As with condenser leaks, it is important to determine whether existing controls detected the ingress in order to apply the proper correction.

Chapter 9

GENERATORS, MOTORS AND SWITCH GEARS

The electrical motor is the most common type of power that is used to start the gas turbines in a combined cycle power plant. There are some applications where the start-up power is provided by steam turbines and diesel engines. Diesel engines are used in plants where black starts are needed. A black start is needed where there is no electrical power available for start-up power. The new large gas turbines use the generator as a motor for start-up requirements. The generator acts as a motor to bring the turbine to a point where the turbine is fired. Once the turbine is fired and reaches a given rpm of about 40% to 60% of design speed, the motor is declutched and then converted to a generator.

The generator is used to convert the rotating mechanical energy of a gas turbine or steam turbine into electrical energy. In the gas turbine, the hot air is expanded through the turbine impinging on the turbine blades causing them to rotate converting the hot gases compressed in the compressor and heated in the combustor into mechanical energy. The high-pressure steam created from the hot exhaust gases from the heat recovery steam generator is channeled to the steam turbine and impinges on the turbine blades causing them to rotate, thus converting the energy in the steam into mechanical energy. The turbine rotors are connected to the generator rotor through a coupling and as the turbine blades rotate, they cause the generator rotor shaft to rotate also. The generator converts the rotating mechanical energy transmitted to it into electrical energy. In majority of the combined cycle power plants, the gas turbine and the steam turbine both have separate generators. There are some turbines where the gas turbine, steam turbine are on the same shaft and thus, there is only one generator, all three components on a single shaft.

Electric motors and generators perform very similarly. Synchronous motors convert electrical power to mechanical power and synchronous generators convert mechanical power to electrical power. This chapter deals with some of the basic characteristics of the electrical motors and generators.

Motors

All electric motors operate on the same basic principle regardless of type or size. When a wire carries electric current in the presence of a magnetic field (at least partially perpendicular to the current), a force on the wire is produced perpendicular to both the current and the magnetic field. In a motor, the magnetic field radiates either toward or outward from the motor axis (shaft) across the air gap, which is the annular space between the rotor and stator. Current-carrying conductors parallel to the axis (shaft) then have a force on them tangent to the rotor circumference. The force on the wire opposes an equal force (or reaction) on the magnetic field. It makes no difference whether the magnetic field is created in the rotor or the stator; the net result is the same: the shaft rotates.

Within these basic principles, there are many types of electric motors. Each has its own individual operating characteristics suited for specific drive applications. When several types are suitable, selection is based on initial installed cost and operating costs (including maintenance and consideration of reliability).

Constant Speed Motors

The majority of drives in a combined cycle power plants are constant speed. Typical applications include:

- Pumps
- Compressors
- Fans.

Alternating Current Squirrel-Cage Induction Motors

These motors are by far the most common constant-speed drives. They are relatively simple in design and, therefore, both low in cost and highly reliable. They vary in price from about US$70/kW for medium size motors (10 to 200 kW) to about $35/kW for larger motors. The power output of a motor is given by the following relationship:

Three Phase:

$$P(\text{kW}) = 0.00173 \times V \times I \times PF \times \eta \qquad (9-1)$$

Single Phase:

$$P(\text{kW}) = 0.001 \times V \times I \times PF \times \eta \qquad (9-2)$$

where
  V = voltage
  I = line current, A
  PF = power factor
  η = motor efficiency

The allowable power factor in any given installation is a function of the reactive components (leakage paths) to the real components.

The typical three-phase squirrel-cage motor has stator windings, which are connected to the power source. The rotor is a cylindrical magnet structure mounted on the shaft with slots in the surface, parallel (or slightly skewed) to the shaft; either bars are inserted into these slots or molten metal is cast in place and connected by a short-circuiting end ring at both ends of the rotor. The name "squirrel-cage" is derived from this rotor-bar construction. In operation, current passing through the stator winding creates a rotating magnetic field, which cuts the rotor winding unless the rotor is turning in exact synchronism with the stator field. This cutting action induces a voltage, and hence a current, in the rotor which in turn reacts with the magnetic field to produce torque.

The typical medium-sized squirrel-cage motor is designed to operate at 2% to 3% (97% to 98% of synchronous speed). The synchronous speed is determined by the power-system frequency and the stator-winding configuration. If the stator is wound to produce one north and one south magnetic pole, it is a two-pole motor; there is always an even number of poles (2, 4, 6, 8, etc.). The synchronous speed (n) is:

$$n = 120f/p \qquad (9-3)$$

where
 f = frequency, Hz
 p = number of poles

The actual operating speed will be slightly less by the amount of slip. Slip depends upon motor size and application. Typically, the larger motor, the less slip; and ordinary 7460 W (10 hp) motor may have 2 $^1/_2$% slip, whereas motors over 746 kW (1000 hp) may have less than $^1/_2$%. High-slip motors (as much as 13% slip) are used for applications with high inertia and requiring high starting torque.

Synchronous Alternating-Current Motors

These motors run in exact clock synchronism with the power system. For most modern power systems, these are truly constant-speed motors. The use of synchronous motors requires a detail calculation of the torsional dynamics on the rotor, since the high acceleration causes some very high torsional torques.

In the conventional synchronous motor, a rotating magnetic field is developed by the stator currents as in induction motors. The rotor, however, is different, consisting typically of pairs of electromagnets (poles) spaced around the rotor periphery. The rotor field corresponds to the field produced by the AC stator having the same number of poles. The rotor or field coils are supplied with DC; the magnetic field is therefore stationary with respect to the rotor structure. Torque is developed by the interaction of the rotor magnetic field and the stator current (in-phase component). Under no-load conditions and with appropriate DC field current, rotor and stator magnetic-field centers coincide. The voltage applied to the stator winding is balanced by an opposing voltage generated in the stator by the rotor field (induced), and no AC power current flows. As load is applied, the rotor tends to decelerate momentarily, causing a shift of rotor position with respect to the AC field. This shift produces a difference between the applied and induced voltage; the voltage difference causes current to flow; the current reacts with the rotor magnetic flux, producing torque.

Synchronous motors should not be started with the DC field applied. Instead, they are started as inductions motors; bars, acting like a squirrel-cage rotor, are embedded in the field-pole surface and connected by end rings at both ends of the rotor. These damper bars also serve to damp out oscillations under normal running conditions. When the motor is at approximately 95% speed (depending upon application and motor design), DC is applied to the field and the motor pulls into step (synchronism). Because the damper bars do not affect the synchronous-speed characteristics, they are designed for starting performance. This provides flexibility in the accelerating characteristics to meet specific application requirements without affecting running efficiency and other synchronous-speed characteristics. The rotor design of a squirrel-cage motor, on the other hand, must be a compromise between starting and running performance. The DC field is usually shorted by a resistor during starting and contributes accelerating torque, particularly near synchronous speed.

Power-Factor Correction

Synchronous motors because of their more complicated design and the necessity for a field power supply are typically applied only in large-horsepower ratings (several hundred horsepower and larger); synchronous motors over 60,000 kW (80,000 hp) have been built. With their latitude in size and characteristics and their important inherent high power factor and efficiency, synchronous motors are applied to a wide variety of drives.

Conventional synchronous-motor power factors are either 100% or 80% leading. Leading-power factor machines are used frequently to correct for the lagging power factor of the remaining plant load. Even 100% power-factor motors can be operated leading at reduced loads. An advantage of synchronous motors over capacitors is their inherent tendency to regulate power-system voltage; as voltage drops, more leading reactive power is delivered to the power system, and conversely as voltage rises, less reactive power, in contrast to capacitors for which the reactive power decreases directly in proportion to the voltage drop squared. The amount of leading reactive power delivered to the system depends on DC field current, which is readily adjustable.

Field current is an important control element. It controls not only the power factor but also the pullout torque (the load at which the motor pulls out of synchronism). For example, field forcing can prevent pullout on anticipated high transient loads or voltage dips. Loads with known high transient torques are driven frequently with 80% power-factor synchronous motors. The needed additional field supplies both additional pullout torque and power-factor correction for the power system. When high pullout torque is required, the leading power-factor machine is often less expensive than a unity-power-factor motor with the same torque capability.

Synchronous speeds are calculated by the same relationship given for induction motors. Speeds above the limits given are obtained through step-up gears. Large high-speed centrifugal compressors are often driven through step-up gears to obtain the desired compressor speed. Two-pole (3600 rpm at 60 Hz) synchronous motors can be built but are uneconomical in comparison with geared drives.

Generator

The generator is used to convert the rotating mechanical energy of a gas turbine or steam turbine into electrical energy. The generator principally comprises of the stator, rotor and mechanical structure including cooling and lubricating systems. The generator rotor windings are supplied with excitation DC from the static excitation system to produce a magnetic field. As the rotor rotates, this magnetic field cuts across the conductors in the generator stator to induce electrical current as seen in Figure 9-1. This electrical current is the output of the generator and is fed through the generator transformer to the grid.

The majority of gas and steam turbines are directly coupled to two-pole generators. For units with ratings below 40 MW, four-pole generators that run at

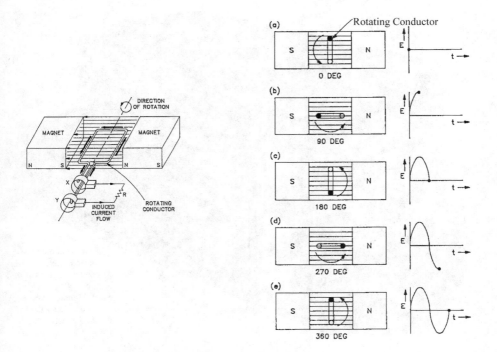

Figure 9-1. Schematic of an A-C Generator System

328 • COGENERATION AND COMBINED CYCLE POWER PLANTS

half speed are more economical. It is advantageous that turbines of this output usually already have a gear so to apply a four-pole generator, only the reduction ratio of the gear must be adapted.

Three main types of generators are used in combined-cycle plants:

- Air-cooled generators are available with open ventilated and with a closed-air circuit or totally enclosed water to air-cooled circuits.
- Hydrogen-cooled generators
- Water-cooled generators.

Generators with open-air circuit cooling are low-cost and have no need for additional cooling, but problems with fouling, corrosive atmosphere and noise can arise.

Generators with closed-air circuit cooling are built for capacities up to 480 MVA as shown in Figure 9-2. These machines are reasonable in cost and provide excellent reliability. The full-load efficiency of modern air-cooled generators is beyond 98%. Air-cooled generators with a closed-air circuit as seen in Figure 9-2 usually use water to cool the air. These machines are well suited for use with single shaft combined-cycle plants where the gas turbine and the steam turbine drive the same generator. In plant condition, this generator can transmit 320 MVA.

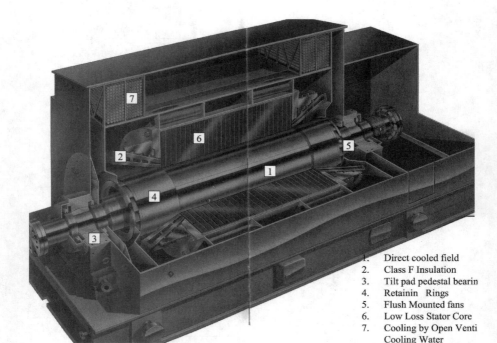

1. Direct cooled field
2. Class F Insulation
3. Tilt pad pedestal bearin
4. Retainin Rings
5. Flush Mounted fans
6. Low Loss Stator Core
7. Cooling by Open Venti Cooling Water

Figure 9-2. Air-Cooled Generator (Courtesy of G. E. Power)

Nowadays, air-cooled generators of up to 480 MVA have now been built and tested due to the overall benefits of air cooling.

A closed circuit air cooling system removes the heat generated by the flow of electrical current through the resistance in the stator windings. Usually, two single-stage axial fans located on the generator shaft circulate air through the cooling circuits. Cooling air losses are replenished with air drawn through mesh filters. The power rating of the generator decreases if the heat removal capability of the air cooling system is in any way reduced.

Hydrogen-cooled generators as seen in Figure 9-3 are used for larger units. They attain an efficiency superior, particularly at part-load, to air-cooled machines. By capitalizing on its low density and high thermal conductivity, they can pack greater efficiency into a smaller generator compared with an air-cooled unit of equal rating. They require additional auxiliaries and monitoring equipment, and are more complex in design. This results in hydrogen-cooled generators being more costly. In general, they do not reach the high reliability of the closed-air circuit-cooled generators.

For generators over 300 MVA, liquid-cooled generators are available. These generators require more effective cooling system, as they rely on hydrogen cooling enhanced by direct water cooling of the stator windings as seen in Figure 9-4. The deionized water, supplied by a closed loop auxiliary system, flows through hollow cooper strands located in the stator windings.

1. Class F Insulation
2. Direct cooled Field
3. Retaining Rings
4. Low Loss stator Core
5. Plate Fin Simplex Coolers

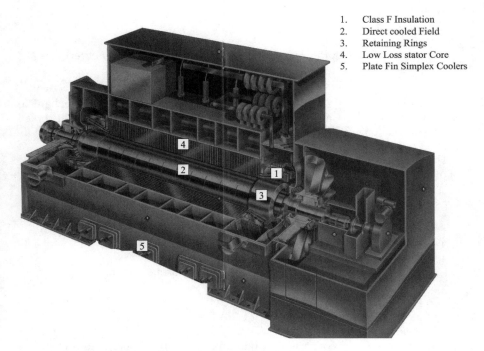

Figure 9-3. Hydrogen-Cooled Generator (Courtesy of G. E. Power)

1. Liquid Cooled Stator Winding
2. Diagonal flow gap pickup rotor cooling
3. Stator Insulation
4. Stator end winding support system
5. Stator Core Support System
6. Stator Winding support
7. Core-end Cooling
8. Retaining Ring
9. Collector arrangement
10. Plate-fin simplex Coolers
11. High Voltage bushings
12. Low-loss stator core

Figure 9-4. Liquid-Cooled Generator (Courtesy of G. E. Power)

Liquid-cooled machines ratings place far greater electrical and mechanical demands on insulation and windings than air, or conventional hydrogen-cooled units. The challenge is to maintain high reliability and efficiency in these highly stressed machines.

The turbine generator and generator transformer system comprises the following major components:

- Generator
- Excitation system
- Automatic voltage regulation (AVR) system
- Isolated phase bus-duct (IPB) system
- Generator transformer
- Generator and transformer protection and metering system.

Design Characteristics

The turbine generator converts the rotational mechanical energy into electrical energy for distribution to the power grid as well as for providing unit excitation power. The generator rotor usually rotates at 3000/3600 rpm and produces three phase, 50/60 Hz electrical power at a nominal terminal voltage (10.5 kV) at the stator. The major components of the generator are:

- Stator (including the stator frame)
- Rotor
- Generator bearings
- Slip-ring assembly.

Stator. The generator stator is the stationary part of the generator and contains the windings in which the generator output current is generated as seen in the schematic in Figure 9-5. The main components of the stator are:

- Stator frame
- Stator magnetic core
- Stator windings.

Stator Frame. The frame of the stator forms the outer casing of the generator and its functions are:

- To support the stator core, bearing assemblies and stator windings terminal box.
- To provide flow paths for the generator cooling air.
- The stator frame is usually made up of a welded construction and supports the laminated core and the winding. In most cases, the air ducts and welded circular ribs provide for the rigidity of the stator frame. End shields usually contain the bearings and are either bolted or welded to the frame end walls.

Feet are welded to the stator end shields to support the stator on the foundation. The stator is firmly anchored to the foundation with anchor bolts through the feet.

The stator frame resembles a tank with centrally located circular openings at its two ends. The generator bearing assemblies, which support the generator rotor shaft, are fitted at these openings. Internal section plates with heavy circumferential bore rings provide strength to support the core and suppress resonant vibration with the magnetic core excitation frequency. The frame also forms an integral part of the cooling air circuit of the generator because its webs and internal sub-divisions form channels through which cooling air flows.

The ends of the generator frame are closed by means of end shields. These are very stiff, reinforced, circular plates designed to support the weight of the rotor and to contain the generator cooling air. The stator end shields are attached to the end flanges of the stator frame. Each end shield has lateral openings to admit cooling air into the generator.

The stator end shields contain the generator bearings. Each bearing sleeve rests on bearing saddles bolted to brackets welded to the end shields. The bearing sleeves are insulated from the saddles. An opening is provided in the exciter end shield to bring out the terminal bushings. Space heaters are often attached to each

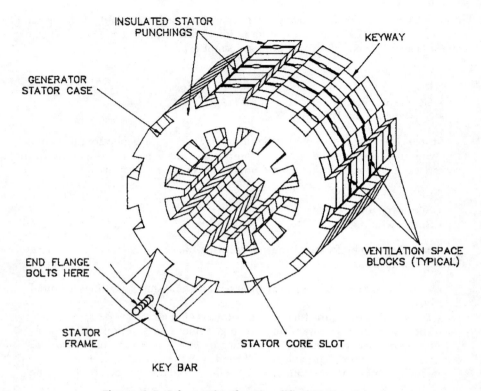

Figure 9-5. Schematic of a Simplified Stator Core

end shield to prevent moisture formation inside the generator when the rotor is at rest or on turning gear. The space heaters are arranged so that the temperature inside the generator is maintained above dew point. Feet welded to the stator end-shields support the stator on the foundation. The stator is firmly anchored to the foundation by grouting bolts threaded through the feet.

Stator Magnetic Core. The stator magnetic core function is to concentrate the rotor field flux around the stator windings. The stator magnetic core is usually stacked from thin insulated electrical silicon-steel laminations with a low loss index and suspended in the stator frame from insulated guide bars. A photograph of such a core is shown in Figure 9-6. The magnetic core laminations are profiled to form parallel surface slots, which contain the stator windings. Spacers placed between groups of laminations form air channels for the generator cooling air. On the core end portions, the cooling air ducts are spaced more closely to account for the higher losses and to ensure more effective cooling of the narrow core sections. Axial compression of the stator core is obtained by clamping fingers, pressure plates, and non-magnetic through-type clamping bolts, which are insulated from the core. The clamping fingers ensure a

Figure 9-6. A Stator Core

uniform clamping pressure, especially within the range of the teeth, and provide for uniform, intensive cooling of the stator core ends. The end sections of the core are stepped to reduce iron losses.

The lamination packets are usually clamped into a solid unified cylindrical core by key bars running through the laminations. Blocks, welded to the ends of the key bars, will hold the core end plates in position. Suspension rings placed around the outer periphery are welded to the key bars to complete a cylindrical frame. The key bars are made of non-magnetic material and are insulated from the core and the pressure plates to prevent short-circuiting and the slow of eddy currents. The core located in the casing by a system of suspension tubes provides positive location and dampens vibration transmission.

Stator Windings. The turbine generator stator contains various sets of windings; each set of windings is called a phase. A separate alternating current (AC) is developed in each of the phases. If there is a three-winding generator it is called a three-phase generator. The stator windings are arranged 120 apart physically, which results in the induced alternating voltage and current in these three windings being 120 out of phase.

The stator winding is a fractional-pitch composed of various layers consisting of individual bars. In order to minimize the stray losses, the bars are composed of separately insulated strands, which are transposed by 360 deg. To minimize the stray losses in the end windings, the strands of the top and bottom bars are separately brazed and insulated from each other. The bars are located in slots of rectangular cross-section, which are uniformly distributed on the circumference of the stator core.

Each generator stator winding has two terminals, the phase and neutral terminals. The electrical output of each generator stator winding is fed to a phase terminal through a bushing mounted in the stator windings terminal box at the top of the generator. The neutral ends of the windings are connected together by a neutral bus and earthed.

After the core has been assembled, the stator winding is formed by inserting insulated bars into the core slots and then joining the ends of the bars to form coils. The stator winding has three phases and two poles. Rectangular section copper coils are often used to form the winding, which is contained in slots in the stator core and retained by wedges. The complete winding is divided into three phases, each phase split into two halves, positioned diametrically opposite each other. The coils are wound so that the alternate groups forming each half phase are in opposition, thus producing two parallel paths per phase.

High-voltage insulation to the bars can be provided by the use of several half-overlapped continuous layers of mica tape applied to the bar. The mica tape is built up from large area mica splittings, which are sandwiched between polyester backed fabric layers with epoxy as an adhesive. The thickness of the insulation depends on the machine voltage. To ensure that the high-voltage insulation obtained is nearly void-free, which ensures excellent electrical and thermal properties and that it is fully waterproof and oil-resistant, the bars are dried under vacuum and impregnated with epoxy resin, which has very good penetration properties due to its low viscosity. After impregnation under vacuum, the bars are subjected to pressure, with nitrogen being used as pressurizing medium (VPI process). The impregnated bars are formed to the required shape in molds and cured in an oven at high temperature.

To minimize corona discharges between the insulation and the slot wall, a final coat of semi-conducting varnish is applied to the surface of all bars within the slot range. In addition, all bars are provided with an end corona protection to control the electric field at the transition from the slot to the end winding and to prevent the formation of creepage spark concentrations.

To protect the stator winding against the effects of magnetic forces due to load and to ensure permanent firm seating of the bars in the slots during operation, the bars are inserted with very small lateral clearances, a hot-curing slot bottom equalizing strip, and a top ripple spring located beneath the slot wedge. In the end windings, the stator winding is firmly lashed to supporting brackets with glass-silk tapes. Spacer blocks arranged between the bars ensure a short-circuit-proof support structure.

The stator winding is connected in the generator interior. The stator winding connections are brought out to bushings located at the exciter end. Current transformers for metering and relaying purposes can be mounted on the bushings.

The stator end windings are the section of the stator bars outside the magnetic core. The end windings are supported by brackets, which are mounted radially on the core end plates. A schematic of such stator end windings is shown in Figure 9-7. A conformable packer is inserted between the end winding and the support bracket. Similar conformable bindings separate the two layers of the end winding, maintaining a space for the through flow of ventilation air. Bindings hold the conductor bars in place and the complete end winding is tied to the radial support brackets. The bindings are finally injected with a resin compound after all the winding components are secured. The winding connections are made by soldering on connectors. Molded covers are fitted over the joints and filled with epoxy resin to provide complete encapsulation. The windings nearest the centerline of the generator are the neutrals, the others being the phase terminals. The phase terminal connects to the isolated phase bus-duct (IPB) section while the neutral leads are earthed.

Rotor. The synchronous rotor is a highly engineered unitized assembly capable of rotating at the designed synchronized speeds with minimal vibration. At the center of the axial rotor assembly is the rotor core embodying the magnetic poles as seen in Figure 9-8. Turbine generators, usually have round rotors, this is called

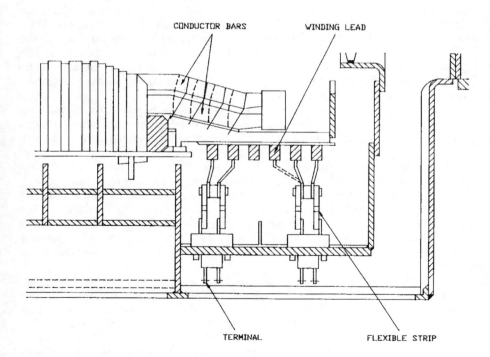

Figure 9-7. Schematic of Stator End Winding

Figure 9-8. A Rotor Core of a Generator

"non-salient pole" construction. The generator rotor has two or four magnetic poles; windings are supplied with excitation DC to produce a magnetic field. This excitation current is provided by a static excitation system comprising an excitation transformer and thyristor bridge rectifiers. As the generator rotor and its magnetic field rotate, electrical currents are induced in the windings of the generator stator. These currents are the electrical output of the generator. The strength of the magnetic field is determined by the amount of current supplied by the static excitation system to the rotor windings. The magnitude of the excitation current is controlled by the automatic voltage regulation.

The rotor shaft is a single-piece solid forging manufactured from a vacuum casting. Slots for insertion of the field winding are milled into the rotor body. The longitudinal slots are distributed over the circumference so that two solid poles are obtained.

To insure that only high-quality forgings are used, strength tests, material analyses, and ultrasonic tests are performed during manufacture of the rotor.

After completion, the rotor is balanced in various planes at different speeds and then subjected to an overspeed test at 120% of rated speed for 2 minutes.

The rotor winding consist of several coils, which are inserted into the slots and series-connected such that two coil groups form one pole. The rotor winding consists of materials such as silver-bearing copper, ensuring an increased thermal stability.

The individual turns of the coils are insulated against each other by interlayers of laminated epoxy glass fiber fabric with a filler to insulate the slot. The slot wedges are made of high-conductivity material and thus act as damper winding. At their ends, the slot wedges are short-circuited through the retaining rings.

The centrifugal forces of the rotor end windings are contained by single-piece rotor retaining rings. The retaining rings are made of non-magnetic high-strength steel in order to reduce stray losses. Each retaining ring with its shrink-fitted insert ring is shrunk onto the rotor body in an overhung position. The retaining ring is secured in the axial position by a snap ring. Many rotors have also built-in integral fans to provide the cooling required by the generator. Figure 9-9 shows the cooling air diagram for a typical generator.

Field Connections and Slip-rings. The field current is supplied to the rotating rotor through carbon brushes and steel slip-rings arranged at the exciter-end shaft end. The electrical connection between the slip-ring and the rotor winding is established by means of radial bolts and insulated semi-circular conductors located in the hollow bore of the rotor at the exciter end.

Bearing Lubrication Oil. The generator is supported on each end on journal bearings called the "driven end bearing DE" closer to the turbine and the "non-driven end NDE" closer to the slip-rings. Lubricating oil to each of these bearings is supplied from the turbine lubrication oil system. Lubricating oil is supplied to the bearing through ports in the bearing pedestal and oil inlets on both sides of the bearing.

Excitation System. The excitation system supplies current to the generator rotor winding to produce the requisite electromagnetic field and provides a means to vary this current to control generator terminal voltage.

The generator produces electrical energy at a voltage (10.5 kV), which is led directly to the LV side of the generator transformer via an Isolated Phase Bus system. The generator transformer steps up this voltage and supplies the Busbars (132 kV) via the main circuit breaker. The gas turbine is usually provided with an extra generator synchronizing breaker; the steam turbine generator is not provided with an extra generator synchronizing breaker, and synchronizing is carried out on the 132-kV side. Generator excitation is supplied through an excitation transformer.

The generator and transformer have protective devices to prevent equipment damage due to phase imbalances, undervoltage, overvoltage, earth faults and other electrical system upsets and malfunctions.

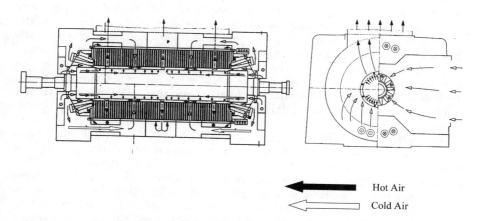

Hot Air
Cold Air

Figure 9-9. Cooling Air Scheme in a Conventional Air-Cooled Generator

The excitation system provides a controlled source of current to the generator rotor field winding. The current flowing through the rotor field winding turns the rotor into an electromagnet with a north and south pole. As the rotor rotates, this magnetic field rotates at the same speed. The excitation system provides means to vary the field winding current to meet the startup, normal operation, and shutdown operation needs of the turbine generator and to connect and disconnect the external source from the rotor field winding. The excitation current is provided by an excitation transformer and rectifier circuit.

Automatic Voltage Regulating System (AVR). The AVR system has several functions:

- Regulate the generator terminal voltage, maintain it within its specified limits, and match it to the grid system voltage.
- Stabilize the generator terminal voltage to keep the unit synchronized with the grid system.
- Protect the generator from excessive thermal damage due to high currents during an internal fault.

The AVR system maintains a constant generator output voltage under varying load conditions by increasing or decreasing the rotor field current. The magnetic field produced by the generator rotor is proportional to the rotor field current.

The grid frequency and voltage can fluctuate constantly, which requires that small adjustments be made to the generator terminal voltage. This is controlled by varying the generator excitation. Just prior to generator synchronization, the automatic voltage regulator (AVR) is placed in service. With the AVR in service, generator terminal voltage is automatically regulated. The AVR controls the excitation current to the field winding of the generator rotor.

Isolated Phase Bus-Duct. The three-phase terminals of the generator are connected to the generator transformer through aluminum conductors mounted on insulators, within cylindrical enclosures called bus-ducts. The bus-ducts are known as isolated phase bus-duct (IPB) since each phase is housed in a separate bus-duct. The cylindrical enclosures are usually supplied with dry low pressure air to prevent moisture buildup and short circuits caused by such moisture. The IPB air supply is provided by air compressors; two compressors are usually provided with one in service at a time. Air from the compressor is routed through driers and pressure reducer before being supplied to the bus-duct enclosure. Branch lines off the bus of each generator phase connect to the excitation transformer and to the potential transformer (PT) cubicles. The PT cubicles contain potential or voltage transformers, which measure the generator output voltage.

Figure 9-10 is a schematic diagram of the generator and generator transformer system and shows the position of the excitation transformer and the isolated phase bus-ducts. The generator and transformer have protective devices to prevent equipment damage due to phase imbalances, undervoltage, overvoltage, earth faults, and other electrical system upsets and malfunctions. Information on generator windings and air temperature, bearing vibration and metal temperature, cooling air dew-point temperature, and generator air cooler water temperatures are usually available on the D-CS system.

Generators, Motors and Switch Gears • 339

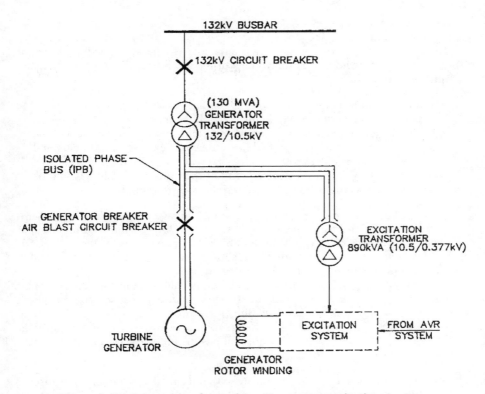

Figure 9-10. Schematic of a Turbine Generator Excitation System

Generator Transformer. The generator transformer steps up the output voltage of the turbine generator from a low voltage (10.5 kV) to the power grid voltage (132 kV). A transformer is a stationary device, which works on the magnetic induction principle to either raise or lower the input voltage it receives. Transformers are essential in large transmission systems since the higher the voltage of transmission, the lower is the current for the same power in the transmission line. The following relationships illustrate this principle:

$$P = V \times I \qquad (9-4)$$

$$V = I \times R \qquad (9-5)$$

$$P = I^2 \times R \qquad (9-6)$$

where
  P = power
  V = voltage

I = current
R = resistance

In other words, the power losses are proportional to the square of the current. If the power were transmitted at the voltage of the generator, there would be many adverse effects such as:

- Transmission losses would be very high.
- Transmission current would be unacceptably high.
- High current means larger conductors and more weight, hence the transmission line supports would need to be bigger. Overall, more capital cost as well as running cost.
- Also, the voltage at the destination end of the transmission line would be very low due to the voltage drop along the transmission line.

For the above reasons, the generator terminal voltage needs to be "stepped-up" to a value sufficient for transmission. Common long distance transmission voltages are 132 kV, 275 kV, 400 kV and 500 kV. In comparison, the voltage for local distribution feeders, where the distance for transmission is appreciably shorter, is lower; typically 11 kV or 33 kV.

The generator transformer comprises low- and high-voltage windings wound around a metallic core. The windings are located inside a tank filled with oil, which provides both an insulation and cooling medium for the winding. For small transformers, natural air cooling is the norm, but for transformers operating at high voltages, oil cooling is used because oil has the following advantages over air as a cooling medium:

- Oil has a larger specific heat than air so that it can absorb larger quantities of heat for the same temperature rise.
- Oil has a greater heat conductivity than air and so enables the heat to be transferred from the windings to the oil more quickly.
- Oil has about six times the breakdown strength of air, which ensures increased reliability at high voltages.

The main parts of a transformer are its primary and secondary windings that are linked by electromagnetic flux contained by a core. The core is a closed magnetic circuit made from laminated magnetic steel having a high permeability.

Figure 9-11 shows the principle of operation of a single phase transformer and illustrates a typical three core-type transformer. The ratio of primary and secondary voltage is the ratio of the primary to secondary winding turns also called the "turns ratio" of the transformer as given below:

$$\frac{V_H}{V_L} = \frac{T_H}{T_L} \qquad (9-7)$$

where
T_H = number of high-voltage winding turns
T_L = number of low-voltage winding turns

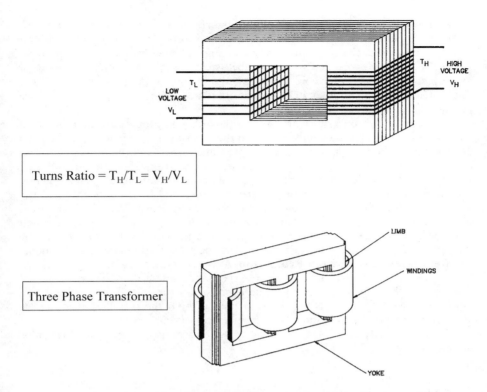

Turns Ratio = $T_H/T_L = V_H/V_L$

Three Phase Transformer

Figure 9-11. A Schematic of the Operation Characteristic of a Transformer

It is a basic requirement of most power transformers that for a constant primary voltage, the secondary voltage can be varied over a specified range. From the transformation formula, it is obvious that a change in secondary voltage can be achieved by changing the turns ratio of the transformer. In practice, the number of primary winding turns is fixed and the secondary winding is "broken up" into several sections with each section having a termination. The turns ratio can then be altered by tapping off the secondary winding at these different sections. Each "tap position" then activates a different number of secondary winding turns. There are two types of tap changer: the on-load tap changer and the off-load or no-voltage tap changer.

The vertical sections of the core are called "limbs" and the horizontal sections "yokes". In a three-phase transformer, the primary and secondary windings of each phase are arranged concentrically on each of the three limbs of the core. Spacers placed between the core and windings form channels through which the transformer oil flows to cool the windings. The core and windings are then placed inside an oil-filled tank for insulation and heat transfer purposes. Insulated terminal bushes are used to enable liquid-tight connections to be made to each of the winding leads through the oil tank.

Magnetic ferrous-alloy material in a transformer core provides an efficient low reluctance path for magnetic fluxes around and through the windings. Without such a magnetic circuit, higher magnetizing currents would be required to establish the fluxes needed to obtain the "stepping-up" or "stepping down" of voltages. The transformer core is laminated and the laminations are electrically insulated from one another to minimize the eddy currents induced in the core material by the main flux.

Substantial eddy currents will result in an increase in supply current, losses, and core overheating. The current supplied to the transformer on load comprises two components; the main power current to be transformed and the no load current, which energizes the core and supplies the core losses. A frame structure surrounding both sides of the core forms an integral part of the core clamping arrangement and, in addition, provides the support for the windings and their connections to the terminals. The clamping framework also provide the means by which the total winding assembly may be lifted and locates the core and windings in the main oil tank. The whole of this frame is insulated from the core and from the tank by intervening insulation. The core is earthed at one point through the main oil tank.

Generator Transformer Protection. A typical transformer package is shown in Figures 9-12a and 9-12b. This generator package shows the cooling system used to protect the transformer coils. These coils can produce large temperature gradients. Usually, there is a blast wall built around the installation to ensure limiting damage due to failure of the transformer system.

The generator protection and metering system comprises the following major components:

- Biased differential
- Restricted earth fault
- Buchholz Relay

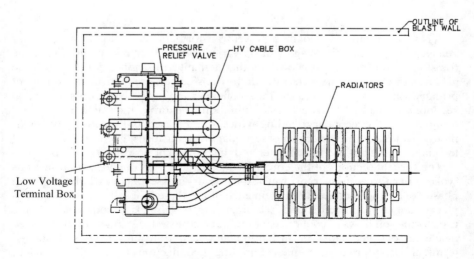

Figure 9-12a. A Drawing of a Typical Transformer System

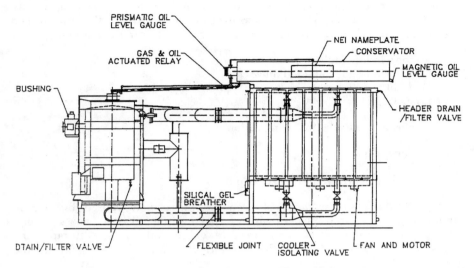

Figure 9-12b. A Drawing of a Typical Transformer System

- Tap changer over pressure
- Instantaneous and time-delayed overcurrent
- Winding temperature second stage.

Biased Differential. This relay protects the generator transformer against internal faults. It is biased to provide stability during large external faults to avoid tripping due to current transformer error. The biased relay also utilize second harmonic restraint to prevent its operation by normal magnetizing inrush currents produced when the transformer is first energized. This relay will not operate for faults external to the generator transformer and activates the following actions:

- Turbine trip
- Field circuit breaker trip
- Generator main circuit breaker trip.

Restricted Earth Fault. This relay detects earth faults on the primary side of the generator transformer including the transformer neutral and connections. Because the generator neutral is grounded through a high impedance, the differential current relaying scheme is not sensitive enough to protect against winding earth faults and at the same time avoid undesirable trippings due to current transformer errors during large external faults. The relay activates the following actions:

- Turbine trip
- Field circuit breaker trip
- Generator main circuit breaker trip.

Buchholz Relay. This relay protects the transformer against build up of gas within the transformer. This gas is formed on breakdown of the transformer oil, which can be caused by one or more of the following condition:

- Minor winding fault
- Minor winding insulation failure
- High oil temperature
- Low oil level.

The relay activates the following actions:

- Turbine trip
- Field circuit breaker trip
- Generator main circuit breaker trip.

Tap Changer Over Pressure. This condition is usually caused by electric arcing within the transformer, which vaporizes the oil and causes a high pressure to develop. This high pressure causes oil to flow through the pressure relief device fitted on the transformer body. The relay activates the following actions:

- Turbine trip
- Field circuit breaker trip
- Generator main circuit breaker trip.

Instantaneous and Time-Delayed Overcurrent. This relay protects the generator transformer against overcurrent conditions on the secondary side of the transformer. The relay has an instantaneous setting and also an inverse definite minimum time current capability. The relay activates the following actions:

- Turbine trip
- Field circuit breaker trip
- Generator main circuit breaker trip.

Winding Temperature. This relay protects the transformer against high winding temperatures, which can be brought about by one or more of the following conditions:

- Transformer cooling fan failure
- Primary or secondary winding phase imbalance, undervoltage or overvoltage
- Transformer overload
- Earth fault.

The relay activates the following actions:

- Turbine trip
- Field circuit breaker trip (through operation of reverse power relay)
- Generator main circuit breaker (through operation of reverse power relay).

Oil Temperature. This relay protects the transformer against high oil temperature. This high oil temperature can be brought about by one or more of the following conditions:

- Transformer cooling fan failure
- Low oil level
- Obstruction in oil flow
- Primary or secondary winding phase imbalance, undervoltage or overvoltage
- Transformer overload
- Earth fault.

The relay activates the following actions:

- Turbine trip
- Field circuit breaker trip (through operation of reverse power relay)
- Generator main circuit breaker trip (through operation of reverse power relay).

Switchgear

The economic benefits of distributed generation offers an opportunity to recover stranded costs from installed standby and emergency power equipment, as well as a means to subsidize the enhanced capability of these systems to meet the demands of modern electrical loads. One of the key considerations in maximizing this opportunity centers on the available switching and control strategies.

The operating mode at the point of interface between the utility derived and on-site power sources is one of the most significant cost determinants in the operating strategy of the on-site powerplant. There are two basic modes to choose from isolated operation of the plants and parallel operation of the plants. There are pros and cons to both, and each provides a suitable return on invested capital while providing critical load backup. In the isolated mode, peaking loads are disconnected from the utility-derived source and connected to the peaking plant. In the parallel mode, both power sources are connected to the load simultaneously. In this latter mode, there is a need for additional control and protective circuits. The function of these additional circuits is to provide adequate safety to personnel and equipment on both sides of the point of common connection.

The issue of confidence in the control strategy of any power system that parallels with a utility derived power source is a universal concern. The electric utility supplier, in any such application, must provide safety to its maintenance staff, as well as continuity and reliability in electric service to its customers. For these reasons, serving utility companies are quite demanding of the synchronizing and paralleling control strategies that they will allow at the point of common coupling.

On the other side of this issue is the increasing demand automation of customer facilities places on continuity of electric service. Increased automation in building environmental and production process control, coupled with reductions in operations staffing, impose less intrusive load transfer strategies on critical electrical loads. Having suffered an interruption in the process upon loss of the utility power source, it is often costly to suffer a second disruption upon restoration to the utility service.

A simplified logic flow chart of the closed transition transfer control strategy is shown in Figure 9-13. Because the closed transition transfer strategy must be capable of both open and closed transition, both logic paths are illustrated. The left-hand path is for closed transition transfer; the other is for open transition. Because the strategy is required to determine when either strategy is to be implemented, it must be capable of both. As an overview, closed transition transfer is initiated by restoration of the preferred source of power to acceptable values for the time set in the respective delay functions or by initiation of either the unit mounted test switch or remote initiating contact. Open transition transfer in automatically initiated when the source to which the load is connected becomes unacceptable and the other source is determined to be acceptable.

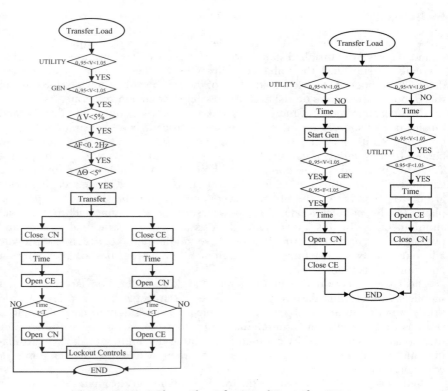

Figure 9-13. A Flow Chart for Load Transfer Strategy

If a closed transition transfer is called for and initiated, the strategy is to continuously check the two sources of power to determine their acceptability. As long as they remain acceptable, the strategy will make the differential determination of voltage, frequency and phase angle. Only when the sources are acceptable, the voltage difference is less than 5%, the frequency difference is less than 0.2 Hz, and the relative phase angle difference between the two sources is less than 5 deg. will the transfer operation initiate. When transferring the load to the normal/preferred source, the strategy is to close the main contacts for the preferred source (CN) contacts and initiate a timing function to permit an overlap time of no more than 100 milliseconds. At the end of this time, the strategy is to open the main contacts for the alternate source (CE). Parallel operation is limited to 100 milliseconds or less because this operation requires no interactive controls between the power sources. Separation of the sources in less than 100 milliseconds will not impact any protective relaying in either of the two power sources.

If the CE contacts fail to open within the time as set in $t<T$, this will reopen the CN contacts and lock out the controls from further attempts at closed transition transfer. When the strategy is successfully completed, the regimen is ended until the next load transfer operation in initiated. As can be seen in the open transition transfer logic stream, the mains of one contact are opened before the mains of the other are closed. This strategy is the typical open transition transfer strategy implemented with a dual operator switching subassembly.

The automatic transfer switch is essentially a double throw device having two positions: closed on the utility source or closed on the peaking source. The device is relatively simple and straightforward and has been around and in this specific service for decades. The control circuitry of the automatic transfer switching equipment (ATSE) used for emergency or legally required standby systems must have a means of initiating load transfer for test purposes. The switch comes with a minimum set of control functions and features mandated by the Art.700 of the NEC and UL1008, Minimum Standard for Safety Automatic Transfer Switches. Among these are voltage sensing of the preferred source; voltage and frequency sensing of the alternate source; means to initiate transfer to the alternate source for testing; timing functions for ride-through of transient voltage sags and outages, preferred source; post-operative power plants cool-down; and stabilization of power source on availability.

The control strategy, when embedded in the ATSE, assures the power system operators that paralleling will only occur when both sources are adequate and within a very narrow window of synchronism. The stability and repetitive accuracies of the control strategies are better than actually required. The control module is also evaluated in the qualification process of the same standards. The control strategies must meet specified repetitive accuracies and stability over a temperature range of -5°C to 40°C (40°F to 104°F). The control strategies include voltage and frequency sensing, time delays and various other control functions. Modern control strategies are typically a microprocessor-based design. The repetitive accuracy is typically ±1% of normal at ambient temperature and the stability is typically ±0.5% of setting across the temperature range. In addition to acceptability of the two power sources, the control strategy measures the difference between the sources in voltage, frequency and phase angle. A typical operation sequence for closed transition transfer is to first determine acceptability of both power sources. Only when both sources are determined to be acceptable and have remained acceptable for a pre-set time will the strategy determine if the differentials are met. When synchronism is determined (synchronism is defined as when the voltage difference between the two sources is less than 5% and the frequency difference is less than 0.2 Hz and the phase angle crosses 5 deg. (electrical) and is going toward zero), the strategy will initiate a closed transition transfer. Upon initiation of transfer, the open contacts are closed. After closure, the initially closed mains are opened. Timing is arranged to limit parallel operation of the sources to less than 100 milliseconds. The control strategy should include automatic recovery should a malfunction occur during the closed transition transfer operation.

The contact structure of the switching device should be evaluated for the intended service. The switch should be qualified for its intended use, as is required by the NEC (NFPA 70) and other electrical building codes. While the intent of qualifying standards is to assure a minimum performance criteria, for application in electric demand abatement, the performance criteria also provides an expectation of service-free longevity. Thus, it addresses controlling operating costs. For a 400-A switch, assuming 250 working days/year, the expectation is 8 years before major contact service is required.

The simplest form of a supplementary load system is when a selected load could be switched from the utility derived sources to the peaking power source using the automated transfer switch. This type of a system is typically used in emergency and standby power systems to transfer critical loads between two power sources as a function of source acceptability.

The most cost effective and simplest scenario for peaking and critical load protection is shown in Figure 9-14. The automatic transfer switch is the only switching device specifically tested and qualified for switching between

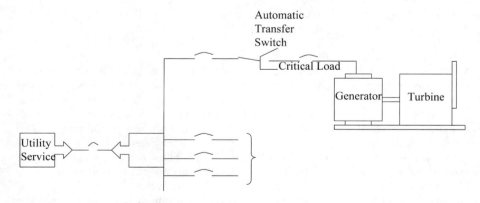

Figure 9-14. A Simple Power System Using the Critical Load as the Peaking Device

unsynchronized power sources. In the circuit of Figure 9-14, the critical load is used as the peaking load as well. The circuitry must have a means of initiating load transfer for test purposes. This same or similar control circuit can be interfaced with a controller that would monitor the facility kilowatt demand and initiate a load transfer to the on-site generator when that demand exceeds a preset value.

Where the critical load is judged to be unsuitable for use as the peaking load, additional ATSE can be installed to provide a peaking load transfer capability. This will more likely be the case for the typical load transfer peaking strategy. Figure 9-15 shows such an installation. In this case, the load can be any load that is operational and constant in demand during the peaking period. In a light industrial or commercial facility, this could be a bulk lighting load. These loads are typically turned on at the beginning of the day and turned off by the cleaning crews at the end of the day. It, therefore, provides a known and constant kilowatt load for peak demand reduction. Figure 9-16 shows a peaking power load transfer using a closed transition transfer technology. In this case, an automatic closed transition transfer switch is used, closed transition transfer is initiated by restoration of the preferred source of power to acceptable values for the time set in the respective delay functions, or by initiation of either the unit mounted test switch or remote initiating contact.

Periodic testing of the complete emergency power system is the only means to instill confidence in its operation. Consequently, upon restoration of the critical load to the normal power source and for test transfer in either direction, closed transition transfer has become the preferred transfer control strategy.

The closed transition strategy without interactive controls between the two power sources will work well where the load being transferred is equal to 25% to 30% of the engine generator set rating. Loading transients for step loading to about 30% of the set rating does not typically cause bothersome transients. Where the switched load represents a higher percentage of the engine generator set rating, consideration for controlled load transition transfer, commonly referred to

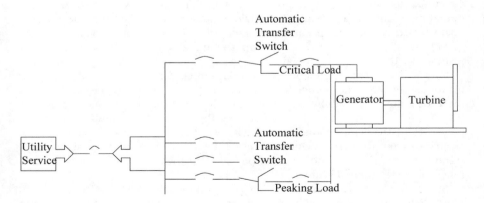

Figure 9-15. A Simple Peaking Power System Using a Non-Critical Load as the Peaking Load

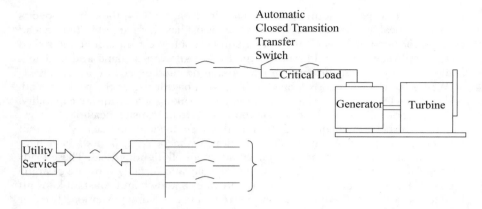

Figure 9-16. A Peaking Power System Using a Closed Transition Transfer Technology

as soft loading, should be given. In the soft load scenario, controls are added to bring the engine generator into synchronism with the utility derived service and hold it there. When synchronism is achieved, the CE contacts are closed. Upon closure, an engine loading control increases fuel flow to the engine at a pre-defined rate. As the fuel flow to the engine increases, it assumes increasing load. Simultaneously with load control, control of generator excitation is initiated so that whatever kilowatt the engine generator is producing, it is produced at a fixed power factor. Since it takes a very short time for the engine generator set to respond to loading control actions, in soft loading, parallel operation can be limited to 30 to 60 seconds.

Optimum flexibility in load and source dispatch occurs with long-term parallel operation of the two sources. Under this operating strategy, engine generator loading can be adjusted to match the real time best economic operating scenario. It should be obvious that this is also the most costly operating strategy from the equipment point of view. However, if the inertia is an add-on to the standby power system, the incremental cost may be readily justified by the savings to be gained.

Given the expansive capabilities of microprocessor-based controls, communications between microprocessor-based strategies is relatively inexpensive and most desirable. The ability to communicate between the engine, generator, transfer switch, load control, etc., conjures the image of a controllable virtual power plant that can be made responsive to changing power demands in the electric distribution system. One could even imagine aggregating several unrelated standby power systems into a virtual power plant that could offer local demand relief for overburdened grids.

More immediately, communications capabilities address the staffing issue. In most distributed power facilities, the primary product of the facility is not electrical energy. Therefore, it is unlikely that there will be competent staff in house to do more than just monitor some points that are readily accessible. As

Generators, Motors and Switch Gears • 351

most facilities contract for outside services to maintain and service special equipment and processes, it would be natural to contract for service and maintenance of the power plant and its associated equipment. The engine generator dealer who furnishes the set can provide appropriate service for it. Similarly, the provider of the electric control and switching system is a ready source of maintenance and service. Incorporating communications into the control scenario enhances this maintenance capability.

Electrical Single Line Diagram

The single line diagram for a combined-cycle power plant for a dual shaft installation is shown in Figure 9-17. When the system is at rest, the breaker

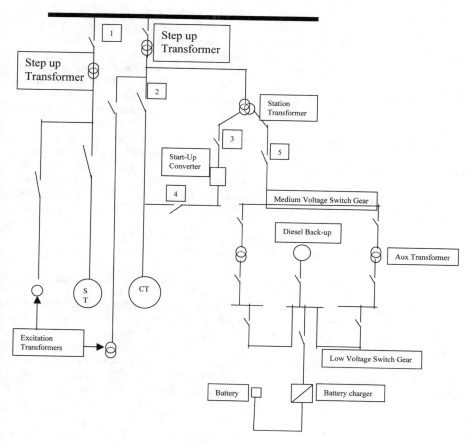

Figure 9-17. A Single Line Electrical Diagram

switches 1, 2, 3, 4, and 5 are open. The auxiliaries such as pumps for the various system are connected to the low voltage bus section. Large refrigeration compressors and other large motors are connected to the medium voltage bus section. There is an uninterruptible power system for the important components of the system. The diesel backup is there for the system in case of grid failure. Auxiliaries are fed from the generator output, which saves a separate in-feed from the grid and allows the plant to run at no load during a grid blackout, ready for being reconnected to the grid if reopened. In this case, the gas turbine generator acts as a motor during startup. However, in many applications especially low power units, either hydraulic, electric, steam, or Diesel engines are available for the start of the gas turbine. In this case, there would be no need for the static frequency converter. The steam turbine is started after the steam can be produced in the HRSG, and thus can lag the gas turbine by about 15 to 20 minutes.

During startup of the gas turbine, the breaker switches 1, 3, 5, and 4 are closed. This enables the turbine generator and its auxiliaries to receive the current through the three winding station transformer and to feed the static frequency converter and enable the generator to act as a motor. When the turbine is at self-sustaining speed, breaker switches 4 and 3 are opened. Once the turbine completes its operation sequence and reaches the full operational speed, the excitation current is fed to the rotor field windings of the generator. At this point, the generator is ready to be synchronized to the grid system by closing its main circuit breaker 2. Before this beaker can be closed, the voltage, frequency, and the phase angle of the generator output and power grid system must be matched.

The generator must be in synchronism with the grid before its main circuit breaker can be closed to avoid large exchanges of electrical power between the generator and grid, severe distortion of generator magnetic fields and the resultant high instantaneous torques on generator and turbine shafts. There are three requirements for power synchronization of the generator to the grid:

- Generator frequency must match power grid running frequency.
- Generator stepped up voltage must be equal in magnitude to the grid system voltage.
- The generator voltage must be in-phase with the grid system voltage.

The phase angle and frequency of the generator output is matched to that of the power grid system with the aid of a synchroscope. Before synchronization, the generator frequency must be slightly higher than the grid frequency to ensure that electricity will flow from the generator to the power grid. If the generator frequency is lower than the grid frequency, the current will flow into the generator causing it to act as synchronous motor and drive the turbine. This phenomenon is called motoring and will cause overheating and damage to the rotor.

Chapter 10

FUELS, FUEL PIPING AND FUEL STORAGE

The major advantage of a combined cycle power plant and a cogeneration plant based around a gas turbine has been its inherent fuel flexibility. Fuel candidates encompass the entire spectrum from gases to solids. Gaseous fuels traditionally include natural gas, process gas, low-Btu coal gas and methanol gas based on compost. "Process gas" is a broad term used to describe gas formed by some industrial process. Process gases include refinery gas, producer gas, coke oven gas, and blast furnace gas among others. Natural gas is usually the basis on which performance for a gas turbine is compared, since it is a clean fuel fostering longer machine life.

Liquid fuels can vary from light volatile naphtha through kerosene to the heavy viscous residuals. The classes of liquid fuels and their requirements are shown in Table 10-1. The light distillates are equal to natural gas as a fuel, and between light distillates and natural gas fuels, 90% of installed units can be counted. Care must be taken in handling liquid fuels to avoid contamination, and the very light distillates like naphtha require special concern in the design of fuel systems because of their high volatility. Generally, a fuel tank of the floating head type with no area for vaporization is employed. The heavy true distillates like No. 2 distillate oil can be considered the standard fuel. The true distillate fuel is a good turbine fuel; however, because trace elements of vanadium, sodium, potassium, lead, and calcium are found in the fuel, the fuel has to be treated. The corrosive effect of sodium and vanadium is very detrimental to the life of a turbine.

Vanadium originates as a metallic compound in crude oil and is concentrated by the distillation process into heavy oil fractions. Sodium compounds are most often present in the form of salt water, which results from salty wells, transport over seawater, or mist ingestion in an ocean environment.

Fuel treatments are costly and do not remove all traces of these metals. As long as the fuel oil properties fall within specific limits, no special treatment is required. Blends are residuals, which have been mixed with lighter distillates to improve properties. Blending can reduce the specific gravity and viscosity. About 1% of total installed machines can operate on blends.

A final fuel group contains high-ash crudes and residuals. These account for 5% of installed units. Residual fuel is the high-ash by-product of distillation. Low cost makes them attractive; however, special equipment must always be added to a fuel system before they can be utilized. Crude is attractive as a fuel, since in pumping applications, it is burned straight from the pipeline. Table 10-2 shows data obtained from a number of users, which indicates a considerable reduction in downtime, depending on the type of service and fuel used. This table also shows that natural gas is by far the best fuel.

The effect of various fuels on the output work of the turbine can be seen in Figure 10-1. Assuming that natural gas is the base line fuel to obtain the same power using diesel fuel the gas turbine would have to be fired at a higher temperature, and for low Btu (400 Btu/cu ft, 14,911 kJ/m^3) gases at the same firing temperature the turbine would produce more power due to the fact that the amount of fuel could be increased by three-fold, thus increasing the overall mass flow through the turbine. The limitation in using low Btu gases is that it takes about 30% of the air for combustion as compared to 10% of the air for natural gas, leaving much less air for cooling the combustor liners. Because of this, for low Btu gases, it is easier to modify annular combustor turbines, which have less of a combustor liner surface area than can annular combustors. Another problem is that in some cases, the extra flow can choke the turbine nozzles. For turbines used in combined cycle application, there is a tendency to keep the same firing temperature at off-load conditions, but the use of the inlet guide vanes vary the airflow rate.

Table 10-1. Comparison of Liquid Fuels for Gas Turbines

Fuel Type → Fuel Characteristics ↓	True Distillates and Naphtha	Blended Heavy Distillates & Low-Ash Crudes	Residual & High-Ash Crudes
Description	High-quality distillate essentially Ash Free	Low-Ash, limited contaminant levels	Low Volatility High Ash
ASTM designation	1-GT, 2-GT, 3-GT	3-GT	4-GT
Turbine Inlet Temperature	Highest	Intermediate	Lowest
Fuel pre-heat	No	Yes	Yes
Base Fuel Cost	Highest	Intermediate	Lowest
Fuel Atomization	Mech/LP air	HP/LP air	HP air
Desalting	No	Some	Yes
Fuel inhibition	Usually none	Limited	Always
Turbine Hot Section Washing	No	Yes except distillate	Yes
Start-up Fuel	With Naphtha	Some fuels	Always

Table 10-2. Operation and Maintenance Life of an Industrial Turbine

Type of Application and Fuel		Firing Temperature Below 1700°F (927°C)			Firing Temperature Above 1700°F (927°C)		
		Comb. Liners	First-Stage Nozzle	First-Stage Blades	Comb. Liners	First-Stage Nozzle	First-Stage Blades
BASE LOAD	Starts/hr	+	+	+			
Nat. gas	1/1000	30,000	60,000	100,000	15,000	25,000	35,000
Nat. gas	1/10	7,500	42,000	72,000	3,750	20,000	25,000
Distillate oil	1/1000	22,000	45,000	72,000	11,250	22,000	30,000
Distillate oil	1/10	6,000	35,000	48,000	3,000	13,500	18,000
Residual	1/1000	3,500	20,000	28,000	2,500	10,000	15,000
Residual	1/10						
SYSTEM PEAKING Normal Max. Load of short duration and daily starts							
Nat. gas	1/10	7,500	34,000	60,000	5,000	15,000	24,000
Nat. gas	1/5	3,800	28,000	40,000	3,000	12,500	18,000
Distillate	1/10	6,000	27,200	53,500	4,000	12,500	19,000
Distillate	1/5	3,000	22,400	32,000	2,500	10,000	16,000
TURBINE PEAKING Operating Above 50–100°F (28–56°C) Firing Temperature							
Nat. gas	1/5	2,000	12,000	20,000	2,000	12,500	18,000
Nat. gas	1/1	400	9,000	15,000	400	10,000	15,000
Distillate	1/5	1,600	10,000	16,000	1,700	11,000	15,000
Distillate	1/1	400	7,300	12,000	400	8,500	12,000

Fuel Specifications

To decide which fuel to use, a host of factors must be considered. The object is to obtain high efficiency, minimum downtime, and the total economic picture. The following are some fuel requirements that are important in designing a combustion system and any necessary fuel treatment equipment:

1. Heating value
2. Cleanliness
3. Corrosivity
4. Deposition and fouling tendencies
5. Availability

The heating of a fuel affects the overall size of the fuel system. Generally, fuel heating is a more important concern in connection with gaseous fuels, since liquid fuels all come from petroleum crude and shows narrow heating value variations. Gaseous fuels, on the other hand, can vary from 1100 Btu/cu ft (41,000 kJ/m^3) for natural gas to 300 Btu/cu ft (11,184 kJ/m^3) or below for process gas. The fuel system will have to be larger for the process gas, since more is required for the same temperature rise.

Cleanliness of the fuel must be monitored if the fuel is naturally "dirty" or can pick up contaminants during transportation. The nature of the contaminants

356 • COGENERATION AND COMBINED CYCLE POWER PLANTS

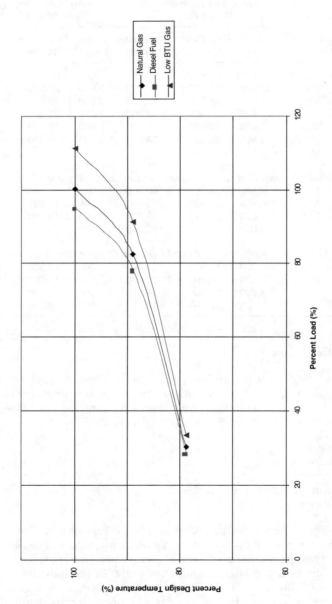

Figure 10-1. Effect of Fuel Various Fuels on Load and Firing Temperature of a Gas Turbine in Combined Cycle Application

depends on the particular fuel. The definition of cleanliness here concerns particulates that can be strained out and is not concerned with soluble contaminants. These contaminants can cause damage or fouling in the fuel system and result in poor combustion.

Corrosion by the fuel usually occurs in the hot section of the engine, either in the combustor or in the turbine blading. Corrosion is related to the amounts of certain heavy metals in the fuel. Fuel corrosivity can be greatly reduced by specific treatments discussed later in this chapter.

Deposition and fouling can occur in the fuel system and in the hot section of the turbine. Deposition rates depend on the amounts of certain compounds contained in the fuel. Some compounds that cause deposits can be removed by fuel treating.

Finally, fuel availability must be considered. If future reserves are unknown, or seasonal variations are expected, dual fuel capability must be considered.

Fuel requirements are defined by various fuel properties. By coincidence, the heating-value requirement is also a property and needs no further mention.

Cleanliness is a measure of the water and sediment and the particulate content. Water and sediment are found primarily in liquid fuels, while particulates are found in gaseous fuels. Particulates and sediments cause clogging of fuel filters. Water leads to oxidation in the fuel system and poor combustion. A fuel can be cleaned by filtration.

Carbon residue, pour point, and viscosity are important properties in relation to deposition and fouling. Carbon residue is found by burning a fuel sample and weighing the amount of carbon left. The carbon residue property shows the tendency of a fuel to deposit carbon on the fuel nozzles and combustion liner. Pour point is the lowest temperature at which a fuel can be poured by gravitational action. Viscosity is related to the pressure loss in pipe flow. Both pour point and viscosity measure the tendency of a fuel to foul the fuel system. Sometimes, heating of the fuel system and piping is necessary to assure a proper flow.

The ash content of liquid fuels is important in connection with cleanliness, corrosion, and deposition characteristics of the fuel. Ash is the material remaining after combustion. Ash is present in two forms: (1) as solid particles corresponding to that material called sediment, and (2) as oil or water-soluble traces of metallic elements. As mentioned earlier, sediment is a measure of cleanliness. The corrosivity of a fuel is related to the amount of various trace elements in the fuel ash. Certain high-ash fuels tend to be very corrosive. Finally, since ash is the fuel element remaining after combustion, the deposition rate is directly related to the ash content of the fuel.

Table 10-3 is a summary of gaseous fuel specifications. The two major areas of concern are heating value with its possible variation and contaminants. Fuels outside a specification can be utilized if some modification is made.

Gas fuels with heating values between 300 and 1100 Btu/cu ft (11,184 to 41,000 kJ/m^3) are in use today; however, future systems may use gas with heating values below 100 Btu/cu ft (3728 kJ/m^3). Although wide ranges of heating values can be accommodated with different fuel systems, the maximum variation that can be used in a given fuel system is ±10%.

Sulfur content must be controlled in units with exhaust recovery systems. If sulfur condenses in the exhaust stack, corrosion can result. In units without exhaust recovery there is no problem since stack temperatures are considerably

Table 10-3. Gaseous Fuel Specifications

Heating value	300–1,200 Btu/cu ft (11,184–41,000 kJ/m^3)
Solid contaminants	<30 ppm
Flammability limits	2.2:1
Composition S, Na, K, Li	<5 ppm
(Sulfur + sodium + potassium + lithium)	(When formed into alkali metasulfate)
H$_2$O (by weight)	<25%

higher than the dew point. Sulfur can, however, promote hot-section corrosion in combustion with certain alkali metals such as sodium or potassium. This type of corrosion is sulfidation or hot corrosion and is controlled by limiting the intake of sulfur and alkali metals. Contaminants found in a gas depend on the particular gas. Common contaminants include tar, lamp black, coke, sand, and lube oil.

Table 10-4 is a summary of liquid fuel specifications set by manufacturers for efficient machine operations. The water and sediment limit is set at 1% by maximum volume to prevent fouling of the fuel system and obstruction of the fuel filters. Viscosity is limited to 20 centistokes at the fuel nozzles to prevent clogging of the fuel lines. Also, it is advisable that the pour point be 20°F (11°C) below the minimum ambient temperature. Failure to meet this specification can be corrected by heating the fuel lines. Carbon residue should be less than 1% by weight based on 100% of the sample. The hydrogen content is related to the smoking tendency of a fuel. Lower hydrogen content fuels emit more smoke than the higher hydrogen fuels. The sulfur standard is to protect from corrosion those systems with exhaust heat recovery.

The ash analysis receives special attention because of certain trace metals in the ash that cause corrosion. Elements of prime concern are vanadium, sodium, potassium, lead, and calcium. The first four are restricted because of their contribution to corrosion at elevated temperatures; however, all these elements may leave deposits on the blading.

Sodium and potassium are restricted because they react with sulfur at elevated temperatures to corrode metals by hot corrosion or sulfurization. The hot corrosion mechanism is not fully understood; however, it can be discussed in general terms. It is believed that the deposition of alkali sulfates (Na$_2$SO$_4$) on the blade reduces the protective oxide layer. Corrosion results from the continual forming and removing of the oxide layer. Also, oxidation of the blades occurs when liquid vanadium is deposited on the blade. Fortunately, lead is not encountered very often. Its presence is primarily from contamination by leaded fuel or as a result of some refinery practice. Presently, there is no fuel treatment to counteract the presence of lead.

Fuel Properties

Natural gas has a Btu content of about 1000 to 1100 Btu/cu ft (37,280 to 41,000 kJ/m^3). By definition, low Btu gases can vary between 100 and 350 Btu/cu ft

Fuels, Fuel Piping and Fuel Storage • 359

(13051 to 41000 kJ/m^3). Presently, little success has been achieved in burning gases with a heating value lower than 200 Btu/cu ft. To provide the same energy as natural gas, a 150 Btu /cu ft low Btu gas must be utilized at the rate of seven times that of natural gas on a volumetric basis. Therefore, the mass flow rate to provide the same energy must be about eight to ten times that of natural gas. The flammability of low Btu gases is very much dependent on the mixture of CH_4, and other inert gases. Figure 10-2 shows this effect by illustrating that a mixture of CH_4-CO_2 of less than 240 Btu/cu ft (8949 kJ/m^3) is inflammable, and a CH_4-N_2 mixture of less than about 150 Btu/cu ft (5594 kJ/m^3) is less inflammable. Low Btu gases near these values have greatly restricted flammability limits when compared to CH_4 in the air. Mixing superheated steam with oil produces vaporized fuel oil gas and then vaporizing the oil to provide a gas whose properties and heating value are close to natural gas.

Important liquid fuel properties for a gas turbine are shown in Table 10-5. The flash point is the temperature at which vapors begin combustion. The flash point is the maximum temperature at which a fuel can be handled safely.

The pour point is an indication of the lowest temperature at which a fuel oil can be stored and still be capable of flowing under gravitational forces. Fuels with higher pour points are permissible where the piping has been heated. Water and sediment in the fuel lead to fouling of the fuel system and obstruction in fuel filters.

The carbon residue is a measure of the carbon compounds left in a fuel after the volatile components have vaporized. Two different carbon residue tests are used,

Table 10-4. Liquid Fuel Specifications

Water and sediment	1.0% (V%) maximum
Viscosity	20 centistokes at fuel nozzle
Pour point	About 20°F (11°C) below minimum ambient
Carbon residue	1.0% (wt) based on 100% of sample
Hydrogen	11% (wt) minimum
Sulfur	1% (wt) maximum
Calcium	10 ppm
Lead	1 ppm
Sodium & Potassium	1
Vanadium	0.5 ppm

Typical Ash Analysis and Specifications

	LEAD	Calcium	Sodium & Potassium	Vanadium
Spec. max. (ppm)	1	10	1	0.5
Naphtha	0–1	0–1	0–1	0–1
Kerosene	0–1	0–1	0–1	0–1
Light distillate	0–1	0–1	0–1	0–1
Heavy distillate (true)	0–1	0–1	0–1	0–1
Heavy distillate (blend)	0–1	0–5	0–20	0.1/80
Residual	0–1	0–20	0–100	5/400
Crude	0–1	0–20	0–122	0.1/80

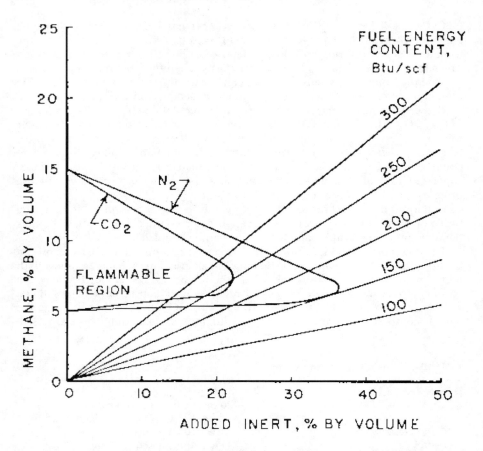

Figure 10-2. Flammable Fuel Mixtures of CH_4-N_2 and CH_4-CO_2 at One Atmosphere Showing Various Energy Levels

one for light distillates, and one for heavier fuels. For the light fuels, 90% of the fuel is vaporized, and the carbon residue is found in the remaining 10%. For heavier fuels, since the carbon residue is large, 100% of the sample can be used. These tests give a rough approximation of the tendency to form carbon deposits in the combustion system. The metallic compounds present in the ash are related to the corrosion properties of the fuel.

Viscosity is a measure of the resistance to flow and is important in the design of fuel pumping systems.

Specific gravity is the weight of the fuel in relation to water. This property is important in the design of centrifugal fuel washing systems. Sulfur content is important in connection with emission concerns and in connection with the alkali metals present in the ash. Sulfur reacting with alkali metals forms compounds that corrode by a process labeled sulfidation.

Table 10-5. Fuel Properties

	Kerosene	Diesel Fuel Burner Fuel			High-Ash Crude Heavy Residual	Typical Libyan Crude	Navy Distillate	Heavy Distillate	Low-Ash Crude
		#2	Oil #2	JP-4					
Flash point °F	130/160	118-220	150/200		175/265		186°F	198	50/200
Pour point °F	−50	−55 to +10	−10/30		15/95	68	10°F		15/110
Visc. CS@100°F	1.4/2.2	2.48/2.67	2.0/4.0	.79	100/1800	7.3	6.11	6.20	2/100
SSU		34.4					45.9		
Sulfur %	.01/.1	.169/.243	.1/.8	.047	.5/4	.15	1.01	1.075	.1/2.7
API gr.		38.1	35.0	53.2			30.5		
Sp. gr. @100°F	.78/.83	.85	.82.88	.7543@60°F	.92/1.05	.84	.874	.8786	.80/.92
Water & ded.			0			.1%wt			
Heating value Btu/lb	19300/19700	18330	19000/19600	18700/18820	18300/18900	18250		18239	19000/19400
Hydrogen %	12.8/14.5	12.83	12/13.2	14.75	10/12.5			12.40	12/13.2
Carbon residue									
10% bottoms	.01/.1	.104	.03/.3			2/10			.3/3
Ash ppm	1/5	.001	0/20		100/1000	36ppm			20/200
Na+K ppm	01.5		0/.1		1/350	2.2/4.5			0/50
V ppm	0/.1		0/.1		5/400	0/1			0/15
Pb ppm	0/.5		0/.1		0/25				
Ca ppm	0/1	0/2	0/2		0/50				

Luminosity is the amount of chemical energy in the fuel that is released as thermal radiation.

Finally, the weight of a fuel, light or heavy, refers to volatility. The most volatile fuels vaporize easily and come out early in the distillation process. Heavy distillates will come out later in the process. What remains after distillation is referred to as residual. The ash content of residual fuels is high.

Catastrophic oxidation requires the presence of Na_2SO_4 and Mo, W, and/or V. Crude oils are high in V; ash will be 65% V_2O_5 or higher. The rate at which corrosion proceeds is related to temperature. At temperatures of more than 1500°F (815.5°C), attack by sulfidation takes place rapidly. At lower temperatures with vanadium-rich fuels, oxidation catalyzed by vanadium pentoxide can exceed sulfidation. The effect of temperature on IN 718 corrosion by sodium and vanadium is shown in Figure 10-3. The corrosive threshold is generally accepted to be in the range of 1100°F to 1200°F (593.3°C to 648.9°C), and this cannot be considered a feasible firing temperature due to losses in efficiency and power output. Figure 10-4 shows the effect of sodium plus potassium and vanadium on life. Allowable limits for 100%, 50%, 20%, and 10% of normal life with uncontaminated fuel at standard firing temperatures are shown.

Natural gas requires no fuel treatment; however, low Btu gas, especially if derived from various coal gasification processes, requires various types of cleaners for use in a gas turbine. These cycles can get very complex as indicated by a typical system, which utilizes a steam bottoming cycle to achieve high efficiency. Vaporized fuel oil gas is already cleansed of its impurities in the vaporization process.

A corrosion inhibiting fuel treatment has been developed for the use of lower grade liquid fuels. Sodium, potassium, and calcium compounds are most often present in fuel in the form of seawater. These compounds result from salty wells and transportation over seawater, or they can be ingested by the compressor in mist form in ocean environments. Methods developed to remove the salt and reduce the sodium, potassium, and calcium relies on the water-solubility of these compounds. Removal of these compounds through their water-solubility is known as fuel washing. A prerequisite for the reduction of sodium content in crude fuel oil is that the oil is washable. A definition of washable fuel is:

- The fuel oil contains no chemically bound sodium.
- The fuel oil does not form an emulsion that can not be broken by the system, and furthermore permits a substantial water separation.

To prevent the formation of an emulsion during the fuel purification and to aid in the separation process an oil-conditioning agent (demulsifier) is added to the fuel. A "Demulsifier" is a chemical agent used to prevent emulsion between water particles and oil in effort to increase separation efficiency. This agent is injected at a rate of approximately 100–200 PPM, based on the sodium and water content and the viscosity of the fuel. Water injection sometimes is necessary when the fuel to be centrifuged has very low quantities of free water or has a high salt content. The water soluble trace metals may not be removed by centrifuging if there is not enough free water for them to bind with. By injecting water up to 5.3 cu.ft/h

Fuels, Fuel Piping and Fuel Storage • 363

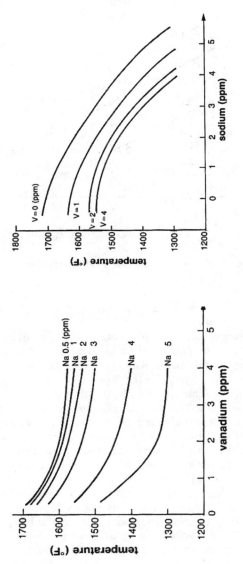

Figure 10-3. The Effect of Temperature on IN 718 Corrosion by Sodium and Vanadium

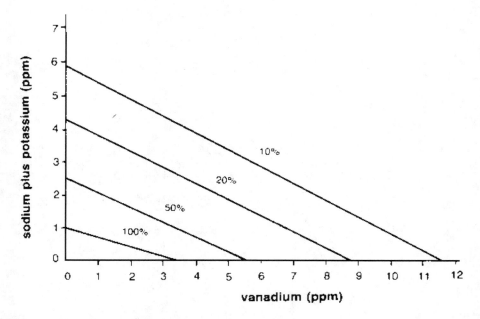

Figure 10-4. Effect of Sodium, Potasssium, and Vanadium on Combustor Life

(0.15 cu.m/h) maximum, the water soluble trace metals can be reduced to acceptable levels by centrifuging.

Fuel washing systems fall into four categories: centrifugal, dc electric, ac electric, and hybrids. The centrifugal fuel cleaning process consists of mixing 5–10% water with the oil plus an emulsion breaker to aid the separation of water and oil. Figure 10-5 shows a typical fuel treatment system. The untreated fuel oil is pumped using a centrifugal pump through a suction strainer, which removes any coarse solids from the fuel. The fuel flow is controlled by modulating the flow control valve in response to signals from the treated crude oil tank. A demulsifier is injected continuously into the oil ahead of the plate heat exchanger through a chemical metering pump in a static in-line mixer. The demulsifier is added to counteract the naturally occurring emulsifying species found in the oil and aid separation of the original water from the oil.

Normal processing temperature is 130°F–140°F (55°C–60°C) and the oil is brought up to this temperature in various stages. The oil is first pre-heated indirectly in the regenerative plate heat exchanger by the hot purified oil leaving the plant. Final heating is carried out by an electric heater. When processing temperature is reached the untreated crude oil is forwarded to the separation modules, which includes the centrifugal separators. Untreated crude oil is fed continuously through the shut-off valve and into the separator bowl where any water and solids are separated from the crude oil by the action of centrifugal force.

Purified crude oil and separated water are discharged continuously through their own outlets and the solids accumulate at the periphery of the bowl. The solids

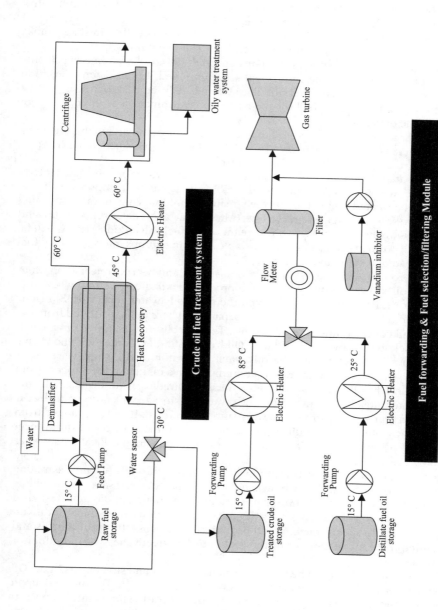

Figure 10-5. A Typical Fuel Treatment System

are discharged periodically before they build up to a point where they would interfere with the separation process. The discharge cycle is initiated at the control panel by either a push button or automatically by the electronic program control system on completion of a set time cycle.

Water and sludge removed by the centrifugal separators collect in sludge tanks and are transferred to a central storage tank by a sludge pump.

An oil and water monitor in the purified crude oil discharge line checks the quality of the oil leaving the centrifugal separators and if there is any deviation from the set-point an alarm is set-off. The system usually consists of a sensor and a monitoring unit. It is designed for the accurate determination of the quantity of water in the oil stream.

Treated fuel oil (Crude) is stored in a treated fuel day tank and maintained at a predetermined temperature below 170°F (77°C). Fuel oil is pumped to the GT valve skid under the control of the Turbine Control System.

The treated fuel will contain some oil soluble trace metals, particularly Vanadium, which is deleterious to the GT hot gas path components if not treated. The treated fuel will be further treated to inhibit Vanadium by injecting a 10% solution of Magnesium Sulfate ($MgSO_4$) at a predetermined ratio and thoroughly mixing it with the fuel upstream of the fuel valve skid. $MgSO_4$ will be pumped via positive displacement, adjustable metering pumps into the treated fuel piping ahead of a static in-line mixer.

Sludge from the fuel wash skid (containing solids, salt bearing reject wash water and some oil) flows by gravity in a batch process to a dedicated sump where it is collected for further processing through the plant oily/water separator. Sludge is pumped from the sump to the oily/water separator where oil is separated from the influent and collected into storage drums. The oil collected is pumped in a batch process back to the raw crude oil tank. Oil free wastewater (water with no visible sheen) flows by gravity from the oily/water separator to the plant wastewater disposal system for discharge. Solids are removed in a manual batch process from the sump and the oily/water separator on an as-required basis for disposal off-site.

Free water must be decanted from the raw crude storage tank automatically (by others) and gravity flows to a wastewater sump (by others). It will be pumped from the sump (by others) to the oily/water separator to remove traces of oil to meet local water discharge criteria before being discharged to the local wastewater receptor.

Fuel chemistry must be continuously monitored through a batch sampling procedure to ensue that the turbine is not contaminated. Grab samples should be gathered from each storage tank periodically throughout the day and tested to determine the level of trace metals, salts and concentrations of chemical additives. An Atomic Photo-spectrometer could be utilized to process the grab samples and determine trace metal content of the fuel and the effectiveness of the inhibitor chemicals.

The GT when fired using heavy fuels usually will start and stop on No 2 Fuel Oil (LFO) to assure that the heavy fuel oil does not coke in the fuel nozzles upon shutdown and that the fuel delivery piping is up to operating temperature on start-up before introducing treated fuel oil. In the case of some light crudes such as the Saudi Crudes starting and stopping has been successful using these crudes for start-up and shutdown. However, in most cases it is recommended that start-up should be with diesel fuel. Typically the GT will operate approximately 30 minutes on LFO during start-up and a similar period of time on shutdown. The valve skid as shown in Figure 10-5 handles the transfer of both fuels as directed by the Turbine Control requirements. The valve skid is located in close proximity to the GT fuel inlet flange and contains the emergency shutoff valve and fuel management instrumentation and control sensors together with a duplex fuel filter. This fuel filter is upstream of the final fuel filter on the GT base skid and serves to assure no foreign particles in the fuel effect the operation of the GT combustion system.

Diesel fuel (LFO), used for starting and stopping the GT, is usually pumped from the LFO storage tank to the valve skid as required. If there are any suspicions that the fuel may have been contaminated with brackish water the diesel fuel should also be treated and the fuel should be sent to a centrifuge before it is injected into the gas turbine, so as to remove the sodium. The Turbine Control will determine the fuel to be delivered to the GT and select the appropriate valve positions. The crude oil treatment plant is controlled, monitored and supervised from its own centralized control console, which is connected to the turbine control system.

In the case of Heavy Fuels where the specific gravity of the fuel is above .96 or viscosity exceeds 3500 SSU @ 100°F (37.8°C), centrifugal separation is impractical, and the specific gravity of one of the components must be increased. Dissolving epsom salt in it can increase water weight. Fuel blending can decrease specific gravity of the fuel. Figure 10-6 shows the relation is linear, and the blend has a specific gravity, which is the average of the constituents. However, viscosity

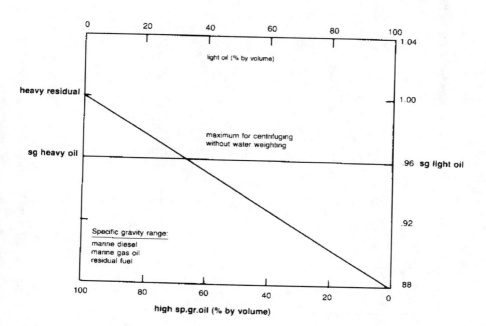

Figure 10-6. Fuel Blending for Specific Gravity Reduction. Specific Gravity Heavy Oil = 1.0. Specific Gravity Light Oil = .88.

368 • COGENERATION AND COMBINED CYCLE POWER PLANTS

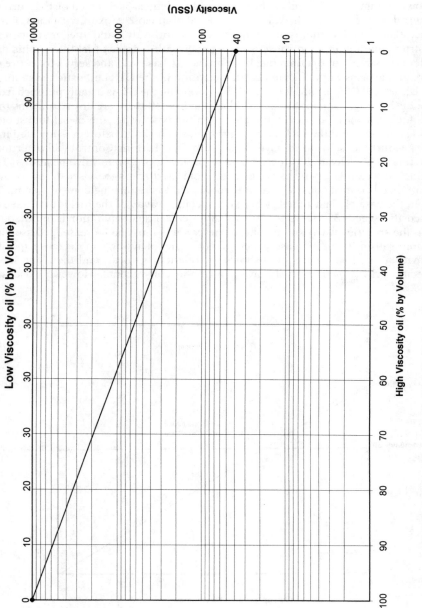

Figure 10-7. Fuel Viscosity Blending Chart. High-Viscosity Oil = 10,000 SSU. Low-Viscosity Oil = 40 SSU

blending is a logarithmic relation as shown in Figure 10-7. To reduce viscosity from 10,000 to 3000 SSU, a 3:1 reduction requires a dilution of only 1:10.

Electrostatic separators operate on a principle similar to centrifugal separation. The salt is first dissolved in the water, and the water is then separated. Electrostatic separators utilize an electric field to coalesce droplets of water for an increase in diameter and an associated increased settling rate. The DC separators are most efficient with light fuels of low conductivity, and AC separators are used with heavier, highly conductive fuels. Electrostatic separators are attractive because of safety considerations (no rotating machinery) and maintenance (few overhauls). However, sludge removal is more difficult. Water washing systems are summarized in Table 10-6.

Vanadium originates as a metallic compound in crude oil and is concentrated by the distillation process into the heavy-oil fractions. Blade oxidation occurs when liquid vanadium is deposited onto a blade and acts as a catalyst. Vanadium compounds are oil-soluble and are thus unaffected by fuel washing. Without additives, vanadium forms low melting temperature compounds which deposit on a blade in a molten slag state that causes rapid corrosion. However, by the addition of a suitable compound (magnesium, for example), the melting point of the vanadates is increased sufficiently to prevent them from being in the liquid state under service conditions. Thus, slag deposition on the blades is avoided. Calcium was initially selected as the inhibiting agent, as tests indicated it was more effective at 1750°F (954.4°C). Subsequent tests showed magnesium gave better protection at 1650°F (898.9°C) and below. However, at temperatures of 1750°F (954.4°C) and over, magnesium no longer inhibits but rather *accelerates* corrosion. Magnesium also provides more friable deposits than calcium inhibitors. A magnesium/vanadium ratio of 3:1 reduces corrosion by a factor of six between temperatures of 1550°F (843.3°C) and 1400°F (760°C).

The particular magnesium compound selected for inhibition is dependent upon fuel characteristics. For low vanadium concentrations (below 50 ppm), an oil-soluble compound such as magnesium sulfate is added in the correct proportion to the vanadium present. The cost of oil soluble inhibitors becomes prohibitive above concentrations of 50 ppm.

At higher concentrations of vanadium, magnesium sulfate or magnesium oxide is used as an inhibitor. Both are approximately equal in material cost, but magnesium sulfate has proven itself, while magnesium oxide is still under study. Magnesium sulfate requires by far the most capital cost, as it must be first dissolved, then adjusted to a known concentration. It is mixed with oil and an

Table 10-6. Selection of Fuel Washing Systems

Fuel	Washing System
Distillate	Centrifugal or DC electrostatic desalter
Heavy distillates	Centrifugal or AC electrostatic desalter
Light-medium crudes	Centrifugal or AC electrostatic desalter
Light residual	Centrifugal or AC electrostatic desalter
Heavy crudes	Centrifugal desalter and hybrid systems
Heavy residuals	Centrifugal desalter and hybrid systems

emulsifying agent to form an emulsion to suspend in the fuel. Two different injection procedures are used. One method is to mix the solution with desalted fuel in a dispersion mixer just prior to the combustion chamber. The inhibited oil is burned quickly, usually within a minute after mixing, because the solution has a tendency to settle out. Also, the solution can be dispersed in the fuel prior to the service tanks. To avoid settling out of the solution, the tanks are re-circulated through distribution headers. Since a magnesium-to-vanadium ratio of 3.25±0.25:1 is used in practice, the second dispersion method is the standard practice as the tanks can be certified "within specification" before burning. Adequate knowledge of contaminants is essential for successful inhibition.

Heavy Fuels

With heavy fuels, the ambient temperature and the fuel type must be considered. Even at warm environmental temperatures, the high viscosity of the residual could require fuel preheating or blending. If the unit is planned for operation in extremely cold regions, the heavier distillates could become too viscous. Fuel system requirements limit viscosity to 20 centistokes at the fuel nozzles.

The start-up and shut down of the gas turbine for heavy fuels is conducted using distillate oil. The problems usually occur when there is an unplanned shut down. In that case the turbine is shutdown on the heavy fuel, and so does not have the opportunity to cleanse the fuel feed system at the turbine with distillate fuel. The heavy fuel, which is in the fuel feed system, solidifies and blocks the fuel system. This in most cases requires the total dismantling of the fuel system to cleanse it of the solidified fuel.

Fuel system fouling is related to the amount of water and sediment in the fuel. A by-product of fuel washing is the de-sludging of the fuel. Washing rids the fuel of those undesirable constituents that cause clogging, deposition, and corrosion in the fuel system. The last part of treatment is filtration just prior to entering the turbine. Washed fuel should have less than .025% bottom sediment and water.

The combustion of liquid fuel oil in a gas turbine requires an air atomization system. When liquid fuel oil is sprayed into the turbine combustion chambers, it forms large droplets as it leaves the fuel nozzles. The droplets will not burn completely in the chambers and many could go out of the exhaust stack in this state. A low pressure atomizing air system is used to provide atomizing air through supplementary orifices in the fuel nozzle which directs the air to impinge upon the fuel jet discharging from each nozzle. This stream of atomizing air breaks the fuel jet up into a fine mist, permitting ignition and combustion with significantly increased efficiency and a decrease of combustion particles discharging through the exhaust into the atmosphere. It is necessary, therefore, that the air atomizing system be operative from the time of ignition firing through acceleration, and through operation of the turbine.

Atomizing air systems provide sufficient pressure in the air atomizing chamber of the fuel nozzle body to maintain the ratio of atomizing air pressure to compressor discharge pressure at approximately 1.4 or greater over the full operating range of the turbine.

In most cases the atomization air is bled off the compressor discharge of the gas turbine and then is sent through a shell tube heat exchanger, where it is cooled, with minimum pressure drop. The discharge air conditions leaving the Gas Turbine depends on the pressure ratio of the gas turbine compressor. The air leaving in most cases is at a pressure of about 150 psia–300 psia (10–20 bar) and a temperature of 750°F–950°F (400°C–480°C). This air bled from the gas turbine is cooled in a shell and tube cooler to about 250°F (120°C) so that the power required to compress that gas is significantly lowered. The compressor used to compress the bleed air is usually a centrifugal compressor, which is operated of the auxiliary gear box of the turbine or a stand alone electric motor. This compressor usually operates at a very high speed, between 40,000–60,000 rpm. The compressor will delivers a flow of about 120%–150% of the fuel flow and at a pressure ratio of about at 1.4 to 1.7 of the compressor discharge pressure depending on the viscosity of the fuel. Since the output of the main atomizing air compressor, driven by the accessory gear, is low at turbine firing speed, a starting atomizing air compressor provides a similar pressure ratio during the firing and warm-up period of the starting cycle, and during a portion of the accelerating cycle.

Major system components include: the main atomizing air compressor, starting atomizing air compressor and atomizing air heat exchanger. Figure 10-8 is a schematic diagram of the system. Air bled from the gas turbine compressor using the atomizing air extraction manifold of the compressor

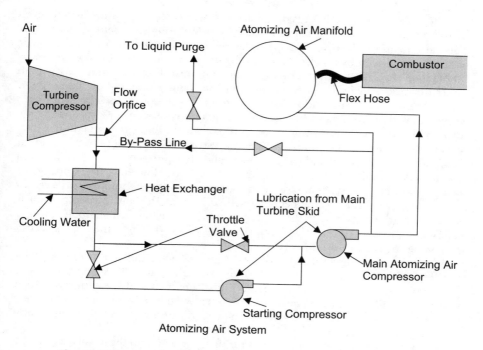

Figure 10-8. Atomization Air System for Liquid Fuel Oil Systems

discharge casing passes through the air-to water heat exchanger (precooler) so as to reduce the temperature of the air. This would maintain a uniform air inlet temperature to the atomizing air compressor, and reduces the power. An alarm is suggested if the temperature of the air from the atomizing air pre-cooler entering the main atomizing air compressor is excessive. Improper control of the temperature will be due to failure of the sensor, the pre-cooler or insufficient cooling water flow. Continued operation above 285°F (140°C) should not be permitted for any significant length of time since it may result in failure of the main atomizing air compressor or in insufficient atomizing air to provide proper combustion.

Compressor discharge air, now cleaned and cooled, reaches the main atomizing air compressor. This is a single-stage, flange mounted, centrifugal compressor driven by an inboard shaft of the turbine accessory gear. It contains a single impeller mounted on the pinion shaft of the integral input speed increasing gear box driven directly the accessory gear. Output of the main compressor provides sufficient air for atomizing and combustion when the turbine is at approximately 70% speed.

Differential pressure switch, located in a bypass around the compressor, monitors the air pressure and annunciates an alarm if the pressure rise across the compressor should drop to a level inadequate for proper atomization of the fuel.

Air, now identified as atomizing air, leaves the compressor and is piped to the atomizing air manifold with "pigtail" piping providing equal pressure distribution of atomizing air to the individual fuel nozzles.

When the turbine is first fired, the accessory gear is not rotating at full speed and the main atomizing air compressor is not outputting sufficient air for proper fuel atomization. During this period, the starting (booster) atomizing air compressor, driven by an electric motor, is in operation supplying the necessary atomizing air. The starting atomizing air compressor at this time has a high pressure ratio and is discharging through the main atomizing air compressor which has a low pressure ratio. The main atomizing air compressor pressure ratio increases with increasing turbine speed, and at approximately 70% speed the slow demand of the main atomizing air compressor approximates the maximum flow capability of the starting atomizing air compressor. The check valve in the air input line to the main compressor begins to open, allowing air to be supplied to the main compressor simultaneously from both the main air line and the starting air compressor. The pressure ratio of the starting atomizing air compressor decreases to one, and when the turbine becomes self-sustaining, the starting compressor is shut down at approximately 95% speed. Now all of the air being supplied to the main compressor is directly from the pre-cooler through the check valve, bypassing the starting air compressor completely.

When water washing the gas turbine's compressor section and turbine section, it is important to keep water out of the atomizing air system. To keep water out of the atomizing air system, the system inlet and discharge should be equipped with a vent valves and an isolation valve. The vent valves are used to avoid completely throttling or deadening the compressors and the drain valves to remove any leakage past the isolation valve.

During normal operation of the gas turbine, the vent and drain valves must be closed and the isolation valve must be open. Before initiating water wash the

Fuels, Fuel Piping and Fuel Storage • 373

isolation valve must be closed to keep water out of the atomizing air system. The vent valve must be opened to allow air to pass through the atomizing air compressors.

Running the atomizing air compressors completely throttled or deadened may result in over heating and damage to the compressor.

At the conclusion of water wash, drain water from the atomizing air piping opening the low point drain valves. When the water has drained from the piping, the drain valves and vent valves must be closed.

Running the atomizing air compressors with water in the piping will result in damage to the atomizing air compressors.

Frequently, no visible smoke and no carbon deposition are design parameters. Smoke is an environmental concern, while excessive carbon can impair the fuel spray quality and cause higher liner temperatures due to the increased radiation emissivity of the carbon particles as compared to the surrounding gases. Smoke and carbon are a fuel related property. The hydrogen saturation influences smoke and free carbon. The less saturated fuels like benzene (C_6H_6) tend to be smokers; the better fuels like methane (CH_4) are saturated hydrocarbons. This effect is shown in Figure 10-9. Boiling temperature is a function of molecular weight. Heavier molecules tend to boil at a higher temperature. Since a less saturated molecule will weigh more (higher molecular weight), one can expect residuals and heavy distillates to be smokers. This expectation is founded in practice. The design

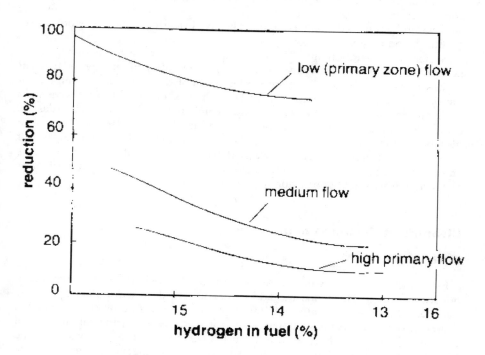

Figure 10-9. Effect of Hydrogen Saturation in Primary Flow on Smoke

374 • COGENERATION AND COMBINED CYCLE POWER PLANTS

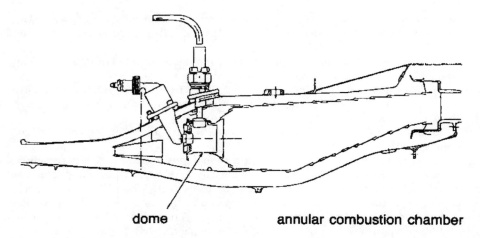

Figure 10-10. Cross Section of an Annular Combustor Showing High Dome Flow Configuration (Courtesy of General Electric Company)

solution pioneered by General Electric on its LM 2500, which has an annular combustor as shown in Figure 10-10, was to increase flow and swirl through the dome surrounding the fuel injector. The increased flow helped to avoid rich pockets and promoted good mixing. The axial swirler achieved a no smoke condition and reduced liner temperature.

Special consideration must be afforded to the combustion chamber walls. Low-grade fuels tend to release a higher amount of their energy as thermal radiation instead of heat. This energy release, coupled with the large diameter of the single can and the formation of carbon deposits, can lead to an overheating problem on the liner. One vendor advocates the use of metallic tiles as combustor liners. The tiles hook into the wall in slots provided for them. The tiles have fine pitched fins cast on the back. The fins form a double wall structure by bridging the gap between the flame-tube wall and the tile. This annulus is fed by air, thus providing a strong cooling action. The standard sheet metal design was abandoned due to warpage.

Cleaning of Turbine Components

Turbine Wash

A fuel treatment system will effectively eliminate corrosion as a major problem, but the ash in the fuel plus the added magnesium does cause deposits in the turbine. Intermittent operation of 100 hours or less offers no problem, since the character of the deposit is such that most of it sheds upon refiring, and no special cleaning is required. However, the deposit does not reach a steady-state value with continuous operation and gradually plugs the first stage nozzle area at a rate of between 5% and 12% per 100 hours. Thus, at present, residual oil use is limited to applications where continuous operation of more than 1000 hours is not required.

If the need exists to increase running time between shutdowns, the turbine can be cleaned by the injection of a mild abrasive into the combustion system. Abrasives include walnut shells, rice, and spent catalyst. Rice is a very poor abrasive, since it tends to shatter into small pieces. Usually, a 10% maximum blockage of the first-stage nozzle is tolerated before abrasive cleaning is initiated. Abrasive cleaning will restore 20% to 40% of the lost power by removing 50% of the deposits. If the frequency of abrasive injection becomes unacceptable and cannot prevent the nozzle blockage from becoming more than 10%, water washing becomes necessary. Water or solvent washing can effectively restore 100% of the lost power. A typical operating plot is shown in Figure 10-11.

The water washing of the hot section of the turbine is required for fuels with high vanadium contents. The addition of magnesium salts to encounter the corrosive action of the vanadium creates ash, which deposits on the blades reducing the flow area. This ash must be removed and in many cases this means that the hot section blades and nozzles must be washed every 100 to 120 hours. This is done by bringing the turbine down and running it on turning gear till the turbine blade temperatures are around 200°F (93.3°C), in most cases this is reached in about 6 to 8 hours). The turbine hot expander section is then blasted by steam and most of the ash is removed. The turbine is then brought up to speed after the turbine blade section is dried. This whole process takes about 20 hours.

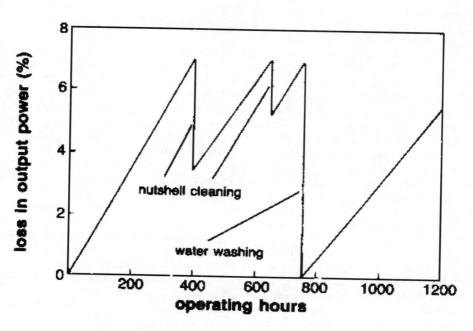

Figure 10-11. Effect of Cleaning on Power Output

Compressor Washing

Compressor washing is also a very important part of turbine operations. Two approaches to compressor cleaning are abrasion and solvent cleaning. The use of abrasive cleaning has diminished due to erosion problems; liquid washing is primarily used. The new high pressure compressor is very susceptible to dirt on the blades, which not only can lead to a reduction in performance but can also lead to compressor surge. Washing efficacy is site-specific due to the different environmental conditions at each plant. There are many excellent techniques and systems for water washing. Operators must often determine the best approach for their gas turbines. This includes what solvents, if any should be used, and the frequencies of wash. Many operators have found that water wash without any solvent is as effective as with the use of solvents. This is a complex technical-economical problem also depending on the service that the gas turbines are in and the plant surroundings. However, the use of non-demineralized water could result in more harm than good.

Water washing (with or without detergents) cleans by water impact and by removing the water-soluble salts. The effect of water cleaning is usually not very effective after the first few stages. It is most important that the manufacturer's recommendations be followed with respect to water wash quality, detergent/water ratio and other operating procedures. Water washing using a water-soap mixture is an efficient method of cleaning. This cleaning is most effective when carried out in several steps, which involve the application of a soap and water solution, followed by several rinse cycles. Each rinse cycle involves the acceleration of the machine to approximately 50% of the starting speed, after which the machine is allowed to coast to a stop. A soaking period follows, during which the soapy water solution may work on dissolving the salt.

A fraction of airborne salt always passes through the filter. The method recommended for determining whether or not the foulants have a substantial salt base is to soap wash the turbine and collect the water from all drainage ports available. Dissolved salts in the water can then be analyzed.

On-line washing is being widely used as a means to control fouling by keeping the problem from developing. Techniques and wash systems have evolved to a point where this can be done effectively and safely. Washing can be accomplished by using water, water-based solvents, petroleum-based solvents or surfactants. The solvents work by dissolving the contaminants while surfactants work by chemically reacting with the foulants. Water-based solvents are effective against salt, but fare poorly against oily deposits. Petroleum-based solvents do not effectively remove salty deposits. With solvents, there is a chance of foulants being re-deposited in the latter compressor stages.

Even with good filtration, salt can collect in the compressor section. During the collection process of both salt and other foulants, an equilibrium condition is quickly reached, after which re-ingestion of large particles occurs. This re-ingestion has to be prevented by the removal of salt from the compressor prior to saturation. The rate at which saturation occurs is highly dependent on filter quality. In general, salts can safely pass through the turbine when gas and metal temperatures are less than 1000°F. Aggressive attacks will occur if the temperatures are much higher. During cleaning, the actual instantaneous rates of salt passage are very high together with greatly increased particle size.

The following are some tips that should be followed by operators during water washes:

- The water used should be demineralized. The use of non-demineralized water would harm the turbine.
- On-line wash should be done whenever compressor performance diminishes by 2% to 3%. It would be imprudent to let foulants build up before commencing water wash.
- Stainless steel for tanks, nozzles and manifolds are recommended to reduce corrosion problems.
- Spray nozzles should be placed where proper misting of the water would occur, and minimize the downstream disturbance of the flow. Care should be taken that a nozzle would not vibrate loosely and enter the flow passage.
- After numerous water washes, the compressor performance will deteriorate and a crank wash will be necessary.

Fuel Economics

Because gas turbine fuel properties are not the ones that determine cost, in some instances the better gas turbine fuel will sell for less than the poorer one. The selection of the most economical fuel depends on many considerations, of which fuel cost is but one. However, users should always burn the most *economical* fuel, which may not be the cheapest fuel.

Fuel properties must be known and economics considered before a fuel is selected. The properties of a fuel greatly affect the cost of a fuel-treatment facility. A doubling of viscosity roughly doubles the cost of desalting equipment, and having a specific gravity of greater than 0.96 greatly complicates the washing system and raises costs. Trying to remove the last trace of a metallic element affects the cost of fuel washing approximately as shown in Table 10-7. The high cost of fuel-treatment systems is the fuel washing system, since the ignition system costs about 10% of that amount. The fuel flow rate as well as the fuel type affects the fuel-treatment system investment cost as shown in Figure 10-12.

Gas turbines, like other mechanical devices, require inspection, maintenance, and service. Maintenance costs include the combustion system, hot gas path, and major inspections. The effect of fuel type on maintenance costs is shown in

Table 10-7. Effect of Washed Fuel Quality on System Cost

Sodium Reduction	Washing System Cost
100 →5 ppm Na	X dollars
100 →2 ppm Na	2 X dollars
100 →1 ppm Na	4 X dollars
100 →1/2 ppm Na	8 X dollars

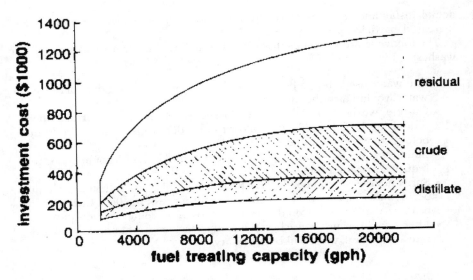

Figure 10-12. Gas Turbine Fuel Treatment Plant Investment Costs (Courtesy of General Electric Company)

Table 10-8. A cost factor is shown using natural gas as unity. The cost of maintenance is subject to great variations. Recognizing the great difficulty in establishing expected maintenance costs for different applications, Table 10-8 should be used as a rough guide in estimating costs. This data is based on actual maintenance costs for heavy-duty gas turbines.

As has been shown, the selection of the most economical fuel can depend on many factors besides cost. Table 10-9 summarizes the major economic considerations in fuel selection.

Heat Tracing of Piping Systems

As mentioned earlier, heavy fuels need to be kept at a temperature where the viscosity of the fuel is limited to 20 centistokes at the fuel nozzles. Heat tracing is

Table 10-8. Maintenance Cost of Various Types of Fuel

Fuel	Maintenance Cost, ¢/kWh	Maintenance Cost Factor
Natural Gas	0.3	1
No. 2 Distillate Oil	0.4	1.25
Crude Oil	0.6	2
No. 6 Residual Oil	1	3.3

Table 10-9. Economic Factors Influencing Fuel Selection

I. Fuel Cost
II. Operation
   1. Power output for given Turbine
   2. Efficiency degradation
   3. Outage (Downtime)
III. Capital Investment
   1. Fuel Washing and Inhibition
   2. Fuel Quality monitoring
   3. Turbine Wash and Cleaning
IV. Duty Cycle
   1. Continuous Duty required
   2. Total Annual Operation
   3. Starts and Stops

used to maintain pipes and the material that pipes contain at temperatures above the ambient temperature. Two common uses of heat tracing are preventing water pipes from freezing and maintaining fuel oil pipes at high enough temperatures such that the viscosity of the fuel oil will allow easy pumping. Heat tracing is also used to prevent the condensation of a liquid from a gas.

A heat-tracing system is often more expensive on an installed cost basis than the piping system it is protecting, and it will also have significant operation costs. A recent study on heat-tracing costs showed installed costs of US$31/ft ($95/m) to $142/ft ($430/m) and yearly operating cost of $1.40/ft (4.35/m) to $16.66/ft ($50/m). In addition to being a major cost, the heat-tracing system is an important component of the reliability of a piping system. A failure in the heat-tracing system will often render the piping system inoperable. For example, with a water freeze protection system, the piping system may be destroyed by the expansion of water as it freezes if the heat-tracing system fails.

The vast majority of heat-traced pipes are insulated to minimize heat loss to the environment. A heat input of 2 to 10 W/ft (6 to 30 W/m) is generally required to prevent an insulated pipe from freezing. With high wind speeds, an uninsulated pipe could require well over 100 W/ft (300 W/m) to prevent freezing. Such a high heat input would be very expensive.

Heat tracing for insulated pipes is generally only required for the period when the material in the pipe is not flowing. The heat loss of an insulated pipe is very small compared to the heat capacity of a flowing fluid. Unless the pipe is extremely long (several thousands of feet or meters), the temperature drop of a flowing fluid will not be significant.

The three major methods of avoiding heat tracing are:

1. Changing the ambient temperature around the pipe to a temperature that will avoid low-temperature problems. Burying water pipes below the frost line or running them through a heated building are the two most common examples of this method.
2. Emptying a pipe after it is used. Arranging the piping such that it drains itself when not in use, can be an effective method of avoiding the need for

heat tracing. Some infrequently used lines can be pigged or blown out with compressed air. This technique is not recommended for commonly used lines due to the high labor requirement.
3. Arranging a process such that some lines have continuous flow can eliminate the need for tracing these lines. This technique is generally not recommended because a failure that causes a flow stoppage can lead to blocked or broken pipes.

Some combination of these techniques may be used to minimize the quantity of traced pipes. However, the majority of pipes containing fluids that must be kept above the minimum ambient temperature are generally going to require heat tracing.

Types of Heat-Tracing Systems

Industrial heat-tracing systems are generally fluid systems or electrical systems. In fluid systems, a pipe or tube called the tracer is attached to the pipe being traced, and a warm fluid is put through it. The tracer is placed under the insulation. Steam is by far the most common fluid used in the tracer, although ethylene glycol and more exotic heat-transfer fluids are used. In electrical systems, an electrical heating cable is placed against the pipe under the insulation.

Steam-Tracing Systems. Steam tracing is the most common type of industrial pipe tracing. In 1960, over 95% of industrial tracing systems were steam-traced. By 1995, improvements in electric heating technology increased the electric share to 30% to 40%, but steam tracing is still the most common system. Fluid systems other than steam are rather uncommon, and account for less than 5% of tracing systems.

Half-inch (12.7 mm) copper tubing is commonly used for steam tracing. Three-eighths-inch (9.525 mm) tubing is also used, but the effective circuit length is then decreased from 150 ft (50 m) to about 60 ft (20 m). In some corrosive environments, stainless steel tubing is used, and occasionally standard carbon steel pipe (1/2 to 1 in.) is used as the tracer.

In addition to the tracer, a steam-tracing system as seen in Figure 10-13, consists of steam supply lines to transport steam from the existing steam lines to the traced pipe, a steam trap to remove the condensate and hold back the steam, and in most cases, a condensate return system to return the condensate to the existing condensate return system. In the past, a significant percentage of condensate from steam tracing was simply dumped to drains, but increased energy cost and environmental rules have caused almost all condensate from new steam-tracing systems to be returned. This has significantly increased the initial cost of steam-tracing systems.

Applications requiring accurate temperature control are generally limited to electric tracing. For example, chocolate lines cannot be exposed to steam temperatures or the product will degrade, and if caustic soda is heated above 150°F (65.6°C), it becomes extremely corrosive to carbon steel pipes.

For some applications, either steam or electricity is simply not available and this makes for a decision. It is rarely economic to install a steam boiler just for

Fuels, Fuel Piping and Fuel Storage • 381

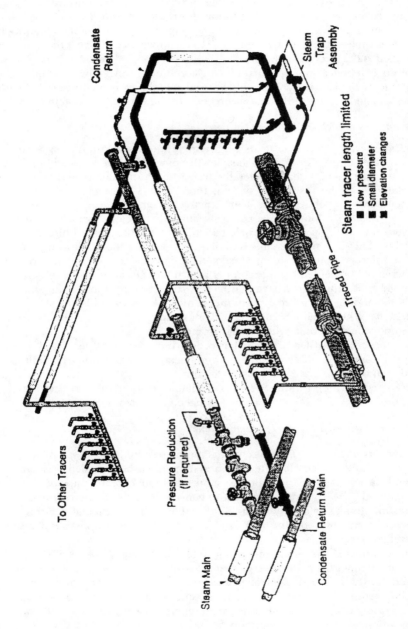

Figure 10-13. Steam Tracing System

tracing. Steam tracing is generally considered only when a boiler already exists or is going to be installed for some other primary purpose. Additional electric capacity can be provided in most situations for reasonable costs. It is considerably more expensive to supply steam from a long distance than it is to provide electricity. Unless steam is available close to the pipes being traced, the automatic choice is usually electric tracing.

For most applications, particularly in combined cycle power plants, either steam tracing or electric tracing could be used, and the correct choice is dependent on the installed costs and the operation costs of the competing systems.

Advantages of a Steam-Tracing System

1. *High heat output.* Due to its high temperature, a steam-tracing system provides a large amount of heat to the pipe. There is a very high heat transfer rate between the metallic tracer and a metallic pipe. Even with damage to the insulation system, there is very little chance of a low temperature failure with a steam-tracing system.
2. *High reliability.* Many things can go wrong with a steam-tracing system but very few of the potential problems lead to a heat-tracing failure. Steam traps fail, but they usually fail in the open position, allowing for a continuous flow of steam. Leaks that can cause wet insulation are generally prevented from becoming heat-tracing failures by the extremely high heat output of a steam tracer. Also, a tracing tube is capable of withstanding a large amount of mechanical abuse without failure.
3. *Safety.* While steam burns are fairly common, there are generally fewer safety concerns than with electric tracing.
4. *Common usage.* Steam tracing has been around for many years and many operators are familiar with the system. Because of this familiarity, failures due to operator error are not very common.
5. *Cost effective operation.* Steam is available in a combined cycle power plant especially at the low pressures needed for steam tracing.

Disadvantages of a Steam-Tracing System

1. *High installed costs.* The incremental piping required for the steam supply system and the condensate return system must be installed, insulated, and, in the case of the supply system, additional steam traps are often required. The tracer itself is not expensive, but the labor required for installation is relatively high. Studies have shown that steam-tracing systems typically cost from 50% to 150% more than a comparable electric-tracing system.
2. *Energy inefficiency.* A steam-tracing system's total energy use is often more than 20 times the actual energy requirement to keep the pipe at the desired temperature. The steam tracer itself puts out significantly more energy than required. The steam traps use energy even when they are properly operating and waste large amounts of energy when they fail in the open position, which is the common failure mode. Steam leaks waste large amounts of energy, and both the steam supply system and the condensate return system use significant amounts of energy.

3. *Poor temperature control.* A steam-tracing system offers very little temperature control capability. The steam is at a constant pressure and temperature of 50 psig (3.45 bar) and 300 F (148.9°C), respectively, usually well above that desired for the pipe. The pipe will reach an equilibrium temperature somewhere between the steam temperature and the ambient temperature. However, the section of pipe against the steam tracer will effectively be at the steam temperature. This is a serious problem for temperature-sensitive fluids such as food products. It also represents a problem with fluids such as bases and acids, which are not damaged by high temperatures but often, become extremely corrosive to piping systems at higher temperatures.
4. *High maintenance costs.* Leaks must be repaired and steam traps must be checked and replaced if they have failed. Numerous studies have shown that, due to the energy lost through leaks and failed steam traps, an extensive maintenance program is an excellent investment. Steam maintenance costs are so high that for low-temperature maintenance applications, total steam operating costs are sometimes greater than electric operating costs, even if no value is placed on the steam.

Electric Tracing. An electric-tracing system as seen in Figure 10-14 consists of an electric heater placed against the pipe under the thermal insulation, the supply of electricity to the tracer, and any control or monitoring system that may be used (optional). The supply of electricity to the tracer usually consists of an electrical panel and electrical conduit or cable trays. Depending on the size of the tracing system and the capacity of the existing electrical system, an additional transformer may be required.

Types of Electric Tracing. Self-Regulating Electric Tracing as seen in Figure 10-14 is by far the most popular type of electric tracing. The heating element in a

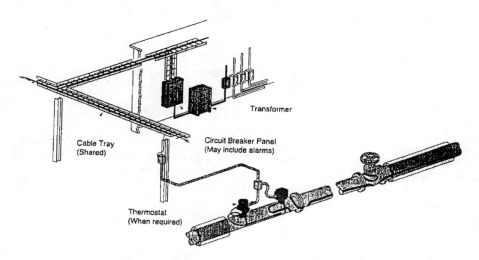

Figure 10-14. Electric Heat Tracing System

self-regulating heater is a conductive polymer between the bus wires. This conductive polymer increases its resistance as its temperature increases. The increase in resistance with temperature causes the heater to lower its heat output at any point where its temperature increases. This self-regulating effect eliminates the most common failure mode of constant wattage electric heaters, which is destruction of the heater by its own heat output.

Because self-regulating heaters are parallel heaters, they may be cut to length at any point without changing their power output per unit of length. This makes them much easier to deal with in the field. They may be terminated, teed, or spliced in the field with hazardous-area-approved components.

MI Cables. Mineral insulated cables are the electric heat tracers of choice for high-temperature applications. High-temperature applications are generally considered to maintain temperatures above 250°F (121.1°C) or exposure temperatures above 420°F (215.6°C), where self-regulating heaters cannot be used. MI cables consist of one or two heating wires, magnesium oxide insulation (from whence it gets its name), and an outer metal sheath. Today, the metal sheath is generally Inconel. This eliminates both the corrosion problems with copper sheaths and the stress cracking problems with stainless steel.

MI cables can maintain temperatures up to 1200°F and withstand exposure to up to 1500°F. The major disadvantage of MI cable is that it must be factory-fabricated to length. It is very difficult to terminate or splice the heater in the field. This means pipe measurements are necessary before the heaters are ordered. Also, any damage to an MI cable generally requires a completely new heater. It's not as easy to splice in a good section as with self-regulating heaters.

Polymer-insulated constant wattage electric heaters are slightly cheaper than self-regulating heaters, but they are generally being replaced with self-regulating heaters due to inferior reliability. These heaters tend to destroy themselves with their own heat output when they are overlapped at valves or flanges. Since overlapping self-regulating heaters is the standard installation technique, it is difficult to prevent this technique from being used on the similar-looking constant-wattage heaters.

Skin-Effect Current Tracing (SECT). This is a special type of electric tracing employing a tracing pipe, usually welded to the pipe being traced, that is used for extremely long lines. With SECT tracing circuits, up to 10 miles can be powered from one power point. All SECT systems are custom-designed by heat-tracing vendors.

Impedance Tracing. This type of tracing uses the pipe being traced to carry the current and generate the heat. Less than 1% of electric heat-tracing systems use this method. Low voltages and special electrical isolation techniques are used. Impedance heating is useful when extremely high heat densities are required, like when a pipe containing aluminum metal must be melted from room temperature on a regular basis. Most impedance systems are specially designed by heat-tracing vendors.

Advantages of Electric Tracing

1. *Lower installed and operating costs.* Most studies have shown that electric tracing is less expensive to install and less expensive to operate. This is true

of most applications. However, for some applications, the installed costs of steam tracing are equal to or less than electric tracing.
2. *Reliability.* In the past, electric heat tracing had a well-deserved reputation for poor reliability. However, since the introduction of self-regulating heaters in 1971, the reliability of electric heat tracing has improved dramatically. Self-regulating heaters cannot destroy themselves with their own heat output. This eliminates the most common failure mode of polymer-insulated constant wattage heaters. Also, the technology used to manufacture mineral-insulated cables, high-temperature electric heat tracing, has improved significantly, and this has improved their reliability.
3. *Temperature control.* Even without a thermostat or any control system, an electric-tracing system usually provides better temperature control than a steam-tracing system. With thermostatic or electronic control, very accurate temperature control can be achieved.
4. *Safety.* The use of self-regulating heaters and ground leakage circuit breakers has answered the safety concerns of most engineers considering electric tracing. Self-regulating heaters eliminate the problems from high-temperature failures, and ground leakage circuit breakers minimize the danger of an electrical fault to ground, causing injury or death.
5. *Monitoring capability.* One question often asked about any heat-tracing system is, "How do I know it's working?". Electric tracing now has available almost any level of monitoring desired. The temperature at any point can be monitored with both high- and low-alarm capability. This capability has allowed many users to switch to electric tracing with a high degree of confidence.
6. *Energy efficiency.* Electric heat tracing can accurately provide the energy required for each application without the large additional energy use of a steam system. Unlike steam-tracing systems, other parts of the system do not use significant amounts of energy.

Disadvantages of Electric Tracing

1. *Poor reputation.* In the past, electric tracing has been less than reliable. Due to past failures, some operating personnel are unwilling to take a chance on any electric tracing.
2. *Design requirements.* A slightly higher level of design expertise is required for electric tracing than for steam tracing.
3. *Lower power output.* Since electric tracing does not provide a large multiple of the required energy, it is less forgiving to problems such as damaged insulation or below design ambient temperatures. Most designers include a 10% to 20% safety factor in the heat-loss calculation to cover these potential problems. Also, a somewhat higher than required design temperature is often specified to provide an additional safety margin. For example, many water systems are designed to maintain 50 °F to prevent freezing.

Economics of Steam Tracing Versus Electric Tracing. The question of the economics of various tracing systems has been examined thoroughly. All of these papers have concluded that electric tracing is generally less expensive to install and significantly less expensive to operate. Electric tracing has significant cost advantages in terms of installation because less labor is required than steam tracing. However, it is clear that there are some special cases where steam tracing is more economical.

The two key variables in the decision to use steam tracing or electric tracing are the temperature at which the pipe must be maintained and the distance to the supply of steam and the source of electric power.

Table 10-10 shows the installed costs and operating costs for 400 ft of 4-in. pipe, maintained at four different temperatures, with supply lengths of 100 ft for both electricity and steam and $25/hour labor.

Choosing the Best Tracing System

Some applications require either steam tracing or electric tracing regardless of the relative economics. For example, a large line that is regularly allowed to cool and needs to be quickly heated would require steam tracing because of its much higher heat output capability. In most heat-up applications, steam tracing is used with heat-transfer cement, and the heat output is increased by a factor of up to 10. This is much more heat than would be practical to provide with electric tracing. For example, a 1/2-in. copper tube containing 50 psig (3.45 Bar) steam with heat transfer cement would provide over 1100 Btu/hr/ft (3806 kJ/hr/m) to a pipe at 50°F (10°C). This is over 300 W/ft or more than 15 times the output of a high-powered electric tracer.

Table 10-10 shows that electric tracing has a large advantage in terms of cost at low temperatures and smaller but still significant advantages at higher temperatures. Steam tracing does relatively better at higher temperatures because steam tracing supplies significantly more power than necessary to maintain a pipe at low temperatures. Table 10-10 indicates that there is very little difference between the steam-tracing system at 50°F (10°C) and the system at 250°F (121.1°C). However, the electric system more than doubles in cost between these two temperatures because more heaters, higher powered heaters, and higher temperature heaters are required.

The effect of supply lengths on a 150°F (65.6°C) system can be seen from Table 10-11. Steam supply pipe is much more expensive to run than electrical conduit.

Table 10-10. Installed Costs and Operating Costs for 400ft of 4-in. Pipe

Temperature Maintained	Total Installed Cost		Ratio Steam/Electric	Total Operating Cost		Ratio Steam/Electric
	Steam	Electric		Steam	Electric	
50°F	22,265	7,733	2.88	1,671	334	5.00
150°F	22,265	13,113	1.70	4,356	1,892	2.30
250°F	22,807	17,624	1.29	5,348	2,114	2.53
400°F	26,924	14,056	1.92	6,724	3,942	1.71

Table 10-11. Ratio of Total Installed Cost of Steam/Electric for Maintaining a Temperature of 150°F (66°C)

Steam Supply Length	Electric Supply Length		
	40 ft (13.1 m)	100 ft (32.8m)	300 ft (98.4 m)
40 ft (13.1 m)	1.1	1.0	0.7
100 ft (32.8m)	1.9	1.7	1.1
300 ft (98.4 m)	4.9	4.2	2.9

With each system having relatively short supply lines (40 ft [13 m] each), the electric system has only a small cost advantage (10%, or a ratio of 1.1). This ratio is 2.1 at 50°F (10°C) and 0.8 at 250°F (121.1°C). However, as the supply lengths increase, electric tracing has a large cost advantage.

Storage of Liquids

Atmospheric Tanks

The term atmospheric tank as used here applies to any tank that is designed to be used within plus or minus a few pounds per square foot (a few tenths of a bar) of atmospheric pressure. It may be either open to the atmosphere or enclosed. Minimum cost is usually obtained with a vertical cylindrical shape and a relatively flat bottom at ground level.

AMERICAN PETROLEUM INSTITUTE (API). The institute has developed a series of atmospheric tank standards and specifications. Some of these are:

- API Specifications 12B, Bolted Production Tanks
- API Specifications 12D, Large Welded Production Tanks
- API Specifications 12F, Small Welded Production Tanks
- API Specifications 650, Steel Tanks for Oil Storage

AMERICAN WATER WORKS ASSOCIATION (AWWA). The association has many standards dealing with water handling and storage. A list of its publications is given in the AWWA Handbook (annually). AWWA D100, Standard for Steel Tanks — Standpipes, Reservoirs, and Elevated Tanks for Water Storage — contains rules for design and fabrication.

Although AWWA tanks are intended for water, they could be used for the storage of other liquids.

Underwriters Laboratories Inc. has published the following tank standards:

- UL 58, Steel Underground Tanks for Flammable and Combustible Liquids
- UL 142, Steel Aboveground Tanks for Flammable and Combustible Liquids

UL 58 covers horizontal steel tanks up to 190 m (3) (50,000 gal), with a maximum diameter of 12 ft (3.66 m), and a maximum length of six diameters. Thickness and a number of design and fabrication details are given. UL 142 covers horizontal steel tanks up to 50,000 gal (190 m^3) (like UL 58), and vertical tanks up to 35 ft (10.7 m) height. Thickness and other details are given. The maximum diameter for a vertical tank is not specified.

The Underwriters Standards overlap API, but include tanks that are too small for API Standards. Underwriters Standards are, however, not as detailed as API and therefore put more responsibility on the designer. They do not specify grades of steel other than requiring steels, which are weldable. Designers should also place their own limits on the diameter (or thickness) of vertical tanks. They can obtain guidance from API.

Elevated Tanks

These can supply a large flow when required, but pump capacities need only be for average flow. Thus, they may save on pump and piping investment. They also provide flow after pump failure, an important consideration for fire systems.

Open Tanks

These may be used to store materials that will not be harmed by water, weather, or atmospheric pollution. Otherwise, a roof, either fixed or floating, is required. Fixed roofs are usually either domed or coned. Large tanks have coned roofs with intermediate supports. Since negligible pressure is involved, snow and wind are the principal design loads. Local building codes often give required values.

Fixed Roof Tanks

Atmospheric tanks require vents to prevent pressure changes, which would otherwise result from temperature changes and withdrawal or addition of liquid. API Standard 2000, venting atmospheric and low-pressure storage tanks, gives practical rules for vent design. The principals of this standard can be applied to fluids other than petroleum products. Excessive losses of volatile liquids, particularly those with flash points below 100°F (38°C), may result from the use of open vents on fixed-roof tanks. Sometimes, vents are manifolded and led to a vent tank, or the vapor may be extracted by a recovery system.

An effective way of preventing vent loss is to use one of the many types of variable-volume tanks. These are built under API Standard 650. They may have floating roofs of the double-deck or the single-deck type. There are lifter-roof types in which the roof either has a skirt moving up and down in an annular liquid seal or is connected to the tank shell by a flexible membrane. A fabric expansion chamber housed in a compartment on top of the tank roof also permits variation in volume.

Floating Roof Tanks

These tanks must have a seal between the roof and the tank shell. If not protected by a fixed roof, they must have a drains for the removal of water, and the tank shell must have a "wind girder" to avoid distortion. An industry has developed to retrofit existing tanks with floating roofs. Much detail on the various types of tank roofs is given in manufacturers' literature. Figure 10-15 shows types. These roofs cause less condensation buildup and are highly recommended.

Pressure Tanks

Vertical cylindrical tanks constructed with domed or coned roofs, which operate at pressures above 15 psia (1 bar) but which are still relatively close to atmospheric pressure, can be built according to API Standard 650. The pressure force acting against the roof is transmitted to the shell, which may have sufficient weight to resist it. If not, the uplift will act on the tank bottom. The strength of the bottom, however, is limited, and if it is not sufficient, an anchor ring or a heavy foundation must be used. In the larger sizes, uplift forces limit this style of tank to very low pressures.

As the size or the pressure goes up, curvature on all surfaces becomes necessary. Tanks in this category, up to and including a pressure of 15 psia (1 bar) can be built according to API Standard 620. Shapes used are spheres, ellipsoids, toroidal structures, and circular cylinders with torispherical, ellipsoidal, or hemispherical heads. The ASME Pressure Vessel Code (sec. VIII of the ASME Boiler and Pressure Vessel Code), although not required below 15 psia (1 bar), is also useful for designing such tanks.

Tanks that could be subjected to vacuum should be provided with vacuum-breaking valves or be designed for vacuum (external pressure). The ASME Pressure Vessel Code contains design procedures.

Calculation of Tank Volume

A tank may be a single geometrical element, such as a cylinder, a sphere, or an ellipsoid. It may also have a compound form, such as a cylinder with hemispherical ends or a combination of a toroid and a sphere. To determine the volume, each geometrical element usually must be calculated separately. Calculations for a full tank are usually simple, but calculations for partially filled tanks may be complicated.

To calculate the volume of a partially filled horizontal cylinder, refer to Figure 10-16. Calculate the angle a in degrees. Any units of length can be used, but they must be the same for H, R, and L. The liquid volume

$$V = LR^2 \left(\frac{\alpha}{57.3} - \sin \alpha \cos \alpha \right) \qquad (10-1)$$

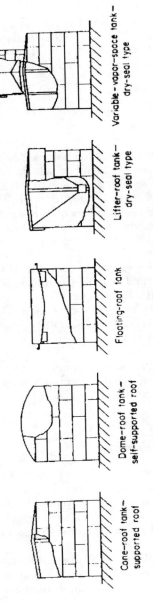

Figure 10-15. Some Types of Atmospheric Storage Tanks

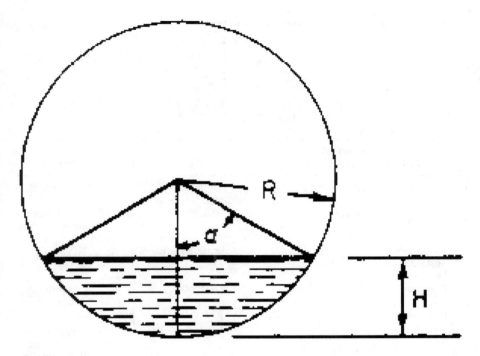

Figure 10-16. Calculation of Partially Filled Horizontal Tanks. H = Depth of Liquid: R = Radius: D = Diameter: L = Length: a = Half of the Included Angle: and $\cos a = 1 - H/R = 1 - 2H/D$

This formula may be used for any depth of liquid between zero and the full tank, provided the algebraic signs are observed. If H is greater than R, $\sin a$, $\cos a$ will be negative and thus will add numerically to $a/57.30$.

The volumes of heads must be calculated separately and added to the volume of the cylindrical portion of the tank. The four types of heads most frequently used are the standard dished head, torispherical or ASME head, ellipsoidal head, and hemispherical head. Dimensions and volumes for all four of these types are given in *Lukens Spun Heads*, Lukens Inc., Coatsville, Pennsylvania. Approximate volumes can also be calculated by the formulas in Table 10-12. Consistent units must be used in these formulas.

A partially filled horizontal tank requires the determination of the partial volume of the heads. The Lukens catalog gives approximate volumes for partially filled (axis horizontal) standard ASME and ellipsoidal heads. A formula for partially filled heads, by Doolittle [Ind. Eng. Chem., 21 (1928) 322-323], is

$$V = 0.215\, H^2 (3R - H) \qquad (10-2)$$

Table 10-12. Approximate Volumes for Four Types of Heads

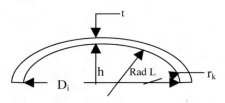

Type of Head	Knuckle Radius, r_k	h	L	Volume	% Error	Remarks
Standard dished	Approx. $3t$		Approx. D_i	Approx. $0.050\,D_i^3 + 1.65\,tD_i^2$	±10	h varies with t
Torispherical or ASME	$0.06L$		D_i	$0.0809\,D_i^3$	±0.1 ±8	r_k must be the larger of $0.06L$ and $3t$
Torispherical or ASME		$3t$	D_i	Approx. $0.513hD_i^2$		
Ellipsoidal				$\pi D_i^2 h/6$	0	
Ellipsoidal		$D_i/4$		$\pi D_i^3 h/24$	0	Standard proportions
Hemispherical		$D_i/2$	$D_i/2$	$\pi D_i^3 h/12$	0	
Conical				$\pi h(D_i^2 + D_i d + d^2)/12$	0	Truncated cone h = height d = diameter at small end

where, in consistent units,
  V = volume
  R = radius
  H = depth of liquid

Doolittle made some simplifying assumptions which affect the volume given by the equation, but the equation is satisfactory for determining the volume as a fraction of the entire head.

When a tank volume cannot be calculated or when greater precision is required, calibration may be necessary. This is done by draining (or filling) the tank and measuring the volume of liquid. The measurement may be made by weighing, by a calibrated fluid meter, or by repeatedly filling small measuring tanks, which have been calibrated by weight.

Container Materials, Insulation and Support

Materials. Storage tanks are made of almost any structural material; steel and reinforced concrete are most widely used. Plastics and glass-reinforced plastics are used for tanks up to 60,000 gal. Resistance to corrosion, light weight and low cost are some of the advantages.

Post-Tensioned Concrete is a material frequently used for tanks to about 15×10^6 gal (57,000 m^3), usually containing water. Their design is treated in detail by Creasy *(Prestressed Concrete Cylindrical Tanks, Wiley, New York, 1961)*. For the most economical design of large open tanks at ground levels, he recommends limiting vertical height to 6 m (20 ft). Seepage can be a problem if unlined concrete is used with some liquids (e.g., gasoline).

Insulation. Tanks containing materials that have to be maintained over the ambient temperatures need insulation to keep the contents at their desired temperature. Common insulating materials such as glass fiber, mineral wool, calcium silicate, and plastic foams are used. Tanks exposed to weather must have jackets of protective coatings, usually asphalt to keep water out of the insulation. Insulation not damaged by moisture is preferable. In some cases, steam heating is provided to keep the fuel at about 20°F (11°C) above the pour point.

Support. Large vertical atmospheric steel tanks may be built on a base of about 6 in. (150 cm) of sand, gravel, or crushed stone if the subsoil has adequate bearing strength. The bearing pressure of the tank must not exceed the bearing strength of the soil. The base may be conical depending on the shape of the tank bottom. Heavy tanks require a foundation ring. Pre-stressed concrete tanks are sufficiently heavy to require foundation rings.

The porosity of the base is good for drainage in case of leaks. A few feet beyond the tank perimeter, the surface should drop about 3 ft (1 m) to create good drainage of the subsoil. API standards 650, Appendix B, and API Standard 620, Appendix C, give recommendations for tank foundations.

Chapter 11

BEARINGS, SEALS AND LUBRICATION SYSTEMS

Bearings

The bearings in gas and steam turbines provide support and positioning for the rotating components. Radial support is generally provided by journal or roller bearings, and axial positioning is provided by thrust bearings. Some engines, mainly aircraft jet engines — use ball or roller bearings for radial support but nearly all-industrial gas turbines use journal bearings.

A long service life, a high degree of reliability, and economic efficiency are the chief aims when designing bearing arrangements. To reach these criteria, design engineers examine all the influencing factors:

1. Load and speed
2. Lubrication
3. Temperatures
4. Shaft arrangements
5. Life
6. Mounting and dismounting
7. Noise
8. Environmental Conditions.

Rolling Bearings

The aeroderivative design, with its low supported-weight rotors - for example, the LM 5000 HP rotor weighs 1230 lb/558 kg, incorporates roller bearings throughout. These do not require the large lube oil reservoirs, coolers and pumps, or the pre- and post-lube cycle associated with journal bearing designs. Roller bearings have proven to be extremely rugged and have demonstrated excellent life in industrial service. Most bearings provide reliable service for over 100,000 hours. In practice, it is advisable to replace bearings when exposed during major repairs, estimated at 50,000 hours for gas generators and 100,000 hours for power turbines.

There are many roller bearing types and they are differentiated according to the direction of the main radial loads (radial bearings), or axial load (thrust bearings), and the type of rolling elements used, balls or rollers. Figure 11-1

Radial ball bearings

Deep groove ball bearing | Angular contact ball bearing single row | Angular contact ball bearing double row | Four-point bearing | Self-aligning ball bearing

Radial roller bearings

Cylindrical roller bearing | Needle roller bearing | Tapered roller bearing | Barrel roller bearing | Spherical roller bearing

Thrust ball bearings

  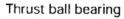

Thrust ball bearing   Angular contact thrust ball bearing double direction

Thrust roller bearings

Cylindrical roller thrust bearing   Spherical roller thrust bearing

Figure 11-1. Types of Rolling Bearings (Courtesy of FAG Bearings)

shows the different types of bearings. The essential difference between ball bearings and roller bearings is that ball bearings have a lower carrying capacity and higher speeds, while the roller bearings have higher load-carrying capacity and lower speeds.

The rolling elements transmit loads from one bearing ring to the other in the direction of the contact lines. The contact angle α is the angle formed by the contact lines and the radial plane of the bearing. α refers to the nominal contact angle, i.e., the contact angle of the load-free bearing as seen in Figure 11-2. Under axial loads, the contact angle of deep groove ball bearings, angular contact ball bearings, etc., increase. Under a combined load, it changes from one rolling element to the next. These changing contact angles are taken into account when calculating the pressure distribution within the bearing. Ball bearings and roller bearings with symmetrical rolling elements have identical contact angles at their inner rings and outer rings. In roller bearings with asymmetrical rollers, the contact angles at the inner rings and outer rings are not identical. The equilibrium of forces in these bearings is maintained by a force component which is directed towards the lip. The pressure cone apex is that point on the bearing axis where the contact lines of an angular contact bearing, i.e., an angular contact ball bearing, a tapered roller bearing or a spherical roller thrust bearing, intersect. The contact lines are the generatrices of the pressure cone apex. In angular contact bearings, the external forces act not at the bearing center but at the pressure cone apex.

Rolling bearings generally consist of bearing rings, inner ring and outer ring, rolling elements which roll on the raceways of the rings, and a cage which surrounds the rolling elements as seen in Figure 11-3. The rolling elements are classified according to their shapes. Into balls, Cylindrical rollers, needle rollers, tapered rollers, and barrel rollers are shown in Figure 11-4.

The rolling elements' function is to transmit the force acting on the bearing from one ring to the other. For a high load-carrying capacity, it is important that as many rolling elements as possible, which are as large as possible, are accommodated between the bearing rings. Their number and size depend on the cross-section of the bearing. It is just as important for load ability that the rolling elements within the bearing are of identical sizes. Therefore, they are sorted according to grades. The tolerance of one grade is very slight. The generatrices of cylindrical rollers and tapered rollers have a logarithmic profile. The center part of the generatrix of a needle roller is straight and the ends are slightly crowned. This profile prevents edge stressing when under load.

The bearing rings comprise of an inner ring and an outer ring to guide the rolling elements in the direction of rotation. Raceway grooves, lips and inclined running areas guide the rollers and transmit axial loads in transverse direction as seen in Figure 11-5. Cylindrical roller bearings and needle roller bearings, which need to accommodate shaft expansions, have lips only on one bearing ring and are commonly known as floating bearings.

The functions of a cage are: to keep the rolling elements apart so that they do not rub against each other; to keep the rolling elements evenly spaced for uniform load distribution; to prevent rolling elements from falling out of separable bearings and bearings, which are swiveled out; and to guide the rolling elements in the unloaded zone of the bearing. The transmission of forces is not one of the cage's functions.

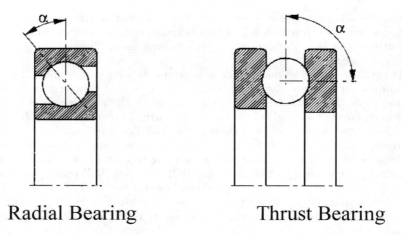

Radial Bearing   Thrust Bearing

Contact Angle

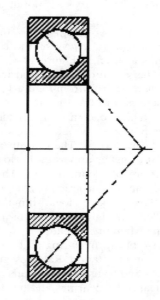

Pressure Cone Apex

Figure 11-2. Rolling Bearing Terminology (Courtesy of FAG Bearings)

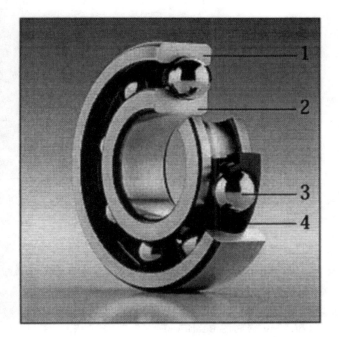

1 Outer ring, 2 Inner ring, 3 Rolling element, 4 Cage

Figure 11-3. Roller Bearing Showing the Various Components
(Courtesy of FAG Bearings)

Cages are classified into pressed cages, machined cages and moulded cages. Pressed cages are usually made of steel but sometimes of brass, too. They are lighter than machined metal cages. Since a pressed cage barely closes the gap between inner ring and outer ring, lubricant can easily penetrate into the bearing.

Machined cages of metal and textile-laminated phenolic resin are made from tubes of steel, light metal or textile laminated phenolic resin, or cast brass rings. To obtain the required strength, large, heavily loaded bearings are fitted with machined cages. Machined cages are also used where lip guidance of the cage is required. Lip-guided cages for high-speed bearings are, in many cases, made of light materials such as light metal or textile-laminated phenolic resin to keep the forces of gravity low.

Moulded Cages using injection moulding techniques can realize designs with an especially high load-carrying capacity. Injection moulding has made it possible to realize cage designs with an especially high load-carrying capacity. The elasticity and low weight of the cages are advantages where shock-type bearing loads, great accelerations and decelerations, as well as tilting of the bearing rings relative to each other have to be accommodated. Polyamide cages feature very good sliding and dry running properties.

400 • COGENERATION AND COMBINED CYCLE POWER PLANTS

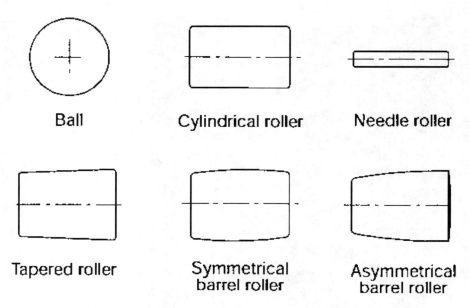

Figure 11-4. Various Types of Rollers Used in Rolling Bearing
(Courtesy of FAG Bearings)

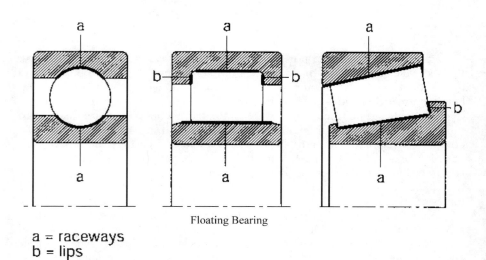

a = raceways
b = lips

Figure 11-5. Raceway Groves and Lips for Typical Roller Bearings
(Courtesy of FAG Bearings)

There are a number of special rolling bearing designs and some series of cylindrical roller bearings without cages. By omitting the cage, the bearing can accommodate more rolling elements. This yields an increased load rating but due to the increased friction the bearing is suitable for lower speeds only.

Load ratings. The load rating of a bearing reflects its load-carrying capacity and is an important factor in the dimensioning of rolling bearings. It is determined by the number and size of the rolling elements, the curvature ratio, the contact angle and the pitch circle diameter of the bearing. Due to the larger contact area between rollers and raceways, the load ratings of roller bearings are higher than those of ball bearings.

The load rating of a radial bearing is defined by radial loads, whereas that of a thrust bearing is defined by axial loads. Every rolling bearing has a dynamic load rating and a static load rating. The terms "dynamic" and "static" refer to the movement of the bearing but not to the type of load.

In all rolling bearings with a curved raceway profile, the radius of the raceway is slightly larger than that of the rolling elements. This curvature difference in the axial plane is defined by the curvature ratio x. The curvature ratio is the curvature difference between the rolling element radius and the slightly larger groove radius.

$$\text{radius curvature ratio} \, x = (\text{groove radius} - \text{rolling element}) / \text{rolling element radius}$$

Thrust ball bearings are used where purely axial loads have to be accommodated. The single direction (= single row) design is designed for loads from one direction, and the double direction one (= double row) for reversing loads. Besides the design with flat washers, designs with spherical housing washers and seating washers are also available which can compensate for misalignment.

Spherical roller thrust bearings can accommodate high axial loads. They are suitable for relatively high speeds. The raceways, which are inclined towards the bearing axis, allow the bearings to accommodate radial loads as well. The radial load must not exceed 55% of the axial load.

The bearings have asymmetrical barrel rollers and compensate for misalignment. As a rule, spherical roller thrust bearings have to be lubricated with oil.

Wear. The life of rolling bearings can be terminated, apart from fatigue, as a result of wear. The clearance of a worn bearing gets too large.

One frequent cause of wear are foreign particles which penetrate into a bearing due to insufficient sealing and have an abrasive effect. Wear is also caused by starved lubrication and when the lubricant is used up.

Therefore, wear can be considerably reduced by providing good lubrication conditions (viscosity ratio $x > 2$ if possible) and a good degree of cleanliness in the rolling bearing. Where $x \leq 0.4$, wear will dominate in the bearing if it is not prevented by suitable additives (EP additives).

The kinematically permissible speed may be higher or lower than the thermal reference speed. The basis of the thermal reference speed for cases where the operating conditions (load, oil viscosity or permissible temperature) deviate from the reference conditions. Decisive criteria for the kinematically permissible speed are, e.g., the strength limit of the bearing parts or the permissible sliding velocity of rubbing seals. Kinematically permissible speeds, which are higher than the thermal reference speeds, can be reached, for example, with specially designed lubrication, bearing clearance adapted to the operating conditions, accurate machining of the bearing seats with special regard to heat dissipation.

The thermal reference speed is a new index of the speed suitability of rolling bearings. It is defined as the speed at which the reference temperature of 70°C is established.

For high-temperature rolling bearings, the steel used for bearing rings and rolling elements is generally heat-treated so that it can be used at operating temperatures of up to +150°C. At higher temperatures, dimensional changes and hardness reductions result. Therefore, operating temperatures over +150°C require special heat treatment.

Journal Bearings

Heavy frame type gas turbines use journal bearings. Journal bearings may either full round or split; the lining may be heavy as those used in large-size bearings for heavy machinery, or thin, as used in precision insert-type bearings in internal combustion engines. (1) Most sleeve bearings are of the split type for convenience in servicing and replacement. Often in split bearings, where the load is entirely downward, the top half of the bearing acts only as a cover to protect the bearing and to hold the oil fittings. Figure 11-6 shows a number of different types of journal bearings. A description of a few of the pertinent types of journal bearings is given here:

1. Plain Journal. Bearing is bored with equal amounts of clearance (on the order of one and one-half to two thousands of an inch per inch of journal diameter) between the journal and bearing.
2. Circumferential grooved bearing. Normally has the oil groove at half the bearing length. This configuration provides better cooling but reduces load capacity by dividing the bearing into two parts.
3. Cylindrical bore bearings. Another common bearing type used in turbines. It has a split construction with two axial oil-feed grooves at the split.
4. Pressure or pressure dam. Used in many places where bearing stability is required, this bearing is a plain journal bearing with a pressure pocket cut in the unloaded half. This pocket is approximately $1/32$ of an inch deep with a width 50% of the bearing length. This groove channel covers an arc of 135 and terminates abruptly in a sharp edge dam. The direction of the rotation is such that the oil is pumped down the channel towards the sharp edge. Pressure dam bearings are for one direction of rotation. They can be used in conjunction with cylindrical bore bearings as shown in Figure 11-6.
5. Lemon bore or elliptical. This bearing is bored with shims at the split line which are removed before installation. The resulting bore shape approximates an eclipse with the major axis clearance. Elliptical bearings are for both directions of rotation.

Bearings, Seals and Lubrication Systems • 403

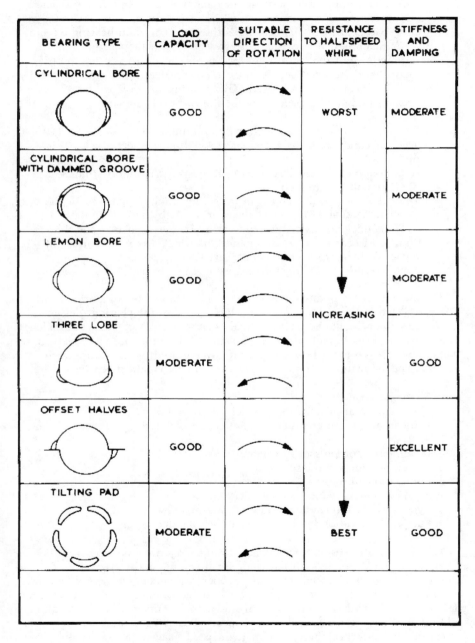

Figure 11-6. Different Types of Journal Bearings

6. Three-lobe bearing. Three-lobe bearing is not commonly used in turbomachines. It has a moderate load-carrying capacity. It is restricted to one direction of rotation.
7. Tilting pad bearings. This bearing is the most common bearing type in today's machines. It consists of several bearing pads posed around the circumference of the shaft. Each pad is able to tilt to assume the most effective working position. Its most important feature is self-alignment when spherical pivots are used. This bearing also offers the greatest increase in fatigue life because of the following advantages:

- Self-aligning for optimum alignment and minimum limit.
- Thermal conductivity backing material to dissipate heat developed in oil film.
- A thin Babbit layer can be centrifugally cast with a uniform thickness of about 0.005 of an inch.
- Thick Babbit greatly reduce bearing life. Babbit thickness in the neighborhood of 0.01 reduces the bearing life by more than half.
- Oil film thickness is critical in bearing stiffness calculations. In a tilting-pad bearing, one can change this thickness in a number of ways: (a) change in the number of pads, (b) load can be directed on or in-between the pads, (c) changing axial length of pad.

The previous list contains some of the most common types of journal bearings. They are listed in the order of growing stability. All of the bearings designed for increased stability is obtained at higher manufacturing costs and reduced efficiency. The anti-whirl bearings all impose a parasitic load on the journal, which causes higher power losses to the bearings, and in turn requires higher oil flow to cool the bearing. The following are some of the factors which that affect bearing design:

1. Shaft speed range
2. Maximum shaft misalignment which can be tolerated
3. Critical speed analysis, and the influence of bearing stiffness on this analysis
4. Loading the compressor impellers
5. Oil temperatures and viscosity
6. Foundation stiffness
7. Axial movement, which can be tolerated
8. Type of lubrication system and its contamination
9. Maximum vibration levels, which can be tolerated.

Tilting Pad-Bearings. Normally, the tilting pad journal bearing is considered when shaft loads are light because of its inherent ability to resist oil whirl vibration. However, this bearing, when properly designed, has a very high load-carrying capacity. It has the ability to tilt to accommodate the forces being developed in the hydrodynamic oil film, and therefore operates with an optimum oil film thickness for the given load and speed. This ability to operate over a large range of load is especially useful in high-speed gear reductions with various combinations of input and output shafts.

Another important advantage of the tilting-pad journal bearing is its ability to accommodate shaft misalignment. Because of its relatively short length-to-diameter ratio, it can accommodate minor misalignment quite easily.

As shown earlier, bearing stiffness varies with the oil-film thickness so that the critical speed is directly influenced to a certain degree by oil-film thickness. Again, in the area of critical speeds, the tilting-pad journal bearing has the greatest degree of design flexibility. There are sophisticated computer programs that show the influence of various load and design factors on the stiffness of tilting-pad journal bearings. The following variations are possible in the design of tilting-pad bearings:

1. The number of pads can be varied from three to any practical number.
2. The load can be placed either directly on a pad or to occur between pads.
3. The unit loading on the pad can be varied by either adjusting the arc length or the axial length of the bearing pad.
4. A parasitic preload can be designed into the bearing by varying the circular curvature of the shaft.
5. An optimum support point can be selected to obtain a maximum oil-film thickness.

On a high-speed rotor system, it is necessary to use tilting-pad bearings because of the dynamic stability of these bearings. A high-speed rotor system operates at speeds above the first critical speed of the system. It should be understood that a rotor system includes the rotor, the bearings, the bearing support system, seals couplings and other items attached to the rotor. The systems natural frequency is therefore on the stiffness and damping effect of these components.

Commercial multipurpose tilting-pad bearings are usually designed for multidirectional rotation so that the pivot point is at pad midpoint. However, the design criteria generally applied for producing maximum stability and load-carrying capacity locates the pivot at two-thirds of the pad arc in the direction of rotation.

Bearing load is another important design criterion for tilting-pad bearings. Bearing preload is bearing assembly clearance divided by machined clearance

$$\text{Preload Ratio} = \frac{C'}{C} = \frac{\text{Concentric Pivot Film Thickness}}{\text{Machined Clearance}}$$

A preload of 0.5–1.0 provides for stable operation because a converging wedge is produced between the bearing journal and the bearing pads.

The variable C is an installed clearance and is dependent upon radial pivot position. The variable C is the machine clearance and is fixed for a given bearing. Figure 11-7 shows two pads of a five-pad tilting-pad bearing where the pads have been installed such that the preload ratio is less than one, and Pad 2 has a preload ratio of 1.0 (10). The solid line in Figure 11-7 represents the position of the journal in the concentric position. The dashed line represents the journal in a position with a load applied to the bottom pads.

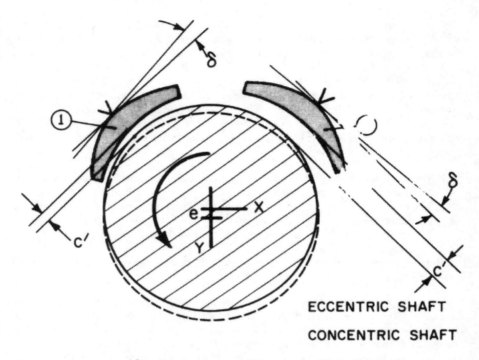

PAD 1 C'/C < 1.0 CONVERGING CLEARANCE
PAD 2 C'/C > 1.0 DIVERGING CLEARANCE

Figure 11-7. Tilting Pad Bearing Characteristics

Unloaded pads are also subject to flutter, which leads to a phenomenon known as "Leading-edge lockup". Leading-edge lockup causes the pad to be forced against the shaft, and is then maintained in that position by the frictional interaction of the shaft and the pad. Therefore, it is of prime importance that the bearings be designed with preload, especially for low viscosity lubricants. In many cases, manufacturing reasons and the ability to have two-way rotation cause many bearings to be produced without preload.

Bearing designs are also affected by the film from a laminar to a turbulent region. The transition speed (N_t) can be computed using the following relationship:

$$N_t = 1.57 \times 10^3 \, \frac{\nu}{\sqrt{DC^3}}$$

where
    ν = viscosity of fluid
    D = diameter, in.
    C = diametrical clearance, in.

Turbulence creates more power absorption, thus increasing oil temperature that can lead to severe erosion and fretting problems in bearings. It is desirable to keep the oil discharge temperature below 170°F (76.7°C), but with high-speed bearings, this ideal may not be possible. In those cases, it is better to monitor the temperature difference between the oil entering and leaving as shown in Figure 11-8.

Bearing Materials. In all the time since Isaac Babbitt patented his special alloy in 1839, nothing has been developed that encompasses all of its excellent properties as an oil-lubricated bearing surface material. Babbitts have excellent compatibility and non-scoring characteristics and are outstanding in embedding dirt and conforming to geometric errors in machine construction and operation. They are, however, relatively weak in fatigue strength, especially at elevated temperatures and when the babbitts exceed about 0.015 in. (0.38 mm) thick as seen in Figure 11-9. In general, the selection of a bearing material is always a compromise, and no single composition can include all desirable properties. Babbitts can tolerate momentary rupture of the oil film, and may well minimize shaft or runner damage in the event of a complete failure. Tin Babbitts are more desirable than lead-based materials, because they have better corrosion resistance, less tendency to pick up on the shaft or runner, and are easier to bond to a steel shell.

Application practices suggest a maximum design temperature of about 300°F (148.9°C) for babbitt, and designers will set a limit of about 50°F (10°C) less. As temperatures increase, there is a tendency for the metal to creep under the softening influence of the rising temperature. Creep can occur with generous film thickness and can be observed as ripples on the bearing surface where flow takes place. With tin babbitts, observation has shown that creep temperature ranges from 375°F (190.5°C) for bearing loads below 200 psi (13.8 bar) to about 260°F-270°F (126.7°C-132.2°C) for steady loads of 1000 psi (69 bar). This range will be improved by using very thin layers of babbitt such as those in automotive bearings.

Bearing and Shaft Instabilities. One of the most serious forms of instability encountered in journal bearing operation is known as "half-frequency whirl". "Whirl" is the phenomenon where the shaft center moves in a circular motion in the bearing cavity. There are many types of "Whirling motion," most of which are in the direction of rotation except the "Coulomb Whirl" which is caused by the shaft contacting the bearing surface. The "Oil Whirl" is a self-excited vibration and characterized by the shaft center orbiting around the bearing center at a frequency of approximately half of the shaft rotational speed, as shown in Figure 11-10.

As the speed is increased, the shaft system may be stable until the "whirl" threshold is reached. When the threshold speed is reached, the bearing becomes unstable, and further increase in speed produces more violent instability until eventual seizure results. Unlike an ordinary critical speed, the shaft cannot "pass through," and the instability frequency will increase and follow that half ratio as the shaft speed is increased. This type of instability is associated primarily with high-speed, lightly loaded bearings. At present, this form of instability is well understood, can be theoretically predicted with accuracy, and can be avoided by altering the bearing design.

408 • COGENERATION AND COMBINED CYCLE POWER PLANTS

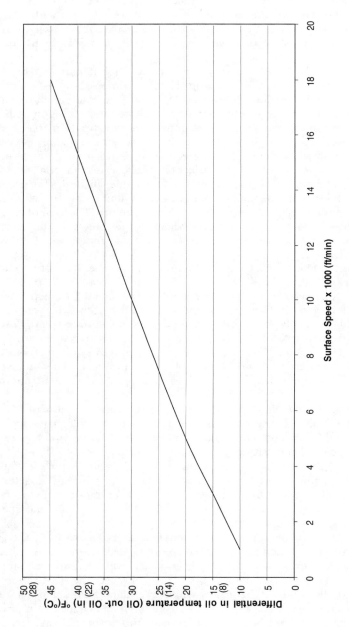

Figure 11-8. Oil Discharge Characteristics

Bearings, Seals and Lubrication Systems • 409

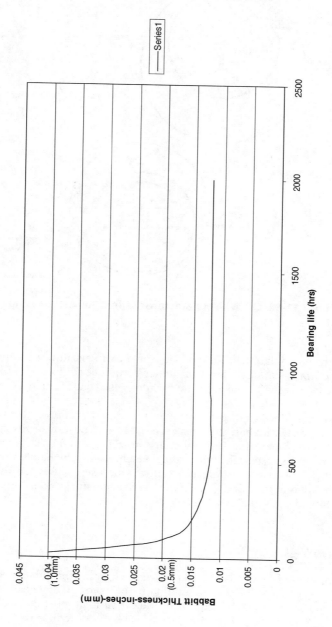

Figure 11-9. Babbitt Fatigue Characteristics

410 • COGENERATION AND COMBINED CYCLE POWER PLANTS

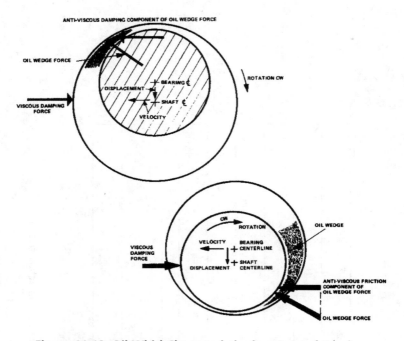

Figure 11-10. Oil Whirl Characteristics in a Journal Bearing

It should be noted that the tilting-pad journal bearing is almost completely free from this form of instability. However, under certain conditions, the tilting pads themselves can become unstable in the form of shoe (pad) flutter, as mentioned previously.

All rotating machines vibrate during operation, but the failure of the bearings is mainly caused by their inability to resist cyclic stresses. The level of vibration a unit can tolerate is shown in the severity charts in Figure 11-11. These charts are modified by many users to reflect the critical machines in which they would like to maintain much lower levels. Care must always be exercised when using these charts, since different machines have different size housings and rotors. Thus, the transmissibility of the signal will vary.

Thrust Bearings

The most important function of a thrust bearing is to resist the unbalanced force in a machine's working fluid and to maintain the rotor in its position (within prescribed limits). A complete analysis of the thrust load must be conducted. As mentioned earlier, compressors with back-to-back rotors reduce this load greatly on thrust bearings. Figure 11-12 shows a number of thrust bearing types. Plain, grooved thrust washers are rarely used with any continuous load, and their use

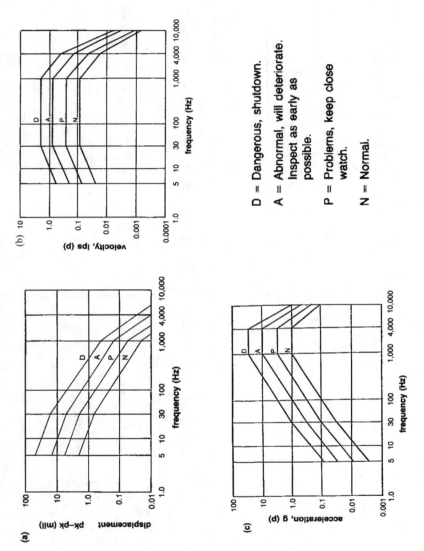

Figure 11-11. Severity Charts for Frame Type Machines

BEARING TYPE		LOAD CAPACITY	SUITABLE DIRECTION OF ROTATION	TOLERANCE OF CHANGING LOAD/SPEED	TOLERANCE OF MISALIGNMENT	SPACE REQUIREMENT
PLAIN WASHER 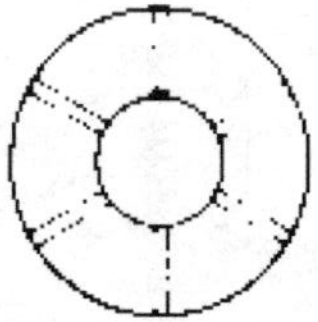		POOR	 ↻↺	GOOD	MODERATE	COMPACT
TAPER LAND 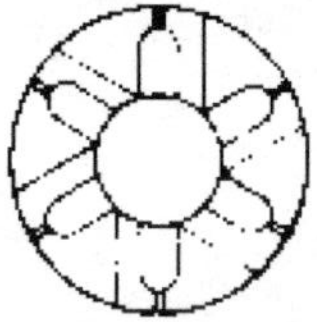	BIDIRECTIONAL	MODERATE	 ↻	POOR	POOR	COMPACT
	UNIDIRECTIONAL	GOOD	 ↻	POOR	POOR	COMPACT
TILTING PAD 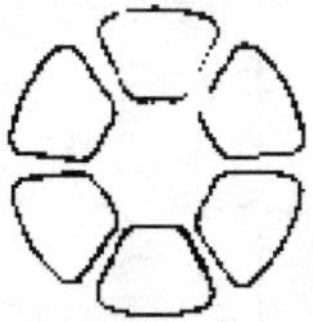	BIDIRECTIONAL	GOOD	 ↻	GOOD	GOOD	GREATER
	UNIDIRECTIONAL	GOOD	↻	GOOD	GOOD	GREATER

Figure 11-12. Comparison of Various Types of Thrust Bearings

tends to be confined to cases where the thrust load is of very short duration or possibly occurs at standstill or low speeds only.

Occasionally, this type of bearing is used for light loads (less than 50 lb/in.), and in these circumstances, the operation is probably hydrodynamic due to small distortions present in the nominally flat bearing surface.

When significant continuous loads have to be taken on a thrust washer, it is necessary to machine into the bearing surface of a profile to generate a fluid film. This profile can be either a tapered wedge or, occasionally, a small step.

The tapered-land thrust bearing, when properly designed, can take and support a load equal to a tilting-pad thrust bearing. With perfect alignment, it can match the load of even a self-equalizing tilting-pad thrust bearing that pivots on the back of the pad along a radial line For variable-speed operation, tilting pad thrust bearings, as shown in Figure 11-13, are advantageous when compared to conventional taper-land bearings. The pads are free to pivot to form a proper angle for lubrication over a long speed range. The self-leveling feature equalizes individual pad loadings and reduces the sensitivity to shaft misalignments which may occur during service. The major drawback of this bearing type is that standard designs require more axial space than a non-equalizing thrust bearing.

Factors Affecting Thrust Bearing Design. The principal function of a thrust bearing is to resist the thrust unbalance developed within the working elements of a turbomachine and to maintain the rotor position within tolerable limits.

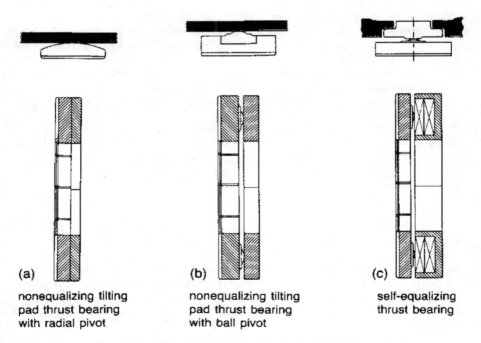

(a) nonequalizing tilting pad thrust bearing with radial pivot

(b) nonequalizing tilting pad thrust bearing with ball pivot

(c) self-equalizing thrust bearing

Figure 11-13. Various Types of Tilting Pad Thrust Bearings

After an accurate analysis has been made of the thrust load, the thrust bearing should be sized to support this load in the most efficient method possible. Many tests have proven that thrust bearings are limited in load capacity by the strength of the babbitt surface in the high load and temperature zone of the bearing. In normal steel-backed, babbitted tilting-pad thrust bearings, this capacity is limited to between 250 psi (17.24 bar) and 500 psi (34.5 bar) average pressure. It is the temperature accumulation at the surface and pad crowning that cause this limit.

The thrust-carrying capacity can be greatly improved by maintaining pad flatness and by removing heat from the loaded zone. By the use of high thermal conductivity backing materials with proper thickness and proper support, the maximum continuous thrust limit can be increased to 1000 psi or more. This new limit can be used to increase the factor of safety and improve the surge capacity of a given size bearing or reduce the thrust bearing size and consequently the losses generated for a given load.

Since the higher thermal conductivity material (copper or bronze) is a much better bearing material than the conventional steel backing, it is possible to reduce the babbitt thickness to 0.010-0.030 in. Embedded thermocouples and RTDs will signal distress in the bearing if properly positioned. Temperature monitoring systems have been found to be more accurate than axial position indicators, which tend to have linearity problems at high temperatures.

In a change from steel-backing to copper-backing, a different set of temperature-limiting criteria should be used. Figure 11-14 shows a typical set of curves for the two backing materials. This chart also shows that drain oil temperature is a poor indicator of bearing operating conditions because there is very little change in drain oil temperature from low load to failure load.

Thrust Bearing Power Loss. The power consumed by various thrust bearing types is an important consideration in any system. Power losses must be accurately predicted so that turbine efficiency can be computed and the oil supply system properly designed.

Figure 11-15 shows the typical power consumption in thrust bearings as a function of unit speed. The total power loss is usually about 0.8% to 1.0% of the total rated power of the unit. New vectored lube bearings that are being tested show preliminary figures of reducing the horsepower loss by as much a 30%.

Seals

Seals are very important and often critical components in turbomachinery, especially on high-pressure and high-speed equipment. This chapter covers the principal sealing systems used between the rotor and stator elements of turbomachinery. They fall into two categories: (1) non-contacting seals and (2) face seals.

Since these seals are an integral part of the rotor system, they affect the dynamic operating characteristics of the machine; for instance, both the stiffness and the

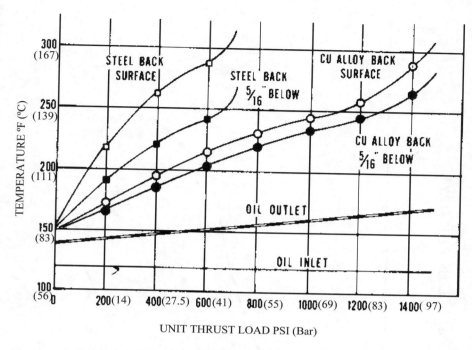

Figure 11-14. Thrust Bearing Temperature Characteristics

damping factors will be changed by seal geometry and pressures. Hence, these effects must be carefully evaluated and factored in during the design of the seal system.

Non-contacting Seals

These seals are used extensively in high-speed turbomachinery and have good mechanical reliability. They are not positive sealing. There are two types of non-contacting seals (or clearance seals): labyrinth seals and ring seals.

Labyrinth Seals. The labyrinth is one of the simplest of sealing devices. It consists of a series of circumferential strips of metal extending from the shaft or from the bore of the shaft housing to form a cascade of annular orifices. Labyrinth seal leakage is greater than that of clearance bushings, contact seals, or film-riding seals. Consequently, labyrinth seals are utilized when a small loss in efficiency can be tolerated. They are sometimes a valuable adjunct to the primary seal.

In large gas turbines, labyrinth seals are used in static as well as dynamic applications. The essentially static function occurs where the casing parts must

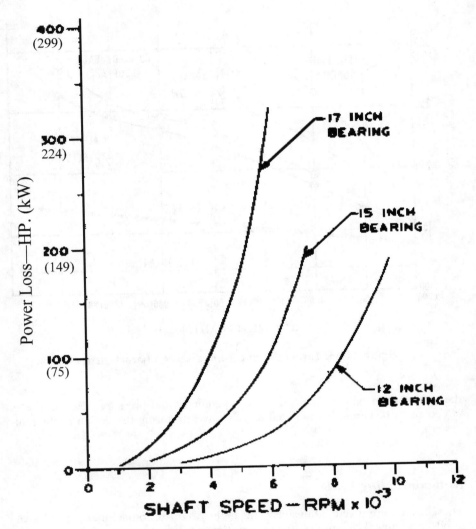

Figure 11-15. Total Power Loss in Thrust Bearings

remain unjoined to allow for differences in thermal expansion. At this junction location, the labyrinth minimizes leakage. Dynamic labyrinth applications for both turbines and compressors are interstage seals, shroud seals, balance pistons, and end seals.

The major advantages of labyrinth seals are their simplicity, reliability, tolerance to dirt, system adaptability, very low shaft power consumption, material selection flexibility, minimal effect on rotor dynamics, back diffusion reduction, integration of pressure, lack of pressure limitations, and tolerance to gross thermal variations. The major disadvantages are the high leakage, loss of machine efficiency, increased buffering costs, tolerance to ingestion of particulates with resulting damage to other critical items such as bearings, the possibility of the cavity clogging due to low gas velocities or back diffusion, and the inability to provide a simple seal system that meets OSHA or EPA standards. Because of some of the foregoing disadvantages, many machines are being converted to other types of seals.

Labyrinth seals are simple to manufacture and can be made from conventional materials. Early designs of labyrinth seals used knife-edge seals and relatively large chambers or pockets between the knives. These relatively long knives are easily subject to damage. The modern, more reliable labyrinth seals consist of sturdy closely spaced lands. Some labyrinth seals are shown in Figure 11-16. Figure 11-16a is the simplest form of the seal. Figure 11-16b shows a grooved seal is more difficult to manufacture but produces a tighter seal. Figures 11-16c and 11-16d are rotating labyrinth-type seals. Figure 11-16e shows a simple labyrinth seal with a buffered gas for which pressure must be maintained above the process gas pressure and the outlet pressure (which can be greater than or less than the atmospheric pressure). The buffered gas produces a fluid barrier to the process gas. The eductor sucks gas from the vent near the atmospheric end. Figure 11-16f shows a buffered, stepped labyrinth. This step labyrinth gives a tighter seal. The matching stationary seal is usually manufactured from soft materials such as babbitt or bronze, while the stationary or rotating labyrinth lands are made from steel. This composition enables the seal to be assembled with minimal clearance. The lands can therefore cut into softer materials to provide the necessary running clearances for adjusting to the dynamic excursions of the rotor.

To maintain maximum sealing efficiency, it is essential that the labyrinth lands maintain sharp edges in the direction of the flow. This requirement is similar to that in orifice plates. A sharp edge provides for maximum venacontracta effect, as seen in Figure 11-17, and hence maximum restriction for the leakage flows.

High fluid velocities are generated at the throats of the constrictions, and the kinetic energy is dissipated by turbulence in the chamber beyond each throat. Thus, the labyrinth is a device wherein there is a multiple loss of velocity head. With a straight labyrinth, there is some velocity carryover that results in a loss of effectiveness, especially if the throats are closely spaced. To maximize the aerodynamic blockage effect of this carryover, the diameters can be stepped or staggered to cause impingement of the expanding orifice jet on a solid transverse surface. The leakage is approximately inversely proportional to the square root of the number of labyrinth lands. Thus, if leakage is to be cut in half at a four-point labyrinth, the number of lands would have to be increased to 16, the leakage formula can be modified and written as

$$\dot{m}_1 = 0.9A \left[\frac{\frac{g}{V_0}(P_0 - P_n)}{n + \ln \frac{P_n}{P_0}} \right]^{1/2}$$

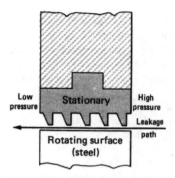

a. Simplest design. (Labyrinth materials: aluminum, bronze, babbitt or steel)

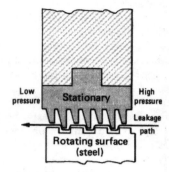

b. More difficult to manufacture but produces a tighter seal. (Same material as in a.)

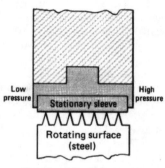

c. Rotating labyrinth type, before operation. (Sleeve material: babbitt, aluminum, nonmetallic or other soft material)

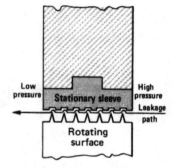

d. Rotating labyrinth, after operation. Radial and axial movement of rotor cuts grooves in sleeve material to simulate staggered type shown in b.

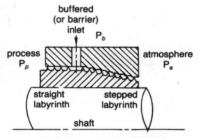

e. buffered combination labyrinth

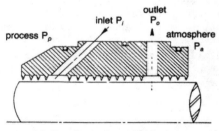

f. buffered-vented straight labyrinth

Figure 11-16. Various Configurations of Labyrinth Seals

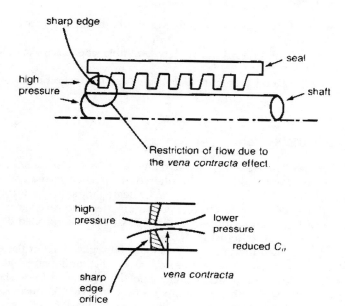

Figure 11-17. Theory Behind the Knife Edge Arrangement

For staggered labyrinths, the equation can be written as

$$\dot{m}_l = 0.75 A \left[\frac{\frac{g}{V_0}(P_0 - P_n)}{n + \ln \frac{P_n}{P_0}} \right]^{1/2}$$

where
   $\dot{m}_l$ = leakage, lb/sec
   A = leakage area of single throttling, ft^2
   P_0 = absolute pressure before the labyrinth, ft^3/lb
   V_0 = specific volume before the labyrinth, lb(m)/ft^2
   P_n = absolute pressure after the labyrinth, lb(f)/ft^2
   n = number of lands

The leakage of a labyrinth seal can be kept to a minimum by providing: (1) minimum clearance between the seal land and the seal sleeve, (2) sharp edges on the lands to to reduce the flow discharge coefficient, and (3) grooves or steps in the flow path for feducing dynamic head carryover from stage to stage.

Operational tests should:

1. Detect and correct all leaks
2. Determine relief pressures and check for proper operation of each relief valve

3. Accomplish a filter cooler changeover without causing startup of the standby pump
4. Demonstrate the control valves have suitable capacity, response, and stability
5. Demonstrate the oil pressure control valve can control oil pressure.

Lubrication Oil System

A typical lubrication oil system is shown in Figure 11-18. Oil is stored in a reservoir to feed the pumps and is then cooled, filtered, distributed to the end users, and returned to the reservoir. The reservoir can be heated for start-up purposes and is provided with local temperature indication, a high-temperature alarm and high-low-level alarm in the control room, a sight glass, and a controlled dry nitrogen purge blanket to minimize moisture intake.

The reservoir shown in Figure 11-19 should be separate from the equipment base plate and sealed against the entrance of dirt and water. The bottom should be sloped to the low drain point, and the return oil lines should enter the reservoir away from the oil pump suction to avoid disturbances of the pump suction. The working capacity should be at least 5 minutes based on normal flow. Reservoir retention time should be 10 minutes, based on normal flow and total volume below minimum operating level. Heating for the oil should also be provided. If thermostatically controlled electrical emersion heating is provided, the maximum watt density should be 15 W/in.^{2} (2.3w/cm^2). When steam heating is used, the heating element should be external to the reservoir.

The rundown level, which is the highest level the oil in the reservoir may reach during system idleness, is computed by considering the oil contained in all components, bearing and seal housings, control elements, and furnished piping that drains back to the reservoir. The rundown capacity should also include a 10% minimum allowance for the interconnecting piping.

The capacity between the minimum and the maximum operating levels in an oil system that discharges seal oil from the unit should be enough for a minimum operation of 3 days with no oil being added to the reservoir. The free surface should be a minimum of 0.25 sq ft/gpm (0.023m^2/gpm) of normal flow.

The reservoir interior should be smooth to avoid pockets and provide an unbroken finish for any interior protection. Reservoir wall-to-top junctions may be welded from the outside by utilizing full-penetration welds.

Each reservoir compartment should be provided with two three-quarter-inch (19 mm) minimum size plugged connections above the rundown oil level. These connections may be used for such services as purge gas, makeup, oil supply, and clarifier return. One connection should be strategically located to ensure an effective sweep of purge gas toward the vents.

The oil system should be equipped with a main oil pump, a standby, and, for critical machines, an emergency pump. Each pump must have its own driver, and check valves must be installed on each pump discharge to prevent reverse flow through idle pumps. The pump capacity of the main and standby pumps should be

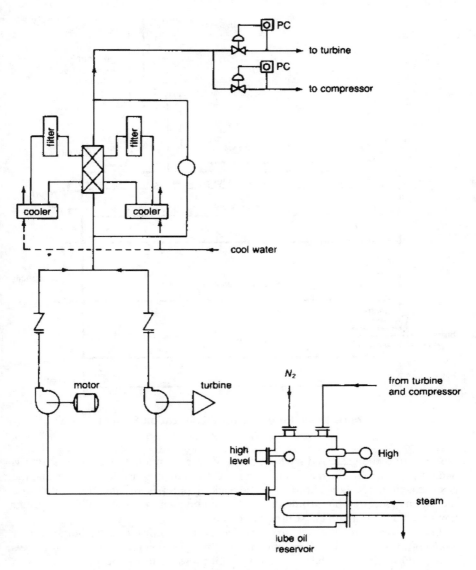

Figure 11-18. A Typical Lubrication System

10% to 15% greater than maximum system usage. The pumps should be provided with different prime movers.

The main pumps are usually steam turbine driven with an electric motor-driven back-up pump. A small mechanical-drive turbine is highly reliable as long as it is running, but it is undependable for starting automatically after long idle periods. A

422 • COGENERATION AND COMBINED CYCLE POWER PLANTS

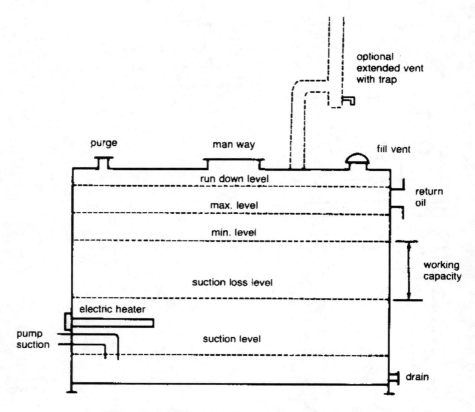

Figure 11-19. Reservoir for a Lubrication System

motor is thus the preferred back-up pump driver. A "ready-to-run" status light is usually provided for the motor in the control room to give visible evidence that the electrical circuit is viable. Starting of the back-up pump is initiated by multiple and redundant sources. The turbine drivers should be maintained for failure by either low-speed or low-steam chest pressure or both.

Low oil pressure switches are provided on the pumps and discharge header ahead of the coolers and filters, sometimes after the cooler and filters, and always at the end of the line where the reduced oil pressure feeds the various users. A signal from any of these should start the motor-driven pump and all alarms should be activated in the control room. The emergency oil pump can be driven with an AC motor but from a power source that is different to the standby pump. When DC power is available, DC electric motors can also be used. Process gas or air-driven turbines and quick-start steam turbines are often used to drive the emergency pumps.

The pump capacities for lube and control oil systems should be based upon the particular system's maximum usage (including transients) plus a minimum of 15%. The pump capacity for a seal oil system should be based upon the system's maximum usage plus either 10 gpm or 20%, whichever is greater. Maximum system usage should include allowance for normal wear. Check valves should be provided on each pump discharge to prevent reverse flow through the idle pump.

The pumps can be either centrifugal or positive displacement types. The centrifugal pumps should have a head curve continuously rising toward the shut-off point. The standby pump should be piped into the system in a manner that permits checking of the pump while the main pump is in operation. To achieve this, a restriction orifice is required with a test bleeder valve piped to the return oil line or the reservoir.

Twin oil coolers, as shown in Figure 11-20, should be provided and piped in parallel using a single multiport transflow valve to direct the oil flow to the

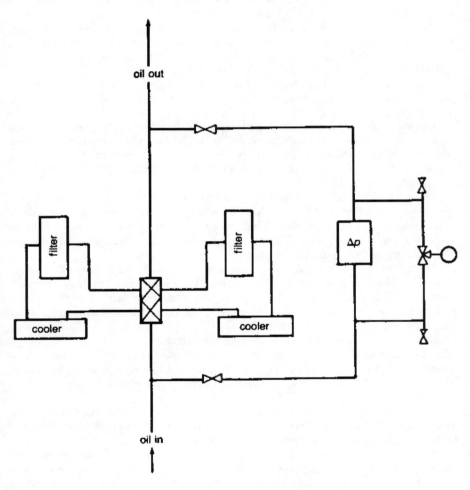

Figure 11-20. Cooler Filter Arrangement

coolers. The water should be on the tube side and the oil on the shell side. The oil-side pressure should be greater than the water-side pressure. This ratio is no assurance that water will not enter the system in the event of a tube leak, but it does reduce the risk. The oil system should be equipped with twin full-flow oil filters located downstream from the oil coolers. Since the filters are located downstream from the oil coolers, only one multiport transflow valve is required to direct the oil flow to the cooler-filter combinations. Do not pipe the filters and coolers with separate inlet and outlet block valves. Separate block valves can cause loss of oil flow from the possible human error of flow blockage during a filter switching operation.

Filtration should be 10 μm nominal. For hydrocarbon and synthetic oils, the pressure drop for clean filters should not exceed 5 psi at 100°F operating temperature at normal flow. Cartridges will have a minimum collapsing differential pressure of 50 psi (3.4 bar).

The system should have an accumulator to maintain sufficient oil pressure while the standby pump accelerates from an idle condition. An accumulator becomes a must if a steam turbine drives the standby pump. Overhead tanks are specified by many users to assure flow to critical machinery components. The sizing of the tanks varies depending on the application. In some gas turbine applications, the bearings reach maximum temperature as long as 20 minutes after shutdown.

The oil coolers and filters are controlled by a local temperature control loop with remote control room indication and high/low alarm. The coolers and filters also have an indicating differential pressure alarm. These usually feed into a common high alarm to pre-warn a need for switching and filter element replacement.

To ensure the required constant pressure, a local pressure control loop is provided on each system-turbine lube oil, compressor lube oil, and control oil. Each oil pressure system should be recorded in the control room to provide troubleshooting information. The success of the oil system depends upon not only the instrumentation, but upon proper instrument location.

The minimum alarms and trips recommended for each major driver and driven machine should be a low oil pressure alarm, a low oil pressure trip (at some point lower than the alarm point), a low oil level alarm (reservoir), a high oil filter differential pressure alarm, a high bearing metal temperature alarm, and a metal chip detector.

Each pressure - and temperature-sensing switch should be in separate housings. The switch type should be single-pole, double-throw, furnished as "open" (de-energized) to alarm and "close" (energize) to trip. The pressure switches for alarms should be installed with a "T" connection pressure gauge and bleeder valve for testing the alarm.

Thermometers should be mounted in the oil piping to measure the oil at the outlet of each radial and thrust bearing and into and out of the coolers. It is also advisable to measure bearing metal temperatures.

Pressure gauges should be provided at the discharge of the pumps, the bearing header, the control oil line, and the seal oil system. Each atmospheric oil drain line should be equipped with steel non-restrictive, bull's-eye-type flow indicators positioned for viewing through the side. View ports in oil lines can be very useful in providing a visual check for oil contamination.

In the piping arrangement and layout, it is very important to eliminate air pockets and trash collectors. Before starting a new or modified oil system, every foot of the entire system — right up to the final connection at the machine — should be methodically cleaned, flushed, drained, refilled and all instruments thoroughly checked.

Lubricant Selection

A good turbomachinery oil must have a rust and oxidation inhibitor, good demulsibility and correct viscosity, and be both non-sludging and form-resistant. Besides lubrication, the oil has to cool bearings and gears, prevent excessive metal-to-metal contact during starts, transmit pressure in control systems, carry away foreign materials, reduce corrosion, and rest, degradation.

For gas turbines, especially the more advanced turbines, synthetic oil should be used since synthetic oils have a high flash point. Mineral oils can be used for steam turbines. It is not uncommon to have two types of oil in a plant. Mineral oil costs much less than synthetic oil.

The selection of the correct lubricant must begin with the manufacturer. Refer to the operator's instruction manual for the oil required and the recommended viscosity range. The local environmental conditions should be seriously considered, including exposure to outside element conditions, acid gas, or steam leaks. As a general rule, most turbornachines are lubricated with premium-quality turbine-grade oil. However, under certain environmental conditions, it may be advantageous to consider another oil. For example, if a machine is subject to exposure to low concentrations of chlorine or anhydrous hydrochloric acid gases, it may be better to select another oil that will outperform the premium turbine oil. Good results have been recorded using oil containing alkaline additives. Certain automotive or diesel engine oils contain the optimum amount and type of alkaline additives to protect the base oil from reaction with chlorine and HCL. In services where the attack on the lubricant by the gas is unknown, laboratory tests are suggested.

Oil Sampling and Testing

Oils from turbomachinery should be tested periodically to determine their suitability for continued use; however, visual inspection of the oil can be useful in detecting contaminated oil when the appearance and odor are changed by the contaminant.

An oil sample should be withdrawn from the system and analyzed in the laboratory. The usual tests of the used oil include: (1) viscosity, (2) pH and neutralization number, and (3) precipitation. The test results will indicate changes from the original specifications and, depending on how extensive these changes are, whether the oil can or cannot be used in the machines.

Oil Contamination

Oil contamination in a turbomachine is one of the major problems that maintenance crews face. However, while contamination is a continuous problem, the levels of contamination are what is causing the most concern.

The greatest source of contamination is an extraneous matter. Atmospheric dirt, for example, is always a serious threat. It can enter the oil system through vents, breathers, and seals. Its primary effect is equipment wear, but plugging of oil lines and ports and reduced oxidation stability of the oil are also serious effects.

Metal particles from wear and rust particles from reservoir and oil piping corrosion can lead to premature equipment failure and oil deterioration. It is important to provide suitable filtering equipment to remove these particles from the system.

Water contamination is a constant threat, especially in the steam turbine. The sources of water are many: atmospheric condensation, steam leaks, oil coolers, and reservoir leaks. Rusting of machine parts and the effects of rust particles in the oil system are the major results of water in oil. In addition, water forms an emulsion and, combined with other impurities, such as wear metal and rust particles, acts as a catalyst to promote oil oxidation.

To remove insoluble contaminants, various types of full-flow filters can be used. Two general types are usually selected: surface filters and depth filters. Both types of filters are effective for the removal of particulate matter.

Surface filters, if manufactured from the correct material, will not be affected by water in the oil. Water-resistant pleated paper elements have much greater surface areas than the depth-type element and yield a much lower differential pressure when used as replacement elements in filters originally equipped with depth-type elements. Pleated paper elements, which will remove particle sizes down to a nominal 0.5 μm, are available.

The depth-type filter elements are used when oil is free from water, and when particle sizes to be removed are in the 5-μm-and-greater range. Generally, the depth-type element is water-sensitive, and when oil is contaminated with moisture, this element type will absorb the water and produce a rapid increase in differential pressure across the filter. The desired maximum differential pressure across a filter with clean elements is 5 psig at normal operating temperature.

Filter Selection

The filter elements should remove particles of 5 μm, must be water-resistant, have a high flow rate capability with low pressure drop, possess high dirt-retention capacity, and be rupture-resistant. The clean pressure drop should not exceed 5 psig (0.34 bar) at 100°F (37.8°C). The elements must have a minimum collapse differential pressure of 50 psig (3.4 bar). Pleated paper elements are preferred provided they meet these requirements. Usually, the pleated paper element will yield the five psig clean drop when used in a filter that was sized to use depth-type elements. This result is due to the greater surface area of the pleated element, more than twice the area of a conventional stacked disc-type or other depth-type elements.

A differential pressure switch set to alarm when the pressure drop reaches a predetermined point protects against the loss of oil flow. In addition to the differential pressure switch, a two-way, three-port valve with a pressure gauge is piped in parallel with the differential pressure switch for accurate indication of inlet and outlet oil filter pressure. When a single transflow valve is used with a cooler filter installation, the differential pressure switch and pressure gauge assembly should span the cooler filter system.

Water contamination in the oil system can cause serious damage to turbomachinery, and every reasonable effort should be made to, first, prevent its entrance into the system, and second, provide suitable removal equipment if water cannot be effectively kept out. Experience indicates that designers and equipment operators can be more effective in keeping water out of the system. Since the main sources of contamination are atmospheric condensation, steam leaks, and faulty oil coolers, preventive measures should be taken.

Condensation will occur in the atmospheric vented oil system whenever the temperature in the vapor space areas drops below the dew point. This effect can take place in the return oil piping as well as the reservoir. Consoles installed in unprotected locations are more vulnerable to climatic changes than those installed inside buildings. The outside locations will be adversely affected by temperature cycles between daytime and night operations — also, by showers and sudden temperature drops due to other weather changes, especially in the fall and winter seasons. There has been great success in "drying up" oil systems by making a few simple alterations. The first step is to check the reservoir unit. The vent should be located in the very top of the reservoir. It should be free of baffles that can collect and return condensate to the reservoir, and the length should be kept as short as possible to provide a minimum of surface areas on which condensate can form. If it is necessary to run the vent up and away from the reservoir, a water trap should be provided as close to the reservoir as possible to remove any condensate formed in the vent stack. The next step is to provide and maintain an inert gas or dry-air purge on the reservoir. Only 2-5 cfh (0.06-0.14 cm/hr) is required. The reservoir purge system will not substitute for the elimination of other water sources.

Steam and condensate leaks are the most difficult water sources to prevent in turbomachinery; however, it can be done, and every effort should be made to eliminate these sources. Obviously, the first means of prevention is to maintain the steam packing in perfect condition. Experience has shown that eventually, the steam packings will leak, and steam condensate will enter the system through the bearing seals. There has also been great success in "drying up" a wet oil system. The procedure is to purge the bearing labyrinths with inert gas or dry air. One method is to drill a one-eighth-inch hole through the bearing cap and intersect the labyrinth. A one-quarter-inch diameter tube is connected to the hole in the bearing cap and to a rotometer. The labyrinth is then purged with 15 cfh (0.42 cm/hr) dry air or inert gas.

Another method is to install an external labyrinth with purge provisions on the bearing housings of a machine that has the necessary space to accommodate the external seal.

Removal of free water from oil systems is usually done with centrifuges or coalescer separators. Centrifuging is the costliest method in both capital outlay and operating costs. The centrifuges are usually the conventional disc-type with manual cleaning. The discs must be cleaned at least once each week with 1 hour required per cleaning. The coalescer separators usually require much less attention. Some separators only require element changes once a year while others may require changes at 6 months or 3 months, and in some instances, once a month. The frequency appears to be related to the amount of water in the oil system. In many instances, coalescer element changes have been reduced by the use of a pre-filter in the system. This element removes the particulates (usually rust) that would restrict the 2-μm coalescer element. The time required to change both the pre-filter element and coalescer separator elements is less than 1 hour.

Cleaning and Flushing

Serious mechanical damage to turbomachinery can result from operation with dirty oil systems. It is essential that an oil system be thoroughly cleaned prior to the initial startup of a new machine, and after each overhaul of an existing machine.

Preliminary steps for the initial startup and startup after the overhaul are similar, except for the reservoir and oil requirements on the machine after an overhaul. For an overhauled machine, the oil is drained and tested for condition. If there is no water or metal changes, the oil may be used again.

Inspect the reservoir interior for rust and other deposits. Remove any rust with scrapers and wire brushes, wash down the interior with a detergent solution, and flush with clean water. Dry the interior by blowing the surfaces with dry air, and use a vacuum cleaner to remove trapped liquids.

Install all new 5-μm pleated-paper elements in the filters. Connect steam piping to the water side of the oil coolers for heating the oil during the flush. Remove the orifice and install jumpers at the bearings, coupling, controls, governor, and other critical parts to prevent damage from debris during the flush. Make provisions for 40-mesh telltale screens at each jumper. The conical-shaped screen is preferred, but a flat screen is acceptable. Adjust all control valves in the full-open position to allow maximum flushing flow. The effectiveness of the flush depends to a large extent on high flow velocities through the system to carry the debris into the reservoir and filters. It may be necessary to sectionalize the system to obtain maximum velocities by alternately blocking off branch lines during the flush.

Fill the reservoir with new or clean used oil. Begin the flush without telltale screens by running the pump or pumps to provide the highest possible flow rate. Heat the oil to 160°F with steam on the oil cooler. Cycle the temperature between 110°F and 160°F to thermally exercise the pipe. Tap the piping to dislodge debris, especially along the horizontal sections. Flush through one complete temperature cycle, shut down and install the telltale screens, and flush for an additional 30 minutes. Remove screens and check for amount and type of debris. Repeat the preceding procedure until the screens are clean after two consecutive inspections. Observe the pressure drop across the filters during the consecutive operation. Do not allow the pressure drop to exceed 20 psig. When the system is considered clean, empty the oil reservoir, and clean out all debris by washing with a detergent solution followed by a freshwater rinse. Dry the interior by blowing with dry air, and vacuum any freestanding water. Replace the filter elements. Remove jumpers and replace orifices. Return controls to their normal settings. Refill the oil reservoir with the same oil used in the flush if lab tests indicate that it is satisfactory; otherwise, refill with new oil.

Because of the high flow velocities obtained during the flush, the previous procedure will allow the fastest possible cleanup of the oil system. The objective is to carry the debris into the reservoir and filters. The turbulence from the high flows, along with the thermal and mechanical exercising of the piping, are the main factors necessary for a fast and effective system cleanup.

Coupling Lubrication

Couplings are a very critical part of any turbomachinery. They must be carefully designed and proper lubrication must be applied. The most common methods of lubricating gear-type couplings are: (1) grease-packed, (2) oil-filled, or (3) continuous oil flow.

The grease-packed and oil-filled couplings offer similar advantages and disadvantages. The main advantage is simplicity of operation. They are also economical, easy to maintain, and the grease type resists the entry of contaminants. In addition, high tooth leading can be accommodated, since lubricants with heavy-bodied oil can be used. An important requisite for the oil-filled type is that the coupling must have an adequate static oil capacity to provide the required amount of oil to submerge the teeth when the coupling is in operation. The greatest disadvantage of these couplings is the possible loss of lubricants during operation due to defective flange gaskets, loose flange bolts, lubricant plugs, and flaws in the coupling flanges and spacers.

The lubricant of a gear coupling must withstand severe service from forces in the coupling, which exceed 8000 g's. For grease-filled couplings, special quality grease is required to prevent mating teeth wear while operating at high-g loads in a sliding load environment. This severe operating condition causes grease separation at high speeds and results in excessive wear. Tests indicate that grease separation is a function of g levels and time. Therefore, the grease coupling is not considered suitable for high-speed service, except when approved high-speed coupling grease is used, and then only for up to 1 year of continuous operation. Presently, new greases on the market do not separate at high speeds and may not deteriorate for 3 years of continuous operation.

The continuous oil-flow method is used primarily in high-speed rotating machinery. This method provides the potential for maximum continuous periods of operation at high operating speeds. The oil flow also provides cooling by carrying away heat generated within the coupling. Another important advantage is maximum reliability, since the oil supply is constant, and the loss of oil from within the coupling is not a problem as it is with oil-filled or grease-packed couplings. The main requisites for this method are: (1) provision of adequate oil flow into the coupling, (2) the oil must be absolutely clean, and (3) it must cool to carry away heat.

Some of the disadvantages of the continuous oil-flow coupling are: (1) increased cost, (2) requires supply oil and return oil piping, and (3) the entry of foreign solids with the oil will cause accelerated wear.

Foreign material in the oil is a major problem with the continuous oil-flow coupling. Since high centrifugal forces are developed within the coupling, any induced solids and water will be extracted from the oil and retained in the coupling. Abrasive wear is usually caused by the trapped sludge. In addition to foreign abrasives, the sludge will retain the wear metal and will contribute to the coupling wear rate. Sludging has been reduced within couplings by improved oil filtration. Filters can be equipped with differential pressure alarms so that replacement can be made.

The gear-type coupling on turbomachinery can be successfully lubricated by both oil and grease methods. The grease-packed and oil-packed couplings must be absolutely oil- and grease-tight to prevent the loss of lubricant, and the very best high-speed coupling grease must be used. The continuous-flow type must have absolutely clean oil supplied continuously at the designed flow rate.

Lubrication Management Program

A well-planned and -managed lubrication program is an important factor in the overall maintenance plan of a plant. A lubrication program includes developing a lubrication period maintenance program, sampling and testing oil, and developing specific procedures to apply lubricants. The initial step in developing a comprehensive plant lubrication program is to conduct a plant survey to determine existing lubrication practices. The survey should utilize machine drawings and external machinery inspections.

A detailed list of lubricant types and their points of application can be compiled from the results of the survey. Combining the list of lubrication types and a current schedule, a master plant lubrication schedule can be published.

A monthly lubrication schedule can then be issued to the appropriate maintenance personnel to serve as a reminder. Issuance of the lubrication schedule does not ensure its compliance, and supervisors should check to see that required lubrication is performed.

As a part of the lubrication program, oil should be periodically tested. Testing requires drawing oil from the system for laboratory analysis. The usual tests conducted to determine the condition of oils include viscosity, pH and neutralization number, precipitation, color and odor, and a check for foreign particles in the oil. The results should be reviewed and compared with new oil characteristics to determine the life characteristics of the oil.

A program for evaluating any new lubrication products can be used to indicate the possible replacement of current lubricants. The general characteristics of new lubricants can be obtained from specifications provided by suppliers or from testing of the lubricant. The final selection of new lubricants should be made only after close observation of the lubricant in several typical plant applications. During the monthly inspections, new lubricants should be checked especially closely to ensure that they retain their desired properties. While all lubrication applications are important to machinery health, gear couplings present special critical lubrication problems and require special attention as explained previously.

Operating experience has proven that unless a continuous program of required lubrication is followed, even the most well-designed units are sure to fail. A proper lubrication management program must incorporate a monthly lubrication schedule, an evaluation of new lubrication products, and supervision to ensure that the prescribed procedures are carried out by maintenance personnel.

In the event of failures due to lubrication problems, the failures should be thoroughly analyzed to determine if they were indeed caused by lubricant failure or incorrect maintenance procedures. Once the problem has been isolated, corrective action can be initiated to prevent subsequent similar failures — whether it requires changing lubricants or procedures.

Chapter 12

CONTROL SYSTEMS AND CONDITION MONITORING

Power plants operate over a large range of power; and, in cogeneration systems, the process requirements also control the operations of the plant. The plants are all part of a major grid and they supply the needs to meet the demand of the grid. Control systems are also closely tied up with the distributed control systems (D-CS) and condition monitoring systems (C-MS) with plant optimization software.

The traditional concept of maintenance in the utilities industry has been undergoing a major change to ensure that equipment not only has the best availability but also is operating at its maximum efficiency. There is a consistent trend in the utilities industry throughout the world to improve maintenance strategy from fix-as-fail to total performance-based planned maintenance. In practice, this calls for on-line monitoring and condition management of all major equipment in the plant. To reach the Utopian goal of just-in-time maintenance with minor disruption in the operation of the plant requires a very close understanding of the thermodynamic and mechanical aspects of plant equipment to be able to implement predictive maintenance programs.

The benefits of total performance-based planned maintenance not only ensure the best and lowest cost maintenance program, but also that the plant is operated at its most efficient point. An important supplementary effect is that the plant will be operating consistently within its environmental constraints.

Control Systems

Control systems can be an open loop or closed loop system. The open loop system positions the manipulated variable either manually or on a programmed basis, without using any process measurements. A closed loop control system is one that receives one or more measured process variables and then uses it to move the manipulated variable to control a device. Most combined cycle power plants have a closed loop control system.

Closed loop systems include either a feedback or feedforward, control loop or both to control the plant. In a feedback control loop, the controlled variable is compared to a set point. The difference between the controlled variable to the set point is the deviation for the controller to act on to minimize the deviation. A feedforward control system uses the measured load or set point to position the manipulated variable in such a manner to minimize any resulting deviation.

In many cases, the feedforward control is usually combined with a feedback system to eliminate any offset resulting from inaccurate measurements and calculations. The feedback controller can either bias or multiply the feedforward calculation.

A controller has tuning parameters related to proportional, integrated, derivative, lag, deadtime, and sampling functions. A negative control loop will oscillate if the controller gain is too high; however, if it is too low, it will be ineffective. The controller must be properly related to the process parameters to ensure close loop stability while still providing effective control. This is accomplished first by the proper selection of control modes to satisfy the requirements of the process, and second by the appropriate tuning of those modes. Figure 12-1 shows a typical block diagram for forward and feedback control.

Computers have been used in the new systems to replace analog PID controllers, either by setting set points, or lower level set points in supervisory control, or by driving valves in direct digital control. Single station digital controllers perform PID control in one or two loops, including computing functions such as mathematical operations, with digital logic and alarms. D-CS provide all the functions, with the digital proc shared among many control loops. A high-level computer may be introduced to provide condition monitoring, optimization and maintenance scheduling.

The combined cycle power plant system consists of control systems of the gas turbine, the HRSG system and the steam turbine. These control systems are fully automated, and ensure the startup of the gas turbine and the steam turbine. The gas turbine control system is complex and has a number of safety interlocks to ensure the safe startup of the turbine. The start-up speed and temperature acceleration curves as shown in Figure 12-2 are one such safety measure. If the temperature or speed is not reached in a certain time span from ignition, the turbine will be shut down. During startup, the turbine is on an auxiliary drive, which speeds up the turbine to about 1500 to 2000 rpm at that point the turbine is ignited and the turbine speed and temperature rise very rapidly. The turbine is usually declutched at speeds around 2400 rpm. If the turbine is a two- or three-shaft turbine, as is the case with aero-derivative turbines, the power turbine shaft will "break loose" at a speed of about 60% of the rated speed of the turbine.

In the early days when these acceleration and temperature curves were not used, the fuel, which was not ignited, was carried from the combustor, and then deposited at the first or second turbine nozzle, where the fuel combusted, which resulted in the burnout of the turbine nozzles. After an aborted start, the turbine must be fully purged of any fuel before the next start is attempted. To achieve the purge of any fuel residual from the turbine, there must be about seven times the turbine volume of air that must be exhausted before combustion is once again attempted.

A combined cycle power plant is a complex system. A typical control system with hierarchic levels of automation is shown in Figure 12-3. The combined cycle power plant has, as the basic machinery units, the gas turbine, the HRSG,

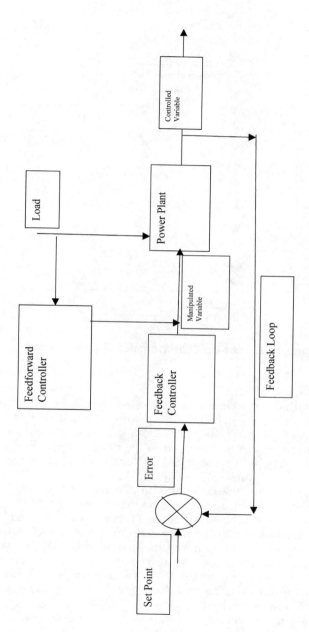

Figure 12-1. Forward and Feedback Control Loop

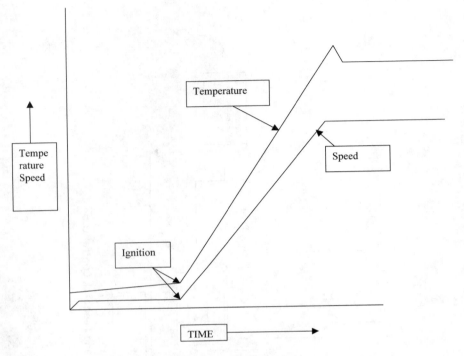

Figure 12-2. Start-Up Characteristics of a Gas Turbine

and the steam turbine. The control system at the plant level consists of a DC-S system, which in many new installations is connected to a condition-monitoring system and an optimization system. The DC-S system is what is considered to be a plant-level system and is connected to the three machine-level systems. It can, in some cases, also be connected to functional level systems such as lubrication systems and fuel-handling systems. In those cases, it would give a signal of readiness from those systems to the machine level systems. The condition-monitoring system receives all its inputs from the DC-S system and from the steam- and gas-turbine controllers. The signals are checked initially for their accuracy and then a full machinery performance analysis is provided. The new performance curves, produced by the condition monitoring system, are then provided to the optimization system. The optimization system receives the load and then sends a signal to the DC-S system, which, in turn, sends the signal to the gas turbine and steam turbine for the best settings of the steam and gas turbine to meet the load.

The gas turbine has a number of systems as it controls, such as the following:

1. Lubrication Skid. The gas-turbine lubrication skid is usually independent of the steam-turbine skid as the lubrication oil is usually synthetic due to

Control Systems and Condition Monitoring • 437

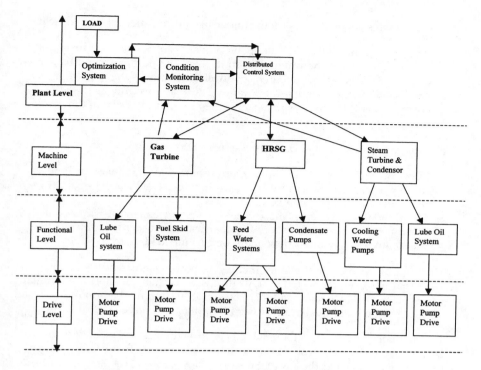

Figure 12-3. Hierarchic Levels of Automation

the high temperatures in the gas turbine. Another reason is due to water contamination of the lubrication oil from the steam turbine is advisable to be totally independent. The gas-turbine lubrication skid would report to the gas-turbine controller. Since the lubrication system is also used for providing cooling, it is usually operated for about 20 minutes after the gas turbine is shutdown. The lubrication skid contains at least three pumps; two pumps which each can provide the head required and a third pump, which is usually recommended to be a DC drive for emergency use. These pumps and their control fall under the drive level hierarchy.

2. The Fuel Skid. This could contain a gas compressor if the fuel gas pressure is low, and a knockout drum for any liquid contamination that the gas may have. The requirement of fuel gas pressure is that it should be operated at a minimum of 50 psi (3.5 bar) above the compressor discharge pressure. The compressor and its motor drive all fall under the drive-level hierarchy. In case of liquid fuels, the skid may also contain a fuel treatment plant, which would have centrifuges, electrostatic precipitators, fuel additive pumps, and other equipment. These could be directly controlled by the D-CS system, which would then report its readiness to the gas turbine controller.

The HRSG has also various systems under its control, such as the following:

1. Water Treatment Plant. This plant consists of various pumps and containers to treat the water. The controls of the pumps should be part of the HRSG hierarchic or can be controlled directly by the D-CS System.
2. Feedwater Pumps Condensate Pumps. These are also under the HRSG control and hierarchy. In some plants, they are placed as part of the steam turbine control system.

The steam turbine has also various systems under its control:

1. Lubrication System. The lubrication system for the steam turbine is usually a separate system from the gas turbine. The lower temperatures mean that mineral oil can be used.
2. The Condenser System. The condenser system can be a water-cooled system or an air-cooled system. In either way, the system is controlled by the steam turbine and is dependent on turbine load.

The control system for the combined cycle power plant works to satisfy the demand that has been requested of it. As discussed earlier, the most predominant control system used in a power plant is a closed loop system. The important closed control loops of a combined cycle power plant falls into three control loops:

1. The main plant control loop, which controls the gas and steam turbines.
2. The secondary control loops, which controls the important process parameters such as the gas turbine firing temperature, the Steam turbine inlet temperature, pressure, and flow.
3. The auxiliary control loops, which maintains the fuel injection pressure, the Lubrication oil pressure and temperature, the various feedwater and condensate pumps.

The gas and steam turbine control loops are the most important in a combined cycle power plant. In the combined cycle power plant, the plant output is controlled by the gas turbine; the steam turbine will always follow the gas turbine by generating power with whatever steam is produced in the HRSG. Figure 12-4 illustrates the control closed loop of a typical combined cycle power plant.

The gas turbine control loop controls the inlet guide vanes (IGV) and the gas turbine inlet temperature (TIT). The gas turbine inlet temperature is defined as the temperature at the inlet of the first-stage turbine nozzle. Presently, 99% of the units the inlet temperature is controlled by an algorithm, which relates the turbine exhaust temperature, or the turbine temperature after the gasifier turbine, the compressor pressure ratio, the compressor exit temperature and the air mass flow to the turbine inlet temperature. New technologies are being developed to measure the TIT directly by the use of pyrometers and other specialized probes, which could last in these harsh environments. The TIT is controlled by the fuel flow and the IGV, which controls the total air mass flow to the gas turbine. In a combined cycle power plant application, the turbine exhaust temperature is maintained at or near a constant down to about 40% of the load.

The steam turbine will adjust its load automatically after a short time, depending on the characteristics of the HRSG. In some cases, the steam turbine could be controlled independently for sudden changes in the load. This would require the steam turbine to be controlled in a continuous throttle position, resulting in poorer efficiencies at full and part loads. Most modern plants rely on the gas turbine control system, as the gas turbine produces about 60% of the power, and also reacts very fast.

Figure 12-4 shows a typical two-gas turbine, HRSG and a single steam turbine configuration which is about 70% of the combined cycle power plants in use today. The load is set into the overall controller, which in conjunction with an optimization program decides on how the load will be shared by each gas turbine. The load of each gas turbine is controlled by the IGV and the fuel flow.

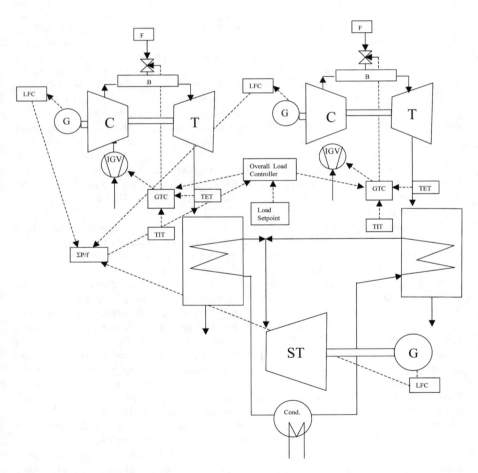

Figure 12-4. Control Closed Loop of a Typical Combined Cycle Power Plant

The entire steam turbine is operated in a sliding pressure mode with fully open steam turbine valves to about 50% steam inlet pressure. Supplementary fired HRSGs would require a separate control system for the steam turbine in a manner very similar to stand-alone steam turbine plants.

All power plants are synchronized to the overall grid and, thus, the operation of the plant at the given frequency is very important. The grid cannot stand much fluctuations of the plant frequency. It is, therefore, very important to operate the plant at its assigned frequency, which is 60 Hz in the U.S. and the Americas as well as many countries in the mid-east. Europe and most of Asia are operating at a 50-Hz frequency. If there is a frequency change, this must be taken care of in seconds.

Frequency response will be needed outside a dead band of ±0.1 Hz. The dead band is essential for stable operation of a plant; otherwise, the plant could oscillate and plant failures have occurred due o a lack of a dead band. Frequency droop is a major problem in plants due to machinery degradation. The standard drop setting is about 5%, which means that a grid frequency droop of 5% would cause an increase of the load by 100%. Gas turbines can easily take swings of 20% to 30%, but large swings cause changes in firing temperature, which places a large strain on the hot section of the turbine. Gas turbines are rated for peak operation to about 10% to 15% of their base load. It is, therefore, suggested that the gas turbine be operated at about 95% of the base load so that there is room for adjustment.

Figure 12-5 shows the behavior of the steam turbine and gas turbine for changes in frequency. The figure shows changes in the gas turbine and the steam turbine in such a plant. The steam turbine in this diagram assumes that there is no steam load control; thus, there is no throttling of the steam so the steam turbine cannot response fast enough to a frequency drop. The falling frequency must be taken up by the gas turbine by a fast change in increasing the load. For an increasing frequency, the gas turbine and the steam turbine both can respond as shown in the figure. The gas turbine (60% load) and the steam turbine (40% load) take their appropriate change in load.

The startup and shutdown of a typical combined cycle power plant is shown in Figures 12-6 and 12-7, respectively. The time and percentages are approximate values and will vary depending upon the plant design, especially for plants where the steam turbine and the gas turbine are on the same shaft.

Studying the behavior of the combined cycle power plant during startup, we need to examine the various components, starting with the gas turbine. The gas turbine during the startup is initially brought to a speed of about 1200 to 1500 rpm when ignition takes place. The turbine temperature, flow and speed increases in a very short time of about 3 to 5 minutes. The gas turbine is then parked at about 80% of its firing temperature to let the HRSG and the steam turbine catch up. Before the steam turbine is started, the condenser must be evacuated and the steam gland system must be in operation. In about 20 minutes after start up of the gas turbine, there is enough steam at an elevated temperature and pressure for the steam turbine to start rolling. The steam turbine is brought up to about 40% of its speed when the system is maintained at a steady condition for the system to soak the heat. This for a large turbine is about 100 minutes from startup and for smaller systems, it can be as low as 50 minutes. At this point, the gas turbine is brought up to its base load and the steam turbine follows and reaches its full load about 20 to 30 minutes after the gas turbine is at full load.

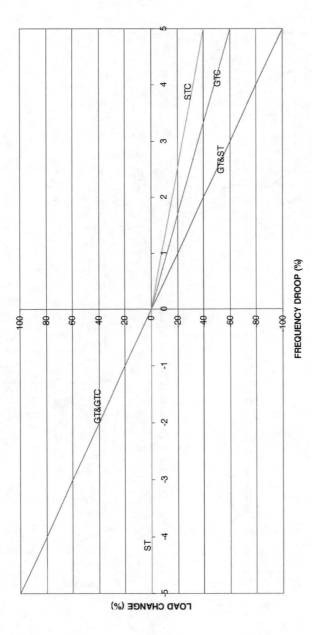

Figure 12-5. Droop Curves for Combined Cycle Power Plants

442 • COGENERATON AND COMBINED CYCLE POWER PLANTS

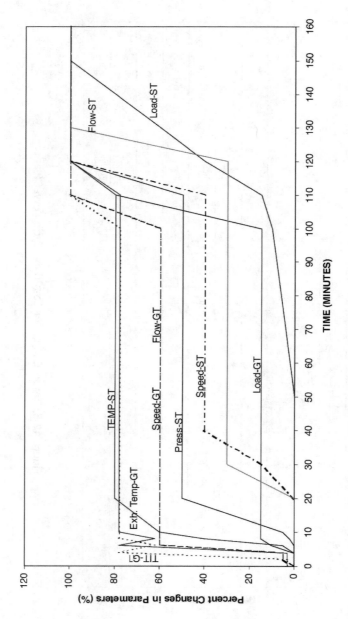

Figure 12-6. A Typical Start-Up Curve for a Combined Cycle Power Plant

Control Systems and Condition Monitoring • 443

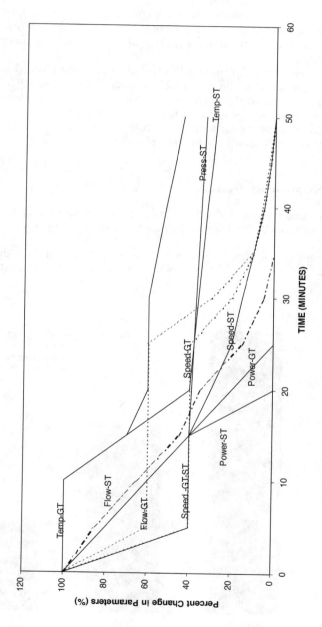

Figure 12-7. A Shutdown Curve for a Combined Cycle Power Plant

444 • COGENERATON AND COMBINED CYCLE POWER PLANTS

If supplementary firing or steam injection for power augmentation is part of the plant system, these should be turned on only after the gas turbine has reached full flow. The injection of steam for power augmentation if done before full load could cause the gas turbine compressor to surge. The supplementary firing should be done only after the steam turbine bypasses are closed and the steam turbine can accommodate.

The shutdown of a typical plant requires the lowering of the gas turbine load to a minimum level about 40%; the steam turbine is shut down. The HRSG and the gas turbine are then unloaded and shut down. The gas turbine and steam turbine must be put on a turning gear to ensure that the turbines do not bow. The lubrication systems must be on so that the lubrication can cool of the various components. This usually takes about 30 to 60 minutes.

Condition Monitoring Systems

Predictive performance-based condition monitoring is emerging, as a major maintenance technique, with large reduction in maintenance costs as shown in Figure 12-8.

The histogram shows that although an approximate one-third reduction in operating and maintenance (O&M) costs was achieved by moving from a

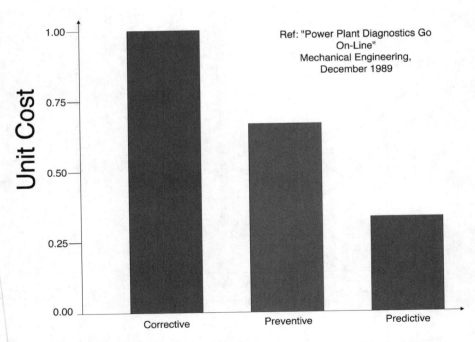

Figure 12-8. Comparison Between Various Maintenance Techniques

"corrective" — more realistically termed a "breakdown" or "fix as fail" — repair strategy to a "preventive" regime, this yielded only approximately half of the maximum cost savings. Although more difficult to introduce than the simple scheduling of traditional maintenance activities required for preventive action, the EPRI research showed that the introduction of "preventive" maintenance strategies could yield a further one-third reduction in O&M costs.

The introduction of the total maintenance condition-monitoring system means the use of composite condition-monitoring systems, which combine mechanical and performance-based analysis with corrosion monitoring. These three components are the primary building blocks that enable the introduction of a comprehensive plantwide condition management strategy. Numerous case studies have shown that many turbomachinery operational problems can only be diagnosed and resolved by correlating the representative performance parameters with mechanical parameters.

In plant health terms, monitoring and measurement both cost money and are only half-way stage on the way to the real objective, which is the avoidance of cost and plant damage. Condition management makes proper use of both activities and exploits information derived from them to generate money for the plant operator. Good plant condition management, therefore, should be the objective of materials and machine health specialists.

The change has further implications: in the past, corrosion and condition monitoring were considered to be service activities, providing only a *reactive* strategy. Condition *management* embodies a *pro-active* stance on plant health. This fundamental understanding should not go unrecognized by the materials and condition-monitoring specialists. Condition management is a huge opportunity for technical specialists to provide the best possible service to clients, whether internal or external. The same specialists also will be able to derive the maximum direct benefit from their expertise.

Conventional alloy selection, coating specification and failure investigation skills will always be required, as will inspection services to confirm the condition of the plant. However, the phenomenon-labeled corrosion should no longer be regarded as a necessary evil as it is only a problem when out of control. The electrochemical behavior characterizing corrosion is also the means by which on-line plant health management can be achieved.

Major power plant complexes contain various types of large machinery. Examples include many types of machinery, in particular, gas and steam turbines, pumps and compressors, and their effect on the heat recovery steam generators (HRSG), condensers, cooling towers, and other major plant equipment. Many combined cycle power trains are on a single shaft. Thus, the logical trend in condition monitoring is to multimachine train monitoring. To accomplish this goal, an extensive database, which contains data from all machine trains, along with many composite multimachine analysis algorithms are implemented in a systematic and modular form in a central system.

Implementation of advanced performance degradation models necessitate the inclusion of advanced instrumentation and sensors, such as pyrometers for monitoring hot section components, dynamic pressure transducers for detection of surge, and other flow instabilities such as combustion especially in the new dry low NO_x combustors. To fully round out a condition-monitoring system, the use of expert systems in determining fault and life cycle of various components is a necessity.

The benefits of total performance-based planned maintenance not only ensure the best and lowest cost maintenance program, but also that the plant is operated at its most efficient point. An important supplementary effect is that the plant will be operating consistently within its environmental constraints.

The new purchasing mantra for the new utility plants is "life cycle cost" and to properly ensure that this is achieved, a "total performance condition monitoring" strategy is unsurpassed.

Identification of Losses in a Combined Cycle Power Plants

Combined cycle power plants is a combination of the Brayton and Rankine Cycles. The two cycles makes a very powerful process, which brings efficient, reliable, and relatively clean power to the next millennium. The high efficiencies that are obtained in this cycle, 50% to 58% far eclipses the efficiencies that were obtained by the steam power plants of the 1980s, which ranged from 30% to 35 %. These high efficiencies are fueled by the increase in the firing temperatures, and the high-pressure ratio, of the new gas turbines. The steam turbines used in this cycle have been more traditional and have not seen the dramatic change that has been seen in the gas turbines. The steam turbine cycle has seen an increase in the pressure and temperature entering the high-pressure steam turbine. This has resulted in the increase of the efficiency of the Rankine Cycle.

To fully understand the performance of the combined cycle, the analysis of the four major components, which make up the combined cycle plant, are:

1. The Gas Turbine
2. Heat Recovery Steam Generator
3. Steam Turbine
4. Condenser.

The Gas Turbine

The new gas turbines are the cornerstone of the rise of the combined cycle as the power source of the new millennium. The new gas turbines have a very high-pressure ratio, a high firing temperature, and, in some case, a reheat burner in the gas turbine. The gas turbines have also new dry low NO_x combustors. The combination of all these components has dramatically increased the thermal efficiency of the gas turbine. The gas turbine since the early 1960s has gone from efficiencies as low 15% to 17% to efficiencies around 45%. This has been due to the pressure ratio increase from around 7:1 to as high as 30:1, and an increase in the firing temperature from about 1500°F (815°C) to about 2500°F (1371°C). With these changes, we have also seen that the efficiency of the major components in the gas turbine increases dramatically. The gas turbine compressor efficiency increased from around 78% to 87%; the combustor efficiency from about 94% to 98%, and the turbine expander efficiency from about 84% to 92%.

The increase in compressor pressure ratio decreases the operating range of the compressor. The operating range of the compressor stretches from the surge line at the low flow end of the compressor speed line to the choke point at the high flow end. As seen in Figure 12-9, the lower-pressure speed line has a larger operational range than the higher-pressure speed line. Therefore, the higher-pressure ratio compressors are subject to fouling, and can result in surge problems or blade excitation problems, which lead to blade failure.

The drop in pressure ratio at the turbine inlet due to filter fouling amounts to a substantial loss in the turbine overall efficiency and the power produced. An increase in the pressure drop of about 25 mm WC, amounts to a drop of about 0.3% reduction in power. Table 12-1 shows the approximate changes that would occur for changes in ambient conditions, the fouling of the inlet filtration system, and the increase in back pressure on the gas turbine in a combined cycle mode. These modes were selected because these are the most common changes that occur on a system in the field. It must be remembered that these are just approximations and will vary for individual power plants.

The gas turbine has to be operated at a constant speed since this is used for power generation and any slight variation in speed could result in major problems for the grid. Thus, the control of the load has to be by controlling the fuel input, therefore, the turbine firing temperature, and the inlet guide vane position, thus

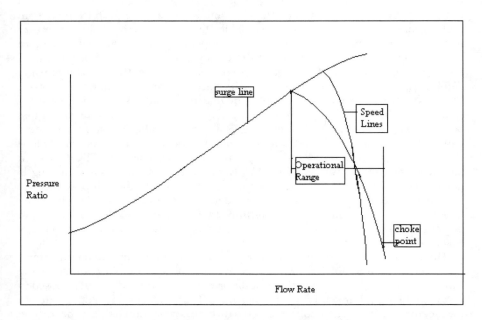

Figure 12-9. Performance Map of an Axial Flow Compressor Used in Most Gas Turbines

448 • COGENERATON AND COMBINED CYCLE POWER PLANTS

Table 12-1. Effect of Various Parameters on the Output and Heat Rate

Parameters	Parameter Change	Power Output	Heat Rate Change
Ambient Temperature	207°F (11°C)	−6.5%	2%
Ambient Pressure	4in. H$_2$O 0.15 psi 10 mbar	0.9%	0.9%
Ambient Relative Humidity	10%	−0.0002%	0.0005%
Pressure Drop in Filter	1in. WC 25mm. WC	−0.5%	−0.3%
Increase in Gas Turbine Back Pressure	1in. WC 25mm. WC	−0.25%	−0.08

controlling the airflow. The effect of this is to try and maintain the exhaust temperature from the gas turbine at a relatively high value since this gas is used in the HRSG, and the effectiveness of the HRSG is dependent on maintaining this temperature.

The effect of compressor fouling is also very important on the overall performance of the gas turbine since it uses nearly 60% of the work generated by the gas turbine. Therefore, a 1% drop in compressor efficiency equates to nearly a 0.5% in the gas turbine efficiency and about a 0.3% drop in the overall cycle efficiency. The cleaning of these blades by on-line water washing is a very important operational requirement. In many plants, this operational procedure has contributed literally hundreds of thousands of dollars to the bottom line of the plant. It has been the experience of many plants that washing using demineralized water is as effective as using a detergent in on-line water wash. The practice of using abrasive cleaning by injecting walnut shells, rice, or spent catalyst is being suspended in most new plants. Whatever practice used, it must be carefully evaluated; rice, for instance, is a very poor abrasive since it shatters and tends to get into seals and bearings and into the lubrication system. Walnut shells should never be used since they tend to collect inside the HRSG system and, in some cases, have been noted to catch on fire. On-line water washing is not the answer to all the problems since after each wash, the full power is not regained; therefore, a time comes when the unit needs to be cleaned off-line. The time for off-line cleaning must be determined by calculating the loss of income in power as well as the cost of labor to do so and equate it against the extra energy costs.

The cleaning of the hot-section turbine nozzles is a major problem in turbines, which use heavy liquid fuels with high vanadium content. To counteract the vanadium, the fuel is treated with the addition of magnesium, which is supposed to mix with the vanadium and results in harmless fly ash. The problem occurs due to the fact that the fly ash gets collected in the turbine nozzles and reduces the turbine nozzle areas. This can be a very major problem since it collects at the rate of 5% to 12% per 100 hours of operation.

The life of the various hot section components of the gas turbine depends on the following operational parameters:

1. Type of fuel. Natural gas is the base fuel against which all other fuels are measured. The use of diesel fuel reduces the average life by about 25%, and the use of residual fuel reduces life by as much as 65%.
2. Type of service. Peaking service tends to reduce life by as much as 20% as compared to base load operation.
3. Number of starts. Each start is equivalent to about 50 hours of operation.
4. Number of full load trips. This is very hard on the turbine and is nearly equivalent to about 400 to 500 hours of operation.
5. Type of material. The properties of the blade and nozzle vanes are very important factors. The new blade materials, which are the single crystal structures, have done much to help the life of these blades in the higher temperatures, which are used in these new turbines. It must be remembered that if more than about 8% of the air is used in cooling, then the advantage of going to higher temperatures is lost. The use of the Larson-Miller parameters describes an alloy's stress rupture characteristics over a wide temperature, life, and stress range. The Larson-Miller parameter is very useful in comparing the elevated temperature capabilities of many alloys.
6. Types of coatings. The use of coatings in both compressor and turbines has extended the life of most of the components. Coatings are also being used on combustor liners. The new overlay coatings are more corrosion-resistant as compared to the old diffusion coatings. The coatings of the compressor are now more prevalent especially since some of the new compressors are operating at very high-pressure ratios, which translate into high exit temperatures from the compressor. Compressor coatings also tend to reduce the frictional losses and can have a very rapid payback.

HRSG

The heat recovery steam generators are key to the high efficiency of the combined cycle. The HRSG transfers the energy exiting the gas turbine to the steam turbine by converting the water entering the HRSG into superheated steam. The new cycles are requiring the steam to have higher pressure and temperatures. The HRSG are operating at very high effectiveness and for this to hold over the wide range of operation, which many plants experience, the gas turbine exit temperature must remain high. This is achieved by adjusting the inlet guide vanes, which reduces the flow through the gas turbine maintaining the high turbine firing temperature.

The water leaving the boiler feed water pump enters the HRSG through the deaerator and feed water heaters. In this region, heat is added to the water and no change in the state of the water is achieved except that the temperature is increased. In some cases, the water is sent through another set of pumps: one for the high-pressure stage and the other for the low-pressure stage. The water from these pumps are sent through an economizer, where the temperature is increased and then sent to the evaporator, where the water is in two phases and comes out as saturated vapor. The saturated vapor is sent to a one- or two-stage superheater where the saturated vapor is superheated.

The effectiveness of the superheater is usually between 90% and 95%. In most plants, this high effectiveness is maintained over a very wide range of operation.

Steam Turbine

The steam turbines in most combined cycle applications are condensing steam turbines so that they can take an advantage of the low temperatures, which can be achieved at these low pressures. The steam in the low-pressure steam turbine is usually in a two-phase (wet) stage. At this point, care must be taken to ensure that the steam has a steam quality of about 85% to 90%. If the quality of the steam is much less than that, then there is a major problem of blade erosion.

The steam leaving the high-pressure stages is often reheated to the initial throttling temperature before it enters the low-pressure stages. The high-pressure steam turbine is usually an impulse type turbine. The impulse turbine (0% reaction) has lower efficiency than the 50% reaction turbines, which are usually the blade design in the intermediate or low-pressure steam turbines. The flow in the lower stages is also usually much higher.

The computation of the power in the various steam turbines stages, especially in the noncondensing stages, is relatively straightforward. The computation of the power of the low-pressure condensing turbine is very difficult since even though the temperature and pressure at the turbine exit can be measured, the quality of the steam cannot be measured. The only way to ascertain the quality of the steam is to model a heat balance between the condenser and the exit conditions at the steam turbine. To conduct the heat balance, the cooling water flow and the inlet and outlet temperatures of the cooling water must be measured. The measurement of the cooling water flow, in most cases, is inaccurate. The cooling water flow being such a large number, small inaccuracies in this measurement lead to large discrepancies in the computation of the quality of the steam at the low-pressure turbine.

Condenser

The steam condensers used in most large power plants are surface condensers whose function is to reduce the exhaust pressure of a prime mover to a pressure below atmospheric pressure and to condense the steam so as to allow the condensate to be reused as boiler feedwater.

The condenser in most of the large plants usually uses water, fresh or salt, depending on the location for condensing the steam. There are number of plants that also use air as the cooling medium. The new condensers use titanium as the tubes carrying the water, which have good heat transfer qualities.

The fouling of the condenser tubes in service must be carefully monitored; otherwise, the fouling of the tubes will decrease the heat transfer and will increase the temperature and, thus, the exit pressure of the steam leaving the low-pressure steam turbine. This, in turn, will reduce the turbine output. In winter, the cold-water inlet temperatures, being low, helps in reducing the turbine exit pressure and increasing the output work.

The cleanliness factor is a term used to express the degree of tube fouling. It is a ratio of the thermal transmittance of tubes in service to the thermal transmittance of new clean tubes all operating under identical operating conditions of circulating water temperature and velocity and the same external steam temperature and flow. The cleanliness factor is a unique value since it applies to only one specific operating condition. To use it as a constant factor of new clean tube, heat transfer for other operating conditions introduces error. To use it over a large operating range, it is necessary to convert the result to thermal resistance and, thus, use it as an additive factor to the overall thermal resistance corresponding to the new clean tube condenser performance. Another approach would be to calculate a set of cleanliness factors over the range of operating conditions under which the plant would operate.

Losses. The losses that are encountered in a plant can be divided into two groups, uncontrollable losses and controllable losses. The uncontrollable losses are usually environmental conditions, such as temperature, pressure, humidity, and the turbine aging. The controllable losses are those that the operator can have some degree of control over and can take corrective actions:

1. *Pressure Drop across the Inlet Filter.* This can be remedied by cleaning or replacing the filter.
2. *Compressor Fouling.* On-line water cleaning can restore part of the drop encountered.
3. *Fuel Lower Heating Value.* In many plants, on-line fuel analyzers have been introduced to not only monitor the turbine performance but to also calculate the fuel payments, which are usually based on the energy content of the fuel.
4. *Turbine Back Pressure.* In this case, the operator is relatively limited since he cannot do anything about the downstream design. If there is some obstruction in the ducting to the HRSG that can be removed or if the duct has collapsed in an area, the duct could be replaced.
5. *HRSG Effectiveness.* A properly instrumented HRSG would have thermocouples placed at intervals through out the HRSG so that effectiveness of each section of the HRSG can be monitored.
6. *Steam Turbine Fouling.* The fouling of the steam turbine can be noted. Most of the fouling will occur in the low-pressure turbine along with some blade erosion problems. By trying to calculate the quality content of the steam, as well as the power output of that section of the turbine, problems can be noted and corrective action can be taken.
7. *Condenser Pressure.* The fouling of the tubes, insufficient water flow through the condenser, the condenser cooling water temperature all lead to increasing the pressure at the low-pressure turbine exit leading to an increase in the exit pressure and temperature.

8. *Condenser Condensate Sub-Cooling.* The subcooling of the condensate can lead to losses, as the condensate has to be heated up so that more energy is spent to do so and the overall energy available in the HRSG is reduced.

Design of a Condition-Monitoring System

A total condition-monitoring system must be designed to provide the operators and rotating equipment engineers with clear insight into machinery performance problems:

- Enhance predictive maintenance capability by diagnostic tools.
- Plant operation with minimum degradation based on optimal washing of the train.
- Integrated condition monitoring utilizing *field proven hardware and software*.
- Provide voltage free contacts at the monitors for machine safeguarding. Data link to and from the plant D-CS system so as to upgrade performance curves in the D-CS system, which control plant processes.

A total condition-monitoring system for a major utilities installation requires a fully integrated system with an extensive database to ensure that it can achieve the following goals:

- High machinery availability
- Maintaining peak efficiency and limiting performance degradation of machine trains
- Extending time between inspections and overhauls
- Corrosion control
- Optimizing the cycle configuration
- Estimating availability
- Evaluating scenarios by means of "What-If" analysis
- Estimating maintenance requirements and life of hot-section components
- Fault identification by expert system analysis.

A condition-monitoring system designed to meet these needs must comprise of hardware and software designed by engineers with experience in machinery and energy system design, operation, and maintenance. Each system needs to be carefully tailored to individual plant and machinery requirements. The systems must obtain real time data from the plant DCS and if required from the gas- and steam-turbine control systems. Dynamic vibration data is taken in from the existing vibration analysis system into a data acquisition system. The system can comprise of several high-performance networked computers depending on plant size and layout. The data must be presented using a graphic user interface (GUI) and include the following:

1. *Aerothermal Analysis* – This pertains to a detailed thermodynamic analysis of the full power plant and individual components. Models are

created of individual components including the gas turbine, steam turbine heat exchangers, and distillation towers. Both the algorithmic and statistical approaches are used. Data is presented in a variety of performance maps, bar charts, summary charts, and baseline plots.

2. *Combustion Analysis* – This includes the use of pyrometers to detect metal temperatures of both stationary and rotating components, such as turbine blades. The use of dynamic pressure transducers to detect flame instabilities in the combustor, especially in the new dry low NO_x applications
3. *Vibration Analysis* – This includes an on-line analysis of the vibration signals, FFT spectral analysis, transient analysis, and diagnostics. A wide variety of displays are available including orbits, cascades, bode and nyquist plots, and transient plots.
4. *Mechanical Analysis* – This includes detailed analysis of the bearing temperatures, lube, and seal oil systems, and other mechanical sub-systems.
5. *Corrosion Analysis* – On-line electrochemical sensors are being used to monitor changes in the corrosivity of flue gases especially in exhaust stacks. The progressive introduction of ever-more stringent regulations to reduce NO_x emissions has resulted in an increase in the risk of water wall tube wastage in large power boilers, refinery process heaters, and municipal waste incinerators.
6. *Diagnosis* – This includes several levels of machinery diagnosis assistance available via expert systems. These systems must integrate both mechanical and aerothermal diagnostics.
7. *Trending and Prognosis* – This includes sophisticated trending and prognostic software. These programs must clearly provide users to clearly understand underlying causes of operating problems.
8. *"What-If" Analysis* – This program should allow the user to do various studies of plant operating scenarios to ascertain the expected performance level of the plant due to environmental and other operational conditions.

Monitoring Software

The monitoring software for every system will be different. However, all software is there to achieve one goal, which is to gather data, ensure that it is correct, and then analyze and diagnose the data. Presentations must be in a convenient form and should be easily understood by plant operational personnel. All priorities must be to the data collection process. This process must not in any manner be hampered since it is the corner stone of the whole system.

A convenient *framework* within which to categorize the software could be as follows:

1. *Graphic User Interface (GUI)* – This consists of screens, which would enable the operator to easily interrogate the system and to visually see where the instruments are installed and their values at any point of time.

By carefully designed screens, the operator will be able to view at a glance the relative positions of all values thus fully understanding the operation of the machinery.
2. *Alarm/System Logs* – To fully understand a machine, we have to have various types of alarms. The following are some of the suggested types of alarms:

   2.1. *Instrument Alarms*: These alarms are based on the instrumentation range.
   2.2. *Value Range Alarm*: These alarms are based on operating values of individual points both measured and calculated points. These alarms should be variable in that they would change with operating conditions.
   2.3. *Rate of Change Alarm*: These alarms must be based on any rapid change in values in a given time range. This type of alarm is very useful to detect bearing problems, surge problems, and other instabilities.
   2.4. *Prognostic Alarms*: These alarms must be based on trends and the prognostics based on those trends. It is advisable not to have prognostics, which project in time more than the time of data that is trended.

3. *Performance Maps* – These are performance maps based on design or initial tests (base lines) of the various machinery parameters. These maps, for example, present how power output varies with ambient conditions, or with properties of the fuel, or the condition of the filtration system; or how close to the surge line a compressor is operating. On these maps, the present value is displayed, thus allowing the operator to determine the degradation in performance occurring in the units.
4. *Analysis Programs* – These include aerothermal and mechanical analysis programs with diagnostics and optimization programs as follows:

   4.1. *Aerothermal Analysis*: Typical aerothermal performance calculations involve evaluation of component unit power, polytropic and adiabatic head, pressure ratio, temperature ratio, polytropic and adiabatic efficiencies, temperature profiles, and a host of other machine specific conditions under steady state as well as during transients — start ups and shut downs. This program must be tailored to individual machinery and to the instrumentation available. Data must be corrected to a base condition so that it can be compared and trended. The base condition can vary from ISO ambient conditions to design conditions of a compressor or pump if those conditions are very different from ISO ambient conditions. To analyze off-design operation, it is necessary to transpose values from the operating points back to the design point for comparison of unit degradation.
   4.2. *Mechanical Analysis*: This program must be tailored to the mechanical properties of the machine train under consideration.

It should include bearing analysis, seal analysis, lubrication analysis, rotor dynamics, and vibration analysis. This includes the evaluation and correlation of bearing metal temperatures, shaft orbits, vibration velocity, spectrum snapshots, waterfall plots, stress analysis, and material properties.

4.3. *Diagnostic Analysis*: This program can be part of an expert system or consists of an operational matrix, which can point to various problems. The program must include comparison of both performance and mechanical health parameters to a machine specific fault matrix to identify if a fault exists. Expert analysis modules can in many cases, aid to faster fault identification but are usually more difficult to integrate into the system.

4.4. *Optimization Analysis*: Optimization programs take into account many variables, such as deterioration rate, overhaul costs, interest, and utilization rates. These programs may also be dependent on more than one machine train if the process is interrelated between various trains.

4.5. *Life Cycle Analysis*: The determination of the effect of the material, the temperature excursions, the number of startups and shutdowns, the type of fuel, all relate to the life of hot-section components.

5. *Historical Data Management* – This includes the data acquisition and storage capabilities. Present-day prices of storage mediums have been dropping rapidly, and systems with 14- to 16-Gigabyte hard disks are available. These disks could store a minimum of 2 years of 1-minute data for most plants. One-minute data is adequate for most steady-state operation, while startups and shutdowns or other nonsteady-state operation should be monitored and stored at an interval of 1 second. To achieve these time rates, data for steady-state operation can be obtained from most plantwide D-CS systems and for unsteady state conditions, data can be obtained from control systems.

Implementation of a Condition-Monitoring System

The implementation of a condition-monitoring system in a major utilities plant requires a great deal of fore thought. A major utilities plant will have a number of varied, large rotating equipment. This will consist usually of various types of prime movers such as large gas turbines, steam turbines, pumps, electric generators, and motors. The following are some of the major steps, which need to be taken to ensure a successful system installation:

1. The first decision is to decide on what equipment should be monitored on-line and what systems should be monitored off-line. This requires an assessment of the equipment in terms of both first cost and operating costs, redundancy, reliability, efficiency, and criticality.

2. Obtain all pertinent data of the equipment to be monitored. This would include details of the mechanical design and the performance design. Some of this information may be difficult to obtain from the manufacturer and will have to be calculated from data being obtained in the field or after installation during commissioning tests in a new installation. Obtaining baseline data is critical in the installation of any condition monitoring system. In most systems, it is the rate of change of parameters that are being trended not the absolute values of these points. It is also important to decide what type of alarms will be attached to the various points. Rate of change alarms must be for bearing metal temperatures, especially for thrust bearings where temperature changes are critical. Prognostic alarms should be applied to critical points. Alarms randomly applied tend to slow down the system and do not provide added protection.

The following are some of the basic data that would be necessary in setting up a system:

   2.1. Type of gasses and fluids used in the various processes. The Equation of State and other thermodynamic relationship, which govern these gases and fluids.
   2.2. Type of fuel used in the prime movers. If the fuel analysis is available including the fuel composition and the heating values of the fuel.
   2.3. Materials used in various hot sections such as combustor liners, turbine nozzles, and blades. This includes stress and strain properties as well as Larson-Miller parameters.
   2.4. Performance maps of various critical parameters, such as power and heat consumption as a function of ambient conditions, pressure drop in filters, and the effect of back pressure. Compressor surge, efficiency, and head maps.

3. Determine the instrumentation, which exists, and its actual location. Location of the instrumentation from the inlet or exit of the machinery is important so that proper and effective compensation may be provided for the various measured parameters. In some cases, additional instrumentation will be needed. Experience indicates that older plants require 10% to 20% more instrumentation depending upon the age of the plant.
4. Once the data points have been decided, limits and alarm must be set. This is a long and challenging task, as the limits on many points are not given in the operation manuals. In some cases, the criticality of the equipment may necessitate that the alarm threshold on certain points be lowered to give early warning of any deterioration of the system. It should be noted that since this is a condition-monitoring system, early alarm warnings are, in most cases, desirable.
5. Types of reports and summary charts should be planned to optimize the data and to present it in the most useful manner to the plant operations and maintenance personnel.

6. The types of D-CS and the control systems available in the plant. The protocol of these systems and their relationships to the condition-monitoring system. The slave or master relationship is important in setting up the protocols.
7. Diagnostics for the system requires noting any unusual characteristics of the machinery, especially in older plants, which have a history of operation inspections and overhauls.
8. Costs of operations, such as fuel costs, labor costs, down time costs, overhaul hours, interest rates, are necessary in computing parameters, such as time of major inspections, off-line cleaning, and overhauls.

Plant Power Optimization

On-line optimization processes for large utility plants are gaining tremendous favor. Plant optimization is gaining importance with combined cycle power plants as these plants are operated over a wide range of power in day-to-day operation. On-line optimization may be defined as the place where economics, operation, and maintenance meet. At first sight, it may be imagined that process integration is not connected to condition management or inspection, and this has been the case in the past. However, there is every incentive for complete integration of all these production-related technologies since the condition monitoring of the various components in a plant are upgraded constantly, thus the operational curves with degradation of each unit are no longer stagnant.

Process integration was developed initially as a means of optimizing the design of chemical and petrochemical process plants. Process optimization is still only a preconstruction or preproduction exercise. This is surprising because many process plants are designed for batch manufacture of a range of products, each of which will require continuously changing optimization parameters. Process optimization and re-optimization "on the fly" can enable companies to meet variations in market demand and maximize production efficiency and overall profitability.

When embodied in a modern integrated plant environment, dynamic plant health assessment, process modeling, and process integration provide the means to augment plant reliability, availability, and safety with maximum capacity and flexibility.

Figure 12-10 shows how on-line systems are configured. The system gathers data in real time. The data is gathered from either the D-CS system or from the control system. Data for startups and transients are needed from the control system since the data from the D-CS is usually updated every 3 to 4 seconds, while the control system can have very rapid loops which are updated as often as 40 times per second. To ensure that performance data is taken at a steady-state condition, since most models of the plant are steady state, the system must observe some key parameters and ensure that they are not varying. In turbine parameters, such as turbine wheel, space temperatures should be observed as constant. This data is then checked for accuracy and the errors are removed. This involves simple checks against instrument operational ranges and system operation parameter

458 • COGENERATON AND COMBINED CYCLE POWER PLANTS

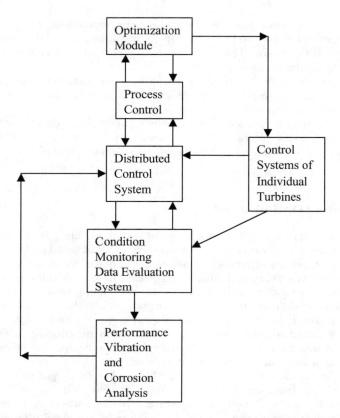

Figure 12-10. A Block Diagram for an On-Line Condition-Monitoring System

ranges. The data is then fully analyzed and various performance data checks are made. New operational and performance maps are then plotted and the system then can optimize itself against an operational model. The operational goal is to maximize the efficiency of the plant at all loads, thus the new performance maps, which show degradation of the plant, are then used in the plant model to ensure that the control is at the right setting for the operation of the plant at any given time. Many maintenance practices are also based on the rate of economic return these operational maintenance practices such as an off-line compressor wash would contribute to the operations of the plant.

Many plants use off-line optimization. Off-line optimization is an open-loop control system. Instead of the closed loop system, which controls the plant settings, data is provided to the operator so that he can make the decisions based on the findings of the operational data. Off-line systems are also used by Engineers to design plants and by maintenance personnel to plan plant maintenance. Comparisons of the on-line systems to off-line systems can be seen in Table 12-2.

Control Systems and Condition Monitoring • 459

Table 12-2. Comparisons of On-Line and Off-Line Plant Optimization System Use

	On-line Systems	Off-line Systems
Objectives	Maximize Economic benefit, operate the plant at its maximum efficiency at all operation points	Maximize Economic benefit, operate the plant at its maximum efficiency at all operation points. Optimize overall facilities design and investment
Target	Existing operating plant	Existing operating plant, New Facilities, Facility Expansion
Prime Use	Process and Maintenance operations	Process and Maintenance Operations, Design Modifications
Users	Operation and Maintenance Engineers	Operation and Maintenance Engineers, Project and Design Engineers

Performance evaluation is also important initially in determining that a plant meets its guarantee points and, subsequently, to ensure it continues to be operated at or near its design operating condition. Maintenance practices are being combined even more closely with operational practices to ensure that plants have the highest reliability with maximum efficiency. When a new plant is built, its cost amounts to only about 7% to 10% of the life cycle cost. Maintenance costs represent approximately 15% to 20% overall. However, operating costs, which in the case of a power plant, for example, consist essentially of energy costs, make up the remainder, and amount to between 70% and 80% of the life cycle costs of the facility. This brings performance monitoring to the forefront as an essential tool in any type of plant condition-monitoring system. Operating a plant as close as possible to its design conditions will guarantee that its operating costs will be reduced. As an illustration of the opportunity cost this represents, large fossil power plants currently being commissioned range from 600 to 2800 MW. The fuel costs for these plants will amount to between US$72 and $168 million/annum. Therefore, savings of 1% to 3% of these costs can amount to an overall cost reduction of upward of $1 million/annum.

A change in approach is clearly necessary in order that the full benefit of integrated plant condition management and control can be recognized and exploited. Improved control and enhanced performance monitoring will enable shutdown intervals to be extended without increasing the risk of premature or unexpected failure. In turn, this will increase the confidence of operations, inspection and management personnel in the effectiveness of unified plant administration.

Life Cycle Costs

The life cycle costs of any machinery are dependent on the life expectancy of the various components, the efficiency of its operation throughout its life. Figure 12-11 shows the cost distribution by the three major categories, initial costs, maintenance costs, and operating or energy costs. This figure indicates that the

460 • COGENERATON AND COMBINED CYCLE POWER PLANTS

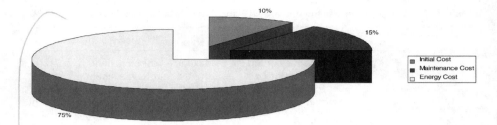

Figure 12-11. Life Cycle Costs for a Combined Cycle Power Plants

new costs are about 7% to 10% of the life cycle costs, while maintenance costs are approximately 15% to 20% of the life cycle costs and operating costs, which essentially consist of energy costs, make up the remainder between 70% and 80 % of the life cycle costs of any major machinery in a utilities plant. It is, therefore, clear why the new purchasing mantra for a utility plant, or for that matter of fact, for any major plant operating large machinery is "life cycle cost".

This brings forth to the forefront performance-monitoring as an essential tool in any type of plant condition-monitoring system. The major costs in a life cycle are the cost of energy. Thus, operating the plant as close to its design conditions guarantees that the plant will reduce its operating costs. This can be achieved by ensuring that the turbine compressor is kept clean and that the driven compressor is operating close to its maximum efficiency, which, in many cases, is close to the surge line. Thus, knowing where the compressor is operating with respect to its surge line is a very critical component in plant operating efficiency.

The life expectancy of most hot-section parts is dependent on various parameters and is usually measured in terms of equivalent engine hours. The following are some of the major parameters that effect the equivalent engine hours in most machinery especially gas turbines:

1. Type of fuel
2. Firing temperature
3. Materials stress and strain properties
4. Effectiveness of cooling systems
5. Number of starts
6. Number of trips.

Maintenance practices are more and more being combined with operational practices to ensure that plants have the highest reliability with maximum efficiency. This has led to the importance of performance-condition monitoring as a major tool in the operation and maintenance of a plant. Life cycle costs, rightly so, now drive the entire purchasing cycle and, thus, the operation of the plant. Life cycle costs, based on a 25-year life, indicate that the following are the major cost parameters:

1. Initial purchase cost of equipment is 7% to 10% of the overall life cycle cost.
2. Maintenance costs are about 15% to 20% of the overall life cycle cost.
3. Energy costs are about 70% to 80% of the life cycle costs.

This distribution in life cycle costs indicates that component efficiency throughout the life period of the plant is the most important factor affecting the cost of a particular machine train. Thus, monitoring the efficiency of the train and ensuring that degradation rates are slowed down ensure that the predicted life cycle costs are achieved. Performance monitoring of the entire train is a must for plants operating on life cycle cost strategies.

Performance-monitoring also plays a major role in extending life, diagnosing problems, and increasing time between overhauls. On-line performance monitoring requires an in-depth understanding of the equipment being measured. Most trains are very complex in nature and, thus, require very careful planning in installation of these types of systems. The development of algorithms for a complex train needs careful planning, understanding of the machinery and process characteristics. In most cases, help from the manufacturer of the machinery would be a great asset. For new equipment, this requirement can be part of the bid requirements. For plants with already installed equipment, a plant audit to determine the plant machinery status is the first step.

To sum up, total performance condition-monitoring systems will help the plant engineers to achieve their goals of:

1. Maintaining high availability of their machinery.
2. Minimizing degradation and maintaining operation near design efficiencies.
3. Diagnosing problems and avoiding operations in regions, which could lead to serious malfunctions.
4. Extending time between inspections and overhauls.
5. Reducing life cycle costs.

Chapter 13

PERFORMANCE TESTING OF COMBINED CYCLE POWER PLANT

The performance analysis of the new generation of gas turbines in combined cycle operation is complex and presents new problems, which have to be addressed. The new units operate at very high turbine firing temperatures. Thus, variation in this firing temperature significantly affects the performance and life of the components in the hot section of the turbine. The compressor pressure ratio is high, which leads to a very narrow operation margin, thus making the turbine very susceptible to compressor fouling. The turbines are also very sensitive to back pressure exerted on them by the heat-recovery steam generators. The pressure drop through the air filter also results in major deterioration of the performance of the turbine.

The performance of the combined cycle is also dependent on the steam-turbine performance. The steam turbine is dependent on the pressure, temperature, and flow generated in the heat-recovery steam generator, which, in turn, is dependent on the turbine firing temperature and the air mass flow through the gas turbine. It is obvious that the entire system is very intertwined and that deterioration of one component will lead to off-design operation of other components, which, in most cases, leads to overall drop in cycle efficiency. Thus, determining component performance and efficiency is the key to determining overall cycle efficiency.

If a life cycle analysis were conducted, the new costs of a plant are about 7% to 10% of the life cycle costs. Maintenance costs are approximately 15% to 20% of the life cycle costs. Operating costs, which essentially consist of energy costs, make up the remainder, between 70% and 80% of the life cycle costs, of any major utility plant. Thus, performance evaluation of the turbine is one of the most important parameters in the operation of a plant.

Total performance monitoring on- or off-line is important for the plant engineers to achieve their goals of:

1. Maintaining high availability of their machinery
2. Minimize degradation and maintain operation near design efficiencies
3. Diagnose problems and avoid operating in regions which could lead to serious malfunctions

Note: Tables on Nomenclature and Greek Symbols may be found at the end of this chapter.

4. Extend time between inspections and overhauls
5. Reduce life-cycle costs.

To determine the deterioration in component performance and efficiency, the values must be corrected to a reference plane. These corrected measurements will be referenced to different reference planes, depending upon the point that is being investigated. Corrected values can further be adjusted to a transposed design value to properly evaluate the deterioration of any given component. Transposed data points are very dependent on the characteristics of the components performance curves. To determine the characteristics of these curves, raw data points must be corrected and then plotted against representative nondimensional parameters. Thus, we must evaluate the turbine train, while its characteristics have not been altered due to component deterioration. If component data was available from the manufacturer, the task would be greatly reduced.

Performance analysis is not only extremely important in determining overall performance of the cycle, but also in determining life cycle considerations of various critical hot-section components.

In this chapter, a detailed technique with all the major equations governing a power plant are presented based on the various ASME test codes. The following five ASME test codes govern the test of a combined cycle power plant:

1. ASME, Performance Test Code on Overall Plant Performance, ASME PTC 46 1996, American Society of Mechanical Engineers, 1996
2. ASME, Performance Test Code on Gas Turbines, ASME PTC 22 1997, American Society of Mechanical Engineers, 1997
3. ASME, Performance Test Code on Gas Turbine Heat-Recovery Steam Generators, ASME PTC 4.4 1981, American Society of Mechanical Engineers, Reaffirmed 1992
4. ASME, Performance Test Code on Steam Turbines, ASME PTC 6 1996, American Society of Mechanical Engineers, 1996
5. ASME, Performance Test Code on Steam Condensing Apparatus, ASME PTC 12.2 1983, American Society of Mechanical Engineers, 1983.

The ASME, performance test code on overall plant performance, ASME PTC 46 was designed to determine the performance of the entire heat cycle as an integrated system. This code provides explicit procedures to determination of power plant thermal performance and electrical output.

The equations and performance parameters for all the major components of a power train must be corrected for ambient conditions and certain parameters must be further corrected to design conditions to accurately compute the degradation. Therefore, to fully compute the performance, as well as degradation of the plant and all its components, the **actual, corrected, and transposed** to reference conditions of critical parameters must be computed.

The overall plant needs the following parameters to be computed. The two most important parameters from an economic point of view are the computation of the Power delivered and the fuel consumed to deliver the power. The following are the parameters that need to be computed to fully understand the macro picture of the plant:

- Overall Plant System
  - Gross Unit Heat Rate
  - Net Unit Heat Rate
  - Gross Output
  - Net Output
  - Auxiliary Power.

The next step is the computation of the various major components that make up the plant. The following are the various calculation modes for a combined cycle power plant and the major parameters computed in each mode. These modes are separated in the four major components which govern a combined cycle power plant:

1. Gas Turbines
2. HRSG
3. Steam Turbine
4. Condenser.

Before starting any performance test the gas turbine shall be run until stable conditions have been established. Stability conditions will be achieved when continuous monitoring indicates the readings have been within the maximum permissible limits. The ASME PTC–22-test code requires that the Performance test will be run as much as possible to the design test conditions as specified in the contract. The maximum permissible variation in a test run shall not vary from the computed average for that operating condition during the complete run by more than the values specified in Table 13-1. If operation conditions vary during any test

Table 13-1. Maximum Permissible Variation in Test Conditions

Variables	Variation of any Station during the Test Run
Power Output (electrical)	±2%
Power factor	±2%
Rotating speed	±1%
Barometric pressure at site	±0.5%
Inlet air temperature	±4.0°F (±2.2°C)
Heat valve–gaseous fuel per unit volume	±1%
Pressure–gaseous fuel as supplied to engine	±1%
Absolute exhaust back pressure at engine	±0.5%
Absolute inlet air pressure at engine	±0.5%
Coolant temperature–outlet [Note (2)]	±5.0°F (±2.8°C)
Coolant temperature–rise [Note (2)]	±5.0°F (±2.8°C)
Turbine control temperature [Note (3)]	±5.0°F (±2.8°C)
Fuel Mass Flow	±0.8%

run by more than the prescribed values in Table 13-1 than the results of that test run shall be discarded. The test run should not exceed 30 minutes and during that time the interval between readings should not exceed 10 minutes. There should be three to four test runs performed, which then could be averaged to get the final guarantee test points.

Correction factors are also provided in ASME PTC Test Code-46. The correction factors for Ambient Temperature, Ambient Pressure and Relative Humidity are presented in this chapter.

Gas Turbine

The gas turbine is the major component of a combined cycle power plant. It produces 60% of the power and its exhaust gas is the medium of energy for the HRSG. Thus, it is important that the calculations of the gas turbine be thorough and in detail. The ASME, performance test code on gas turbines, ASME PTC 22 examines the overall performance of the gas turbine. The ASME PTC 22 only examines the overall turbine and many turbines in the field are better instrumented for computation of the detail characteristics of the gas turbine. Figure 13-1 shows the desired location of the measurement points for a fully instrumented turbine. The following are the various computations required to calculate the gas turbine overall performance based on the code:

- Gas-Turbine Overall Computation
  - Gas-Turbine Output
  - Inlet Air Flow

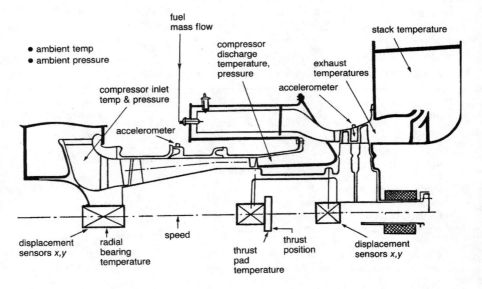

Figure 13-1. Gas Turbine Suggested Measurement Points

- First-Stage Nozzle Cooling Flow Rate
- Total Cooling Flow Rate
- Heat Rate
- Expander Efficiency
- Gas-Turbine Efficiency
- Exhaust Flue Gas Flow
- Specific Heat of Exhaust Flue Gas.

To further analyze the gas turbine must be examined in its four major categories:

1. Air Inlet Filter
2. Compressor
3. Combustor
4. Expander Turbine.

Air Inlet Filter Module

Loss Computation enables the operator to ensure that the filters are clean and that no additional losses than necessary reduce the performance of the gas turbine. The following parameters are necessary to monitor the filter:

- Time to replace each stage of filters
- Filter Plugged Index to monitor the condition of each stages of filters
- Inlet Duct Air Leak.

Compressor Module

The compressor of a gas turbine is one of the most important components of the gas turbine; it consumes between 50% and 65% of the energy produced in a gas turbine. Thus, fouling of the compressor can cause large losses in power and efficiency for the gas turbine. Furthermore, the fouling of the compressor also creates surge problems, which not only affects the performance of the compressor, but also because surge can create bearing problems and flame outs. The following are some of the major characteristics that need to be calculated:

Overall Parameters of the Compressor:

- Efficiency
- Surge Map
- Compressor Power Consumption
- Compressor Fouling Index
- Compressor Deterioration Index
- Humidity Effects on the Fouling
- Stage Deterioration
- Compressor Losses.

These losses are divided into two sections:

- Controllable Losses. Losses which can be controlled by the action of the operator, such as:
  - Compressor fouling
  - Inlet pressure drop.

- Uncontrollable Losses. Losses which cannot be controlled by the operator, such as:
  - Ambient Pressure
  - Ambient Temperature - in cases where refrigerated inlets could be controlled; however, in most applications, it is uncontrolled
  - Ambient Humidity
  - Ageing.

Compressor Wash. When should the compressor be washed on-line, and when should an off-line compressor wash should be considered.

On-Line Wash. This wash is done on many plants as the pressure drop decreases by more than 2%, some plants do it on a daily basis. The water for these washes must be treated.

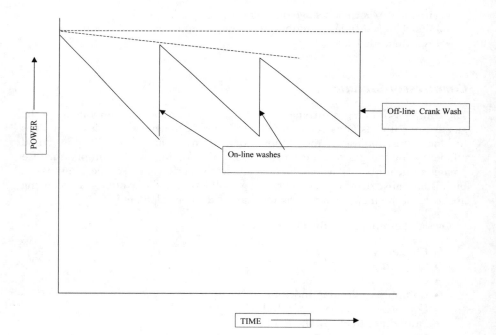

Figure 13-2. Water Wash Characteristics

Off-Line Wash. Figure 13-2 shows that on-line water wash will not return the power to normal; thus, after a number of these washes, an off-line water wash must be planned. This is a very expensive maintenance program and must be fully evaluated before it is undertaken.

Combustor Module

The calculation of the firing temperature is one of the most important calculations in the combined cycle performance computation. The temperature is computed using two techniques: (1) Fuel Heat Rate, (2) Power Balance. The following are the important parameters that need to be computed:

- Combustor Efficiency
- Deterioration of Combustor
- Turbine Inlet Temperature (First-Stage Nozzle Inlet Temperature)
- Flash Back Monitor (for Dry Low NO_x Combustors)
- Specific Fuel Consumption.

Expander Module

The calculation of the turbine expander module depends whether or not this is a single-shaft gas turbine, or a multiple-shaft gas turbine. In aero-derivative turbines, there are usually two or more shafts. In the latest aero-derivative turbines, there are usually two compressor sections: the LP compressor section and the HP compressor section. This means that the turbine has three shafts, the third shaft is the power shaft. The turbines that drive the compressor section are known as the gasifier turbines and the turbine that drives the generator is the power turbine. The gasifier turbine produces the work to drive the compressor.

The parameters which must be computed are:

- Expander Efficiency
- Fouled Expander Parameter
- Eroded Turbine Nozzle Monitor Parameter
- Expander Power Produced
- Deterioration Monitor Parameter
- Plugged Turbine Nozzle Monitor Parameter.

Life Cycle Consideration of Various Critical Hot-Section Components

The life expectancy of most hot-section parts is dependent on various parameters and is usually measured in terms of equivalent engine hours. The following are some of the major parameters that affect the equivalent engine hours in most machinery, especially gas turbines:

1. Type of fuel
2. Firing Temperature
3. Materials stress and strain properties
4. Effectiveness of cooling systems
5. Number of starts
6. Number of trips.

- Expander Losses:
  - Controllable Losses
    - Firing Temperature
    - Back Pressure
    - Turbine Fouling (Combustion Deposits)
  - Uncontrollable (Degradation) Losses
    - Turbine Ageing (Increasing Clearances).

HRSG Calculations Module (Based on ASME PTC 4.4)

The hot gases as they leave the gas turbine enter the HRSG. The heat in these gases has a very large amount of energy associated with them. The computations are based on the properties of the gases on the air side and on the water/steam side. Most of the HRSGs are instrumented on the gas side as well as on the liquid/steam side. The effectiveness of the HRSG at all sections can be computed, as well as the overall effectiveness of the HRSG. In case of supplementary fired HRSGs, the firing of the gas turbine gases must be computed separately. Figure 13-3 is a detailed diagram of a typical Gas-turbine heat-recovery steam-generating unit from the ASME PTC 4.4 code. There are many different types of HRSG as discussed in Chapter 7. However, the basic components are the same. The HRSG is divided into one to three sections, each section having the following three sections:

1. Economizer, where the water is heated
2. Evaporator, where the water is converted to saturated steam
3. Superheater, where the saturated steam is converted to superheated steam.

The HRSG temperature profile is known as the Q-T diagram as seen in Figure 13-4. The diagram indicates the pinch point and the approach point, which must be determined to correctly evaluate the effectiveness of the HRSG. The following are the other parameters which need to be evaluated in an HRSG:

- Overall HRSG Effectiveness
- HRSG work output
- Flue Gas Inlet and Exit Flow and Temperature

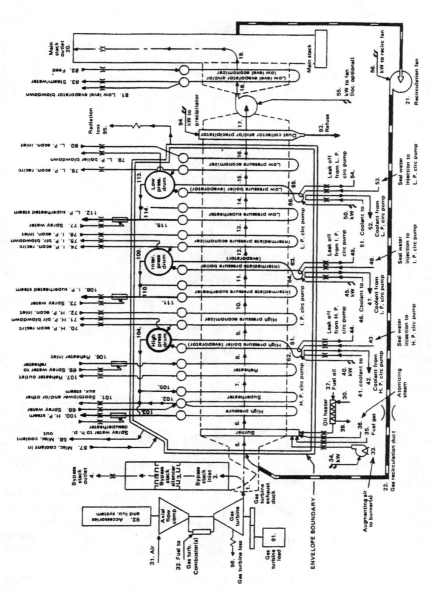

Figure 13-3. Heat Recovery Steam Generating Unit (Courtesy of ASME PTC Code 4.4)

472 • COGENERATION AND COMBINED CYCLE POWER PLANTS

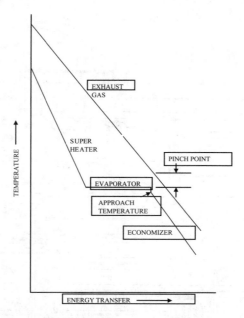

Figure 13-4. Energy Transfer for Diagram for an Heat Recovery Steam Generator

- Flue Gas Enthalpy at the position of each component
- Feed Water Flow
- HP circuit Effectiveness
- IP circuit Effectiveness
- LP circuit Effectiveness.

Steam-Turbine Calculations (ASME PTC 6.1)

The steam turbine in a combined cycle can be either a back-pressure steam turbine or a condensing turbine. The turbine can also be an extraction turbine, a very common type in cogeneration plants. The steam turbines in most combined cycle power plants are of the condensing type and can be three or two sections. In a triple cycle, the steam turbine has three sections, a HP turbine, an IP turbine, and an LP turbine which is a condensing-type turbine. In a dual cycle HRSG, the steam turbine is a two-section turbine with a HP turbine and an LP condensing-type turbine. The accuracy of the LP turbine is dependent on the accuracy of the condenser heat-removal computations, since the measurement of the quality of the steam exiting the turbine is not a measurable quantity. These depend upon accuracy of condenser water-flow measurements in a water-cooled condenser, or the air flow in an air-to-air condenser. Figure 13-5 shows the instrumentation required for a full performance test on a steam turbine that given the ASME

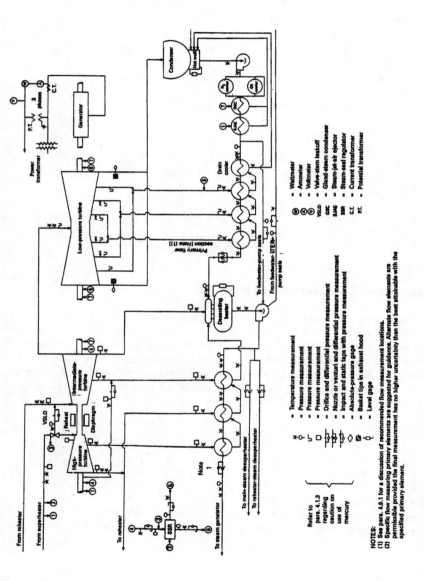

Figure 13-5. Location and Type of Instrumentation for a Steam Turbine Test as per PTC 6 (Courtesy of ASME PTC 6)

performance test codes. The following are the parameters, that are usually computed depending on the instrumentation available:

- Steam-Turbine Output Overall and for each Section
- Heat-Rate Output Overall and for each Section
- HP Steam Flow
- IP Steam Flow
- IP Steam Flow
- LP Steam Flow
- HP Steam-Turbine Efficiency
- HP Steam-Turbine Losses
- IP Steam-Turbine Efficiency
- IP Steam-Turbine Losses
- LP Steam-Turbine Efficiency
- LP Steam-Turbine Losses
- Fouling Monitor Index for LP Steam Turbine
- Deposits Monitor
- Mechanical Damage Monitor
- Flow Reduction
- Erosion Monitor Parameter
- Excessive Clearances Monitor Parameter.

Condenser Calculations (ASME PTC 12.2)

The condenser is a very critical part of a steam turbine. The performance of the ASME test codes PTC 12.2 covers all surface condenser's. The condensers main function is to reduce the exhaust pressure of a prime mover to a pressure at or below atmospheric pressure and to condensate the steam so that the condensate is available for reuse as boiler feed water. The condenser effectiveness depends on the type of the water used, sea or fresh water, and the fouling of the tubes. The following are some of the major parameters computed:

- Condenser Effectiveness
- Heat Reject
- Back Pressure
- Clearance Factor
- Fouling Index
- Air Leakage
- Fouling factors (Actual/Design heat transfer Coefficient)
- Steam Conditions (Steam Quality based on cooling conditions.

Performance Curves

It is very important to form a baseline for the entire power plant. This would enable the operator to determine the section of the plant that is operating below design conditions. The following performance curves should be obtained either from the manufacturer or during acceptance testing so that the in-depth study of the parameters and their interdependency with each other can be defined:

1. Gas Turbine
   - Gas-turbine compressor inlet bell-mouth pressure differential versus air flow rate
   - Gas-turbine output versus compressor inlet temperature
   - Heat rate versus compressor inlet temperature
   - Fuel consumption versus compressor inlet temperature
   - Exhaust temperature versus compressor inlet temperature
   - Exhaust flow versus compressor inlet temperature
   - The NO_x water injection rate for oil firing versus gas-turbine compressor inlet temperature
   - Gas-turbine generator power output and heat-rate correction as result of water injection
   - Effect of water injection on generator output as a function of compressor inlet temperature
   - Effect of water injection rate on heat rate as a function of compressor inlet temperature
   - Ambient humidity corrections to generator output and heat rate
   - Power factor correction
   - Losses in generation due to fuel restriction resulting in operational constraints (e.g., temperature spread, problems on fuel stroke valve, etc.).

2. HRSG
   - Gas-turbine exhaust flow versus HP steam flow as a function of gas-turbine exhaust temperature
   - Gas-turbine exhaust flow versus LP steam flow as a function of gas-turbine exhaust temperature
   - Gas-turbine exhaust flow versus HP super-heater steam temperature as a function of gas-turbine exhaust temperature.

3. Steam Turbine
   - HP steam flow versus Steam-turbine output as a function of HP steam flow
   - HP steam flow versus first-stage shell pressure
   - HP steam pressure versus Steam-turbine output as a function of HP steam pressure
   - LP steam flow versus steam-turbine output as a function of LP steam flow
   - LP steam flow versus first-stage shell pressure
   - LP steam pressure versus steam-turbine output as a function of LP steam pressure.

4. Condenser
   - Condenser vacuum pressure versus steam flow as function of circulating water inlet temperature.

Performance Computations

This section deals with the equations and techniques used to compute and simulate the various performance and mechanical parameters for the combined cycle power plant. The goals have been to be able to operate the entire power plant at its maximum design efficiency and at the maximum power that can be obtained by the turbine without degrading the hot-section life.

The division in power between the gas turbine and the steam turbine in a combined cycle varies considerably with load, as can be seen in Figure 13-6. At the lower load, the steam turbine actually produces more power than the gas turbine. This is because in a utility application, the mechanical speed must remain constant due to unacceptable consequences of frequency fluctuations. The IGV is adjusted to reduce the flow at off-design loads and to maintain the high exhaust gas temperature.

Figure 13-7 shows the typical efficiencies that one could expect from the different major sections of a typical combined cycle power plant. The gas-turbine efficiency drops off quickly at part load as would be expected, as the gas turbine is very dependent on turbine firing temperature and mass flow of the incoming air. Figure 13-8 shows that the plant heat rate increases rapidly at part load conditions; however, the rate of increase of the gas-turbine heat rate is even steeper, agreeing with the data shown in Figure 13-7.

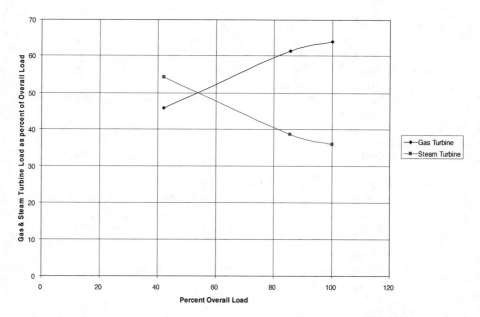

Figure 13-6. Power Sharing Between Gas Turbine and Steam Turbine in a Large Combined Cycle Power Plant

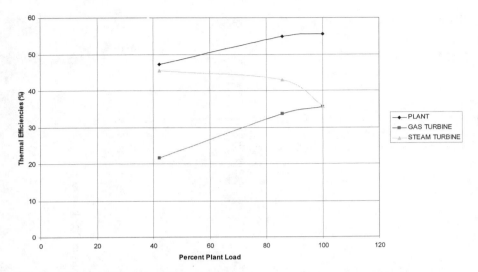

Figure 13-7. Thermal Efficiencies of the Overall Plant, Gas Turbine and Steam Turbine at Part Load Conditions

The plant overall power and the heat rate are very dependent on the inlet conditions as seen in Figure 13-9, which is based on a typical combined cycle plant. The effect of temperature is the most critical component in the ambient condition variations of temperature, pressure, and humidity.

General Governing Equations

The four fundamental equations, which govern the properties of the combined cycle, are the equation of state, conservation of mass, momentum, and energy equations.

Equation of state:

$$\frac{P}{\rho} = Z \frac{R}{MW} T \qquad (13-1)$$

which can also be written as:

$$\frac{P}{\rho^n} = C \qquad (13-1a)$$

where
    n varies from 0 to ∞
    $n = 0$, $P = C$ (constant pressure process)
    $n = 1$, $T = C$ (constant temperature process)

478 • COGENERATION AND COMBINED CYCLE POWER PLANTS

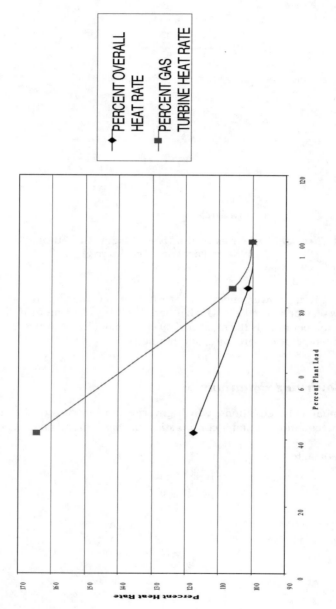

Figure 13-8. Heat Rate Variation with Variation in Plant Load

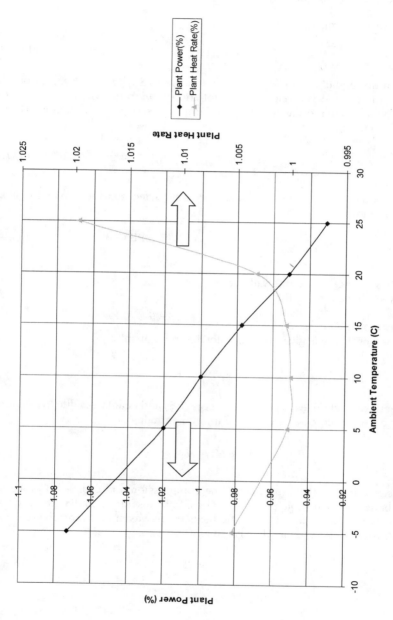

Figure 13-9. Plant Conditions as a Function of Inlet Ambient Temperature

$n = \gamma$ $n = \gamma (\gamma = \frac{c_p}{c_v})$, $S = C$ (constant entropy process)
$n = \infty$ $V = C$ (constant volume process)

Conservation of mass:

$$m = \rho A V \qquad (13-2)$$

Momentum equation; for a caloricaly and thermally perfect gas, and one in which the radial and axial velocities do not contribute to the forces generated on the rotor, the adiabatic energy (E_{ad}) per unit mass is given as follows (Euler Turbine Equation):

$$E_{ad} = \frac{1}{g_c}(U_1 V_{\theta 1} - U_2 V_{\theta 2}) \qquad (13-3)$$

Energy equation; for a caloricaly and thermally perfect gas, the Work (W) can be written as follows:

$$Q_{rad} + \Delta UE + \Delta PV + \Delta KE + \Delta PE = W \qquad (13-4)$$

where
 U = change in the internal energy
 ΔPV = change in the flow energy
 ΔKE = change in kinetic energy
 ΔPE = change in potential energy

The total enthalpy is given by the following relationship:

$$H = U + PV + KE \qquad (13-5)$$

Neglecting the changes in potential energy (PE) and heat losses due to radiation (Q_{rad}), the work is equal to the change in total enthalpy:

$$W = H_2 - H_1 \qquad (13-6)$$

In the gas turbine (Brayton Cycle), the compression and expansion processes are adiabatic and isentropic processes. Thus, for an isentropic adiabatic process, $\gamma = \frac{c_p}{c_v}$; where c_p and c_v are the specific heats of the gas at constant pressure and volume, respectively, and can be written as:

$$c_p - c_v = R \qquad (13-7)$$

where

$$c_p = \frac{\gamma R}{\gamma - 1} \text{ and } c_v = \frac{R}{\gamma - 1} \qquad (13-8)$$

values for air and products of combustion (400% theoretical air) are given in Appendix B.

It is important to note that the pressure measured can be either total or static, however, only total temperature can be measured. The relationship between total and static conditions for pressure and temperature are as follows:

$$T = T_s + \frac{V^2}{2c_p} \qquad (13-9)$$

where
 T_s = static temperature
 V = gas stream velocity

$$P = P_s + \rho \frac{V^2}{2g_c} \qquad (13-10)$$

where
 P_s = static pressure and the acoustic velocity in a gas is given by the following relationship:

$$a^2 = \left(\frac{\partial P}{\partial \rho}\right)_{s=c} \qquad (13-11)$$

for an adiabatic process (s = entropy = constant) the acoustic speed can be written as follows:

$$a = \sqrt{\frac{\gamma g_c R T_s}{MW}} \qquad (13-12)$$

where
 T_s = static temperature
The Mach number is defined as:

$$M = \frac{V}{a} \qquad (13-13)$$

it is important to note that the Mach number is based on static temperature.

The turbine compressor efficiency and pressure ratio are closely monitored to ensure that the turbine compressor is not fouling. Based on these computations, the turbine compressor is water-washed with mineralized water and if necessary adjustment of inlet guide vanes (IGV) is carried out to optimize the performance of the compressor, which amounts to between 60% and 65% of the total work produced by the gas turbine.

The turbine firing temperature, which affects the life, power output, as well as the overall thermal efficiency of the turbine, must be calculated very accurately. To ensure the accuracy of this calculation, the turbine firing temperature is computed using two techniques. These techniques are based, firstly, on the fuel heat input and, secondly, on the turbine heat balance. Turbine expander efficiencies are computed and deterioration noted.

Gas-Turbine Performance Calculation

Increase in pressure ratio and increase in the firing temperature are the two most important factors in the increase of gas-turbine efficiency as can be seen from Figure 13-10. Today, the large gas turbines have pressure ratios ranging from 15:1 to as high as 30:1, and firing temperatures as high as 2500°F (1371°C). These high-pressure ratios lead to a very narrow operational margin in the gas-turbine compressor. The operating margin, between the surge line and the choke region, is reduced with increasing pressure ratio. This means, in a practical sense, that the new compressors on these gas turbines are very susceptible to any fouling of the compressor, indicating that the inlet filters must be very efficient and the turbines must be performance-monitored to ensure maximum operational efficiency.

The overall compressor work is calculated using the following relationship:

$$W_c = (H_{2a} - H_1) = c_{\text{pavg}} T_1 \left\{ \left(\frac{P_2}{P_1}\right)^{\left(\frac{\gamma-1}{\gamma}\right)} - 1 \right\} \qquad (13-14)$$

The work per stage is calculated assuming the energy per stage is equal; this has been found to be a better assumption than assuming the pressure ratio per stage to

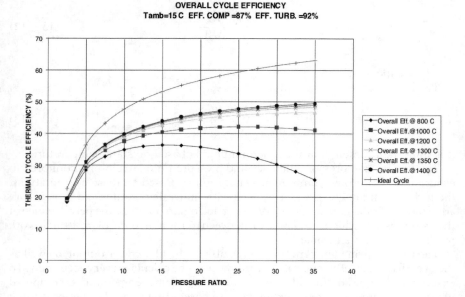

Figure 13-10. Effect of Pressure Ratio and Firing Temperature on the Performance of a Gas Turbine

be equal. It is necessary to know the work per stage if there is inter-stage bleed of the air for cooling or other reasons.

$$w_{stg} = \frac{(H_{2a} - H_1)}{n_{stg}} \qquad (13-15)$$

where
n_{stg} = number of compressor stages

The computation of the compressor total energy requirements can now be computed.

$$\text{Pow}_c = \dot{m}_a w_{stg} n_1 + (\dot{m}_a - \dot{m}_{b1})w_{stg} n_2 + (\dot{m}_a - \dot{m}_{b1} - \dot{m}_{b2})w_{stg} n_3 \ldots \qquad (3-16)$$

The work of the compressor under ideal conditions occurs at constant entropy. The actual work occurs with an increase in entropy; thus, the adiabatic efficiency can be written in terms of the total changes in enthalpy:

$$\eta_{ac} = \frac{\text{Isentropic Work}}{\text{Actual work}} = \frac{(H_{2TI} - H_{1T})}{(H_{2a} - H_{1T})} \qquad (13-17)$$

where
H_{2TI} = total enthalpy of the gas at isentropic exit conditions
H_{2a} = total enthalpy of the gas at actual exit conditions
H_1 = total enthalpy of the gas at inlet conditions for a caloricaly perfect gas whose equation can be written as:

$$\eta_{ac} = \frac{\left[\left(\frac{P_2}{P_1}\right)^{\left(\frac{\gamma-1}{\gamma}\right)} - 1\right]}{\left[\frac{T_{2a}}{T_1} - 1\right]}. \qquad (13-18)$$

The gas-turbine compressor, which produces the high-pressure gas at elevated temperature, uses a very large part of the turbine power produced by the gas turbine, this can amount to about 60% of the total power produced. Figure 13-11 shows the distribution of the gasifier power required as a function of the gas-turbine load, of a typical large gas turbine. The fouling of the compressor, therefore, is a large parasitic load on the gas turbine. Figure 13-12 shows the effect on the compressor efficiency at part load conditions. The flow and the firing temperature affect the turbine expander.

The calculation of the turbine firing temperature (T_{tit}) is based firstly on the fuel injected into the turbine and the fuel's lower heating value (LHV). The lower heating value of the gas is one in which the H_2O in the products has not condensed. The lower heating value is equal to the higher heating value minus the latent heat of the condensed water vapor.

$$H_{tit} = \frac{(\dot{m}_a - \dot{m}_b)H_{2a} + \dot{m}_f \eta_b \text{ LHV}}{(\dot{m}_a + \dot{m}_f - \dot{m}_b)} \qquad (13-19)$$

484 • COGENERATION AND COMBINED CYCLE POWER PLANTS

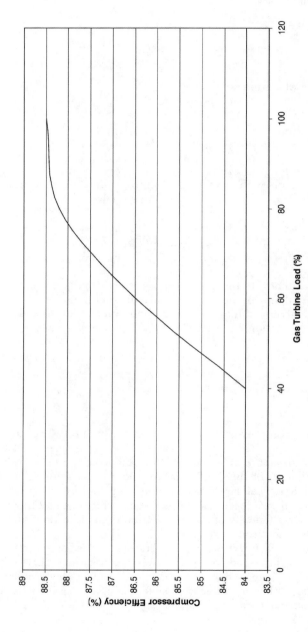

Figure 13-11. Gasifier Power as a Function of Total Gas Turbine Power

Performance Testing of Combined Cycle Power Plant • 485

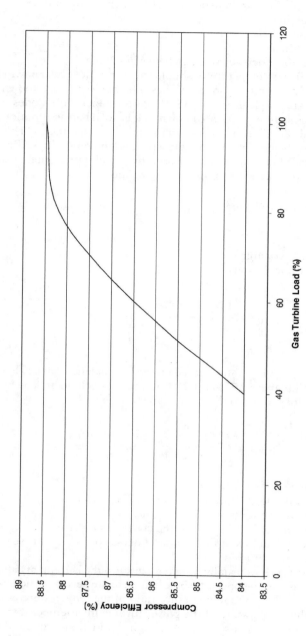

Figure 13-12. Gas Turbine Compressor Efficiency as a Function of Gas Turbine Load

where
- H_{tit} = enthalpy of the combustion gas at the firing temperature
- $\dot{m}_a$ = mass of air
- $\dot{m}_b$ = bleed air
- $\dot{m}_f$ = mass of fuel
- ν_b = combustor efficiency (usually between 97% and 99%)

The turbine firing temperature should be computed by knowing the gas characteristics of the combustion gas. If these characteristics are known, then one can use the equations given in the ASME Performance Test Codes 4.4 (1991). Usually, the gas constituents are not known so it is not a bad assumption to use the 400% theoretical air tables in the Keenan and Kaye gas tables. The following equations for specific heat at constant pressure and the ratio of specific heats have been obtained based on the air tables based on a fuel with a mole weight of the combustion gas to be 28.9553 lb_m/pmole (kg/kg mole).

$$c_p = (-2.76 \times 10^{(-10)} T^2 + 1.1528 \times 10^{(-5)} T + 0.237) \times C1 \qquad (13-20)$$

where
- $C1 = 1.0$ in the US units
- $C1 = 4.186$ in the SI units

and:

$$\gamma = \frac{c_p}{\left(c_p - \frac{R}{\frac{778.16}{MW}}\right)} \qquad (13-21)$$

The turbine firing temperature based on the heat balance can be also computed and must be within about 2°F to 6°F (1°C to 3°C) of each other. The heat balance relationships as they apply to the gas turbine:

$$H_{tit} = \frac{\frac{Pow_c}{\eta_{mc}} + \frac{Pow_g}{\eta_{mt}} + (\dot{m}_a + \dot{m}_f) H_{exit}}{(\dot{m}_a + \dot{m}_f - \sum \dot{m}_b)} \qquad (13-22)$$

where
- Pow_c = work of the gas-turbine compressor, BTU/sec
- Pow_g = generator output
- ν_{mc} = mechanical loss in the turbine compressor drive
- ν_{mt} = mechanical loss in the turbine process compressor drive
- H_{exit} = enthalpy at turbine exit

Split-shaft gas turbines usually have temperature measurements at the gasifier turbine exit and also at the power turbine exit. From experience and also based on theoretical relationships, the temperature ratio of the temperature at the gasifier inlet (T_{tit}) and the temperature of the power turbine inlet temperature (T_{pit}) for a given geometry remains constant even though the load and ambient conditions change. It is because of this that most manufacturers limit the engine based on the power turbine inlet temperature.

$$T_r = \frac{T_{tit}}{T_{pit}} \qquad (13-23)$$

This also enables Eq. (13-19) for the case of a split shaft turbine to be rewritten as:

$$H_{tit} = \frac{\frac{Pow_c}{\eta_{mc}} + (\dot{m}_a + \dot{m}_f - 0.6\dot{m}_b)H_{pit}}{(\dot{m}_a + \dot{m}_f - \dot{m}_b)} \qquad (13-24)$$

where an assumption of 40% of the bleed flow was assumed to have entered the turbine through the cooling mechanisms of the first few stages of the turbine.

To ensure that the heat balance is accurate, the following relationship indicates the accuracy of the computations. This heat balance ratio can be written as follows:

$$HB_{ratio} = \frac{\frac{Pow_c}{\eta_{mt}} + (\dot{m}_a + \dot{m}_f)H_{exit} - \dot{m}_a H_{inlet}}{\dot{m}_f \, LHV} \qquad (13-25)$$

this ratio should be between 0.96 and 1.04.

Figure 13-13 shows the effect of the turbine firing temperature on the turbine expander efficiency. The decrease in firing temperature reduces the absolute velocity, as also does the reduction in the mass flow, both of which occur at part load conditions. Figure 13-14 shows the variation in the firing temperature and the exhaust gas temperature as a function of the load. It is interesting to note that the firing temperature of the turbine is greatly reduced, while the exhaust temperature remains nearly constant, accounting for the steam turbine producing more work at low part loads.

The work produced by the gasifier turbine (W_{gt}) is equal to the gas-turbine compressor work (W_c):

$$Pow_{gt} = \frac{Pow_c}{\eta_{mc}} \qquad (13-26)$$

The gasifier turbine efficiency (η_{gt})

$$\eta_{gt} = \frac{H_{tit} - H_{pita}}{H_{tit} - H_{piti}} \times 100 \qquad (13-27)$$

where
H_{pita} = enthalpy of the gas based on the actual temperature at the exit of the gasifier turbine
H_{piti} = enthalpy of the gas based on the ideal temperature at the exit of the gasifier turbine

488 • COGENERATION AND COMBINED CYCLE POWER PLANTS

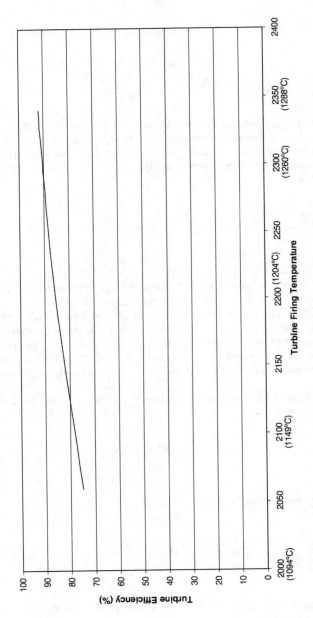

Figure 13-13. Gas Turbine Turbine Expander Efficiency as a Function of Firing Temperature

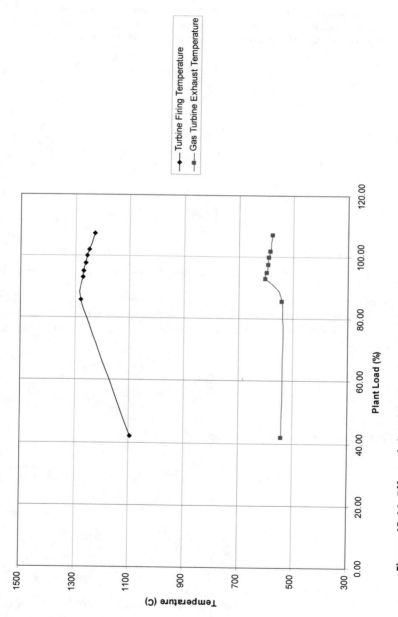

Figure 13-14. Effect of the Plant Load on Turbine Firing Temperature and the Turbine Exhaust

To obtain this ideal enthalpy, the pressure ratio across the gasifier turbine must be known.

The pressure ratio (P_{grt}) across the turbine depends on the pressure drop (ΔP_{cb}) through the combustor. This varies in various combustor designs, where a pressure drop of between 1% and 3% of the compressor discharges pressure.

$$P_{grt} = \frac{P_{dc}(1 - \Delta P_{cb})}{P_{dgt}} \quad (13-28)$$

where
  P_{dgt} = pressure at the gasifier turbine exit

Thus, the ideal enthalpy at the gasifier turbine exit is given by:

$$H_{piti} = \frac{H_{tit}}{\dfrac{c_{p_{tit}}}{c_{p_{tit}}}\left(P_{grt}^{\left(\frac{\gamma-1}{\gamma}\right)}\right)} \quad (13-29)$$

where
  γ = based on an average temperature across the gasifier turbine based on Eq. (13-21)

The power turbine efficiency can be computed using Eqs. (13-27) and (13-29).

The overall thermal efficiency of the gas turbine in a simple cycle (varies between 25% and 45%, depending on the turbine) is computed to determine deterioration of the turbine:

$$\eta_{ovt} = \frac{\dfrac{Pow_g}{\eta_{mt}}}{\dot{m}_f \, LHV} \times 100 \quad (13-30)$$

The heat rate can now be easily computed:

$$HR = \frac{2544.4}{\dfrac{\eta_{th}}{100}} = \frac{Btu}{hp-hr} \quad (13-31)$$

$$= \frac{3600}{\dfrac{\eta_{th}}{100}} = \frac{kJ}{kW-hr}$$

Correction Factors for Gas-Turbine Performance

The performance of the gas turbine is based on the basic equations in the prior section. To relate these relationships to the turbine in concern, and to calculate the deterioration of different sections of the gas turbine, the values obtained must be corrected to design conditions and, in some cases, values would have to be transposed from off-design conditions to the design conditions. The corrected values define the engine-corrected performance values. Geometric similarity, such as blade characteristics, clearances, nozzle areas, and guide vane settings, do not change when geometric similarity is constant. Dynamic similarity, which relates to parameters such as gas velocities, and turbine speeds, when maintained together with the geometric similarity, ensure that these corrected parameters will maintain the engine performance at all operating conditions.

Corrected mass flow:

$$\dot{m}_{\text{acorr}} = \frac{\dot{m}_a \sqrt{\frac{T_{\text{inlet}}}{T_{\text{std}}}}}{\frac{P_{\text{inlet}}}{P_{\text{std}}}} \qquad (13-32)$$

where

$\dot{m}_{\text{acorr}}$ = corrected mass flow of the air entering the gas-turbine inlet

These corrections are from the ambient conditions to usually the ISO conditions (14.7 psia, 60°F, RH = 60%), (1 bar, 15°C, RH = 60%).

The corrected speed for both the gasifier and power turbine defines the corrected engine performance.

Corrected speed:

$$N_{\text{corr}} = \frac{N_{\text{act}}}{\sqrt{\left(\frac{R_a T_a}{(RT)_{\text{std}}}\right)}} \qquad (13-33)$$

Corrected temperature:

$$T_{\text{corr}} = \frac{T_a}{\frac{T_{\text{inlet}}}{T_{\text{std}}}} \qquad (13-34)$$

Corrected fuel flow:

$$\dot{m}_{\text{fcorr}} = \frac{\dot{m}_f}{\left(\frac{P_{\text{inlet}}}{P_{\text{std}}}\right) / \left(\sqrt{\frac{T_{\text{inlet}}}{T_{\text{std}}}}\right)} \qquad (13-35)$$

Corrected power:

$$HP_{\text{corr}} = \frac{HP_{\text{act}} \frac{T_{\text{inlet}}}{T_{\text{std}}}}{\frac{P_{\text{inlet}}}{P_{\text{std}}}} \qquad (13-36)$$

The above relationship has to be further modified to take into account the pressure drop in the inlet ducting, the increase in back pressure due to exhaust ducting, the off-design operation due to decrease in turbine firing temperature and decrease in speed of the power turbine. These modifications are used to calculate the transposed power (HP$_{pt}$) by transposing from the off-design output power at operating conditions of the turbine to the design conditions.

Transpose power output:

$$\text{HP}_{tp} = \text{HP}_{corr} + (\Delta P_c(\text{PW}_i)) + (\Delta P_e(\text{PW}_e)) + (T_{dtit} - T_{atit})c_p(\dot{m}_d - \dot{m}_a)\eta_{at}$$

$$+ \left[1 + 0.45\left(1 - \frac{N_{ptcorr}}{N_{ptdes}}\right)^m\right]\text{HP}_{act} \qquad (13-37)$$

where
 ΔP_c = pressure drop at the inlet due to the filters and evaporator in the inlet ducting
 PW$_i$ = power loss per inch of H$_2$O drop
 ΔP_e = back pressure at the discharge due to the exhaust ducting
 PW$_e$ = power loss per inch of H$_2$O drop

The last term of the equation only applies to split shaft turbines. The power factor (m) to which the speed ratio is raised will vary with turbines; in the case of this turbine, the value was $m = 0.4$.

Heat-Recovery Steam Generator

The Exhaust gas from the gas turbine has a very large amount of energy. The energy converted into power in a typical large combined cycle power plant is about 36% of the energy available at the Gas-Turbine Exhaust.

The HRSG as shown in Figure 13-15 is composed of many sections depending on the design. Figure 13-15 shows a three-section unit, where each section is composed of a preheater or an economizer, an evaporator, and one or two stages of superheaters. The condenser water flow enters the HRSG unit after being pumped by the boiler feedwater pump through the feedwater heater. In this configuration, a dearator heater is also provided. The feedwater heater and the dearator provide some energy to the water, increasing the temperature by about 90°F to 125°F (50°C to 70°C) to the Dearator tank. The boiler feedwater pump increases the pressure from the condenser discharge pressure of about 0.5 to 2.0 psia (0.033 to 0.13 bar), to about 60 to 400 psia (5 to 27 bar), depending on the configuration. In the end, with the LP feedwater pump, the LP Evaporator has a pressure of 450 to 800 psia (30 to 54 bar). The energy transferred is about 80% to the HP loop and the 20% to the LP and the IP loops. The gas entering the HRSG has energy equivalent to the rated power of most of these large plants. This energy can range from about 100 to 360 MW per unit in the large combined cycle plants. The gas temperatures entering the HRSG are between 950°F and 1200°F (510°C to 650°C), and leaving the HRSG stack are around 275°F (135°C).

Economizer. The flow in the economizer or the preheaters on the "steam side" is in a liquid state. This means that all the calculations of the enthalpy are for a

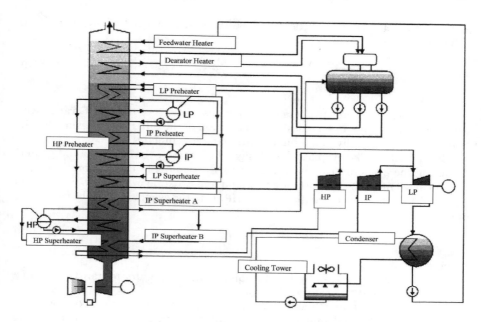

Figure 13-15. Three Section of HRSG

saturated liquid. The flow enters as a liquid, heat is added to the liquid, but the state of the flow remains a liquid. The heat added means that the temperature of the water is increased, but the water remains in the state of a saturated liquid. The computation of the energy increased in the economizer can be calculated as follows:

$$Q_{secon} = \dot{m}_{sl}(H_{fe} - H_{fi}) \qquad (13-38)$$

where

Q_{secon} = energy transferred to the liquid in the economizer
$\dot{m}_{sl}$ = mass flow of the liquid being pumped through the economizer
H_{fe}, H_{fi} = the enthalpy of a saturated liquid at the exit and inlet temperatures, respectively

The energy available on the gas side can be computed if the HRSG is well instrumented with temperature probes at the entrance and exit of each section. The energy available on the gas side can be calculated as follows:

$$Q_{gecon} = (\dot{m}_a + \dot{m}_f)(H_{ge} - H_{gi}) \qquad (13-39)$$

where

Q_{gecon} = energy available in the gas side for the economizer
H_{ge}, H_{gi} = enthalpy of the gas (air and fuel) at the exit and inlet of the HRSG economizer section, respectively

The economizer effectiveness can be computed by the following relationship:

$$h_{eefc} = \frac{Q_{secon}}{Q_{gecon}} \qquad (13-40)$$

The value of the effectiveness of the various sections of the HRSG varies with the gas temperature and the design and location of the different components. The effectiveness is usually in the high eighties and low nineties.

Evaporator. The evaporator is designed to transfer the heated liquid (water) entering the evaporator to a saturated vapor. The flow in the evaporator on the steam side enters the evaporator as a liquid and exits the evaporator as saturated steam. The heat transferred through the HRSG to the evaporator can be calculated as shown:

$$Q_{sevap} = \dot{m}_{sl}(H_{sve} - H_{fe}) \qquad (13-41)$$

where

Q_{sevap} = energy transferred to the liquid in the evaporator to transfer it from a liquid to a saturated vapor
$\dot{m}_{sl}$ = mass flow of the liquid being pumped through the evaporator
H_{sve} = saturated enthalpy at the exit temperature of the steam at the evaporator exit
H_{fe} = enthalpy of the saturated liquid leaving the economizer

The energy available to the evaporator in the HRSG can be computed similarly as the energy available to the economizer. Using Eqs. (13-38) and (13-39) with the corresponding values applicable for the evaporator section, the energy available and the effectiveness of the evaporator section of the HRSG can be computed. The effectiveness is usually slightly higher than the economizer as the gas temperatures are higher.

Superheater. The superheaters in the HRSG transfer the saturated vapor leaving the evaporator to a superheated vapor for injection into the steam turbines. The process in the superheater is a constant pressure process. The temperature in the superheater is increased from the saturated vapor temperature to the superheated temperature; this temperature increase in the superheater can be as much as 400°F (205°C). The superheater is often divided into two or three stages. The high-pressure superheater is usually placed closest to the inlet of the exhaust gases from the gas turbine. The computation of the energy transferred to the superheater through the HRSG is given by the following relationship:

$$Q_{ssupr} = \dot{m}_{ss}(H_{ssupe} - H_{sve}) \qquad (13-42)$$

where

Q_{ssupr} = energy transferred to the steam in the superheater to transfer it from a saturated vapor to a superheated vapor

$\dot{m}_{ss}$ = mass flow of the steam being sent through the superheater
H_{ssupe} = temperature of the superheated vapor at the exit of the superheater
H_{sve} = saturated enthalpy at the exit temperature of the steam at the evaporator exit

The energy available to the superheater in the HRSG can be computed similarly as the energy available to the economizer. Using Eqs. (13-38) and (13-39) with the corresponding values applicable for the superheater section, the energy available and the effectiveness of the superheater section of the HRSG can be computed. The effectiveness is usually the highest in this section as the gas temperatures are higher.

The HRSG effectiveness is important in transferring the energy to the steam turbine cycle (Rankine Cycle). The effectiveness of the HRSG in a combined cycle power plant is in the 90th percentile. Figure 13-16 shows the energy available to the HRSG, and the effectiveness of the HRSG as a function of plant load. The energy available to the HRSG will decrease as the plant load is reduced, since to obtain a reduction in the plant load the air flow to the gas turbine and the firing temperature are reduced. The effectiveness of the HRSG will vary from one design to the other; its variation with plant load will depend on the flue gas characteristics and the number of sections and components the HRSG is divided into.

Steam Turbines

The steam-turbine computation must be divided into two parts: one for steam turbines operating in the superheated vapor region, and the other for steam turbines operating initially in the superheated vapor region and exiting in the liquid-vapor region.

The high-pressure steam turbines in these large combined cycle power plants operate at very high pressure and high temperatures. The HP steam-turbine power can be computed as follows:

$$\text{Pow}_{st} = \dot{m}_{ss}(H_{svi} - H_{sve}) \qquad (13-43)$$

where

H_{svi}, H_{sve} = enthalpy of the superheated vapor at the inlet and exit temperature and pressure, respectively

The efficiency of the high-pressure steam turbine is computed by calculating the ideal power based on the following relationships:

$$\text{Pow}_{ist} = \dot{m}_{ss}(H_{svi} - H_{isve}) \qquad (13-44)$$

where

H_{isve} = ideal enthalpy based on the exit pressure and the entropy based on the inlet condition

496 • COGENERATION AND COMBINED CYCLE POWER PLANTS

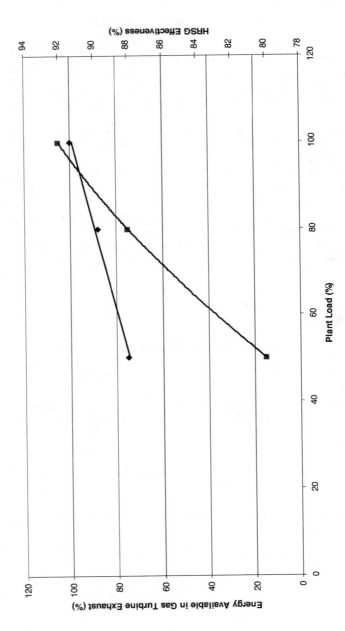

Figure 13-16. The Effect of Energy Available and HRSG Effectiveness as a Function of Plant Load

The calculation of the efficiency is now computed using the following relationships:

$$\eta_{st} = \frac{\text{Pow}_{st}}{\text{Pow}_{st}} \qquad (13-45)$$

where

η_{st} = efficiency of the steam turbine

The low-pressure steam turbine in these large combined cycle power plants operates at its exit in a liquid-vapor region. Thus, though the temperature and pressure are known the exit enthalpy cannot be computed since the quality of the steam is not known. To calculate the work of the LP steam turbine, the heat removed by the condenser, Q_w, must be computed. This is computed by the following relationship:

$$Q_{cw} = \dot{m}_w c_{pw}(T_{2cw} - T_{1cw}) \qquad (13-46)$$

where

T_{2cw}, T_{1cw} = cooling water temperature at the exit and inlet of the water
$\dot{m}_w$ = cooling water flow through the condenser
c_{pw} = specific heat at constant pressure for the water

The calculation of the enthalpy at the LP steam turbine can now be computed by using the following relationship:

$$H_{lv} = Q_{cw} - H_{lc} \qquad (13-47)$$

where

H_{lc} = enthalpy at the liquid state and the exit temperature and pressure of the LP steam turbine

The condenser is sensitive to the cooling water flow and the cooling water temperature. Figure 13-17 shows that with increasing cooling water temperature the back-pressure on the LP steam turbine is increased. The increase in the back-pressure reduces the output of the LP steam turbine. The fouling of the condenser is a cause of great concern in most large power plants, especially plants which use sea water as the cooling water.

The LP steam-turbine power now can be computed using the heat removed from the condenser by the water:

$$\text{Pow}_{slp} = \dot{m}_{slp}\{H_{svi} - (Q_{cw} - H_{lc})\} \qquad (13-48)$$

The efficiency can now be computed for the LP steam turbine based on the relationship obtained in Eq. (13-45), and the computation of the ideal enthalpy based on the entropy at the inlet conditions and the pressure and temperature at the exit conditions of the LP steam turbine.

The steam turbine, in most cases, is divided into two or three sections. The HP section, where most of the work is done, also usually has the highest efficiency. The

498 • COGENERATION AND COMBINED CYCLE POWER PLANTS

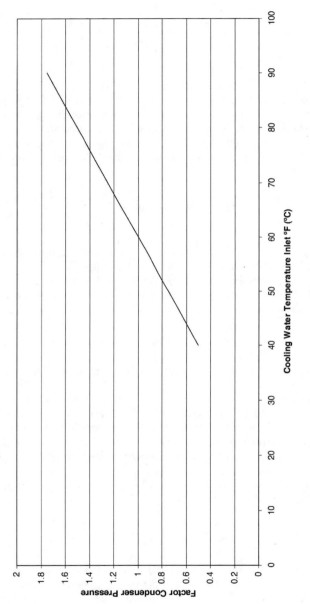

Figure 13-17. Effect of Condenser Inlet Cooling Water on Condenser Vacuum Pressure

calculation of the LP steam-turbine characteristics are difficult to compute since, in most cases, the steam exiting from the LP turbine has some liquid entrained in it. The quality of the steam cannot be measured so the calculation depends on the condenser measurements of the temperatures of the water entering and leaving the condenser and flow through the condenser. Measurements of the flow through the condenser are hard to measure accurately and this results in some inaccuracy in calculating the LP turbine characteristics.

Plant Losses

The losses that are encountered in a plant can be divided into two groups: uncontrollable losses and controllable losses. The uncontrollable losses are usually environmental conditions, such as temperature, pressure, humidity, and the turbine aging. Table 13-2 shows the approximate changes that would occur for these changes. It must be remembered that these are just approximations and will vary for individual power plants.

The controllable losses are those that the operator can have some degree of control over and can take corrective actions:

1. Pressure Drop across the Inlet Filter. This can be remedied by cleaning or replacing the filter.
2. Compressor Fouling. On-line water cleaning can restore part of the drop encountered.
3. Fuel Lower Heating Value. In many plants, on-line fuel analyzers have been introduced not only to monitor the turbine performance, but also to calculate the fuel payments, which are usually based on the energy content of the fuel.
4. Turbine Back Pressure. In this case, the operator is relatively limited since he cannot do anything about the downstream design. If there is some obstruction in the ducting to the HRSG that can be removed or if the duct has collapsed in an area, the duct could be replaced.
5. HRSG Effectiveness. A properly instrumented HRSG would have thermocouples placed at intervals throughout the HRSG so that effectiveness of each section of the HRSG can be monitored.

Table 13-2. Effect of Uncontrollable Losses on the Output and Heat Rate

Parameters	Parameter Change	Power Output	Heat-Rate Change
Ambient Temperature	20°F (11°C)	−6.5%	2%
Ambient Pressure	4 in. H$_2$O (10 mbar) (mmH$_2$O)	0.9%	0.9%
Ambient Relative Humidity	10%	−.0002%	.0005%
Turbine Age	First 10,000 hrs	−0.34%/1000	.5%
Turbine Age	Above 10,000 hrs	−.03%	−.08%

Table 13-3. Effect of Controllable Losses on the Output and Heat Rate

Parameters	Parameter Change	Power Output	Heat-Rate Change
Compressor Fouling	2%	−1.5%	0.65%
Pressure Drop in Filter	1 in. H$_2$O (25 mmH$_2$O)	−0.5%	−0.3%
Increase in Gas Turbine Back Pressure	1 in. H$_2$O (25 mmH$_2$O)	−.25%	.08%
Lower Heating Value	−430 Btu/lb (−1000 kJ/kG)	0.4%	−1.0%
Power Factor	−0.05	−.14%	−.15%

6. Steam-Turbine Fouling. The fouling of the steam turbine can be noted. Most of the fouling will occur in the low-pressure turbine along with some blade erosion problems. By trying to calculate the quality content of the steam, as well as the power output of that section of the turbine, problems can be noted and corrective action is taken.
7. Condenser Pressure. The fouling of the tubes, insufficient water flow through the condenser, the condenser cooling water temperature all lead to increasing the pressure at the low-pressure turbine exit leading to an increase in the exit pressure and temperature.
8. Condenser Condensate Subcooling. The subcooling of the condensate can lead to losses as the condensate has to be heated up and so more energy is spent to do so and the overall energy available in the HRSG is reduced.

Table 13-3 shows the effect of controllable losses in the output and heat rate of a typical combined cycle power plant. The gas turbine has to be operated at a constant speed for power generation and any slight variation in speed could result in major problems for the grid. Thus, the control of the load has to be by controlling the fuel input, i.e., the turbine firing temperature and the inlet guide vane position, thus controlling the airflow. The effect of this is to try and maintain the exhaust temperature from the gas turbine at a relatively high value since this gas is used in the HRSG, and the effectiveness of the HRSG is dependent on maintaining this temperature.

NOMENCLATURE

Symbol	Description	Units, U.S., SI
A	Area	ft^2, m^2
a^2	Acoustic Velocity of a Gas	ft/sec, m/sec
c_{pa}	Specific Heat of Air at Constant Pressure	Btu/lb$_m$ R, kJ/kg K
c_{pg}	Specific Heat of Gas at Constant Pressure	Btu/lb$_m$ R, kJ/kg K
c_{va}	Specific Heat of Air at Constant Volume	Btu/lb$_m$ R, kJ/kg K

Symbol	Description	Units
c_{vg}	Specific Heat of Air at Constant Volume	Btu/lb$_m$ R, kJ/kg K
E	Energy per Unit Mass	ft lb$_f$/lb$_m$v, kJ/kg
F	Force	
g_c	Newtonian Constant	32.2 ft lb$_m$/lb$_f$ sec^2, 9.807 m/sec^2
H	Total Enthalpy	Btu/lb$_m$, kJ/kg
HHV	Lower Heating Value of Fuel	Btu/lb$_m$, kJ/kg
KE	Kinetic Energy	Btu/hr, kJ/hr
LHV	Higher Heating Value of Fuel	Btu/lb$_m$, kJ/kg
M	Mach Number	
$\dot{m}$	Mass Flow	lb$_m$/sec, kg/sec
$\dot{m}_a$	Mass Flow of Air	lb$_m$/sec, kg/sec
$\dot{m}_{b1}, \dot{m}_{b2}, ...\dot{m}_{bn}$	Mass Flow of Bleed Air at Different Bleed Locations	lb$_m$/sec, kg/sec
$\dot{m}_f$	Mass of Fuel Flow	lb$_m$/sec, kg/sec
MW	Mole Weight	lb$_m$/pmole, kg/kg mole
N	Speed	rpm
n	Polytropic Exponent	
$n_1, n_2, ...n_n$	Number of Stages Between Different Bleed Locations	
n_{stg}	Number of Stages	
P	Total Pressure	lb$_f$/ft^2, bar
P_s	Static Pressure	lb$_f$/ft^2, bar
PE	Potential Energy	Btu/lb$_m$, kJ/kg
Pow$_c$	Power Compressor	Btu/hr, kJ/hr or MW
Pow$_g$	Power Generator Output	Btu/hr, kJ/hr or MW
PV	Flow Energy	Btu/lb$_m$, kJ/kg
Q	Heat	Btu/lb$_m$, kJ/kg
Q_{rad}	Radiation Heat Loss	Btu/lb$_m$, kJ/kg
R	Universal Gas Constant	1545 ft lb$_f$/pmole R, 8.3143 kg/kg mole K
RH	Relative Humidity	%
s	Entropy	Btu/lb$_m$ R, kJ/kg K
T	Total Temperature	R, K
T_s	Static Temperature	R, K
U	Blade Velocity	ft/sec, m/sec
V	Absolute Velocity	ft/sec, m/sec
V_θ	Tangential Component of Absolute Velocity	ft/sec, m/sec
W	Work	Btu/lb$_m$, kJ/kg
w_c	Work of Compressor per Stage	Btu/lb$_m$, kJ/kg
W_c	Work of Compressor	Btu/hr, kJ/hr
W_t	Work of Turbine	Btu/hr, kJ/hr
Z	Compressibility Factor	
γ	Ratio of Specific Heats	
ΔP_{cb}	Pressure Drop in Combustor	
η_{ac}	Adiabatic Efficiency of Compressor	%

η_{at}	Adiabatic Efficiency of Turbine	%
η_b	Adiabatic Efficiency of Combustor	%
η_{mc}	Mechanical Efficiency Compressor Section	%
η_{mt}	Mechanical Efficiency Turbine Section	%
η_{ovp}	Overall Adiabatic Efficiency of Plant	%
η_{ovt}	Overall Adiabatic Efficiency of Gas Turbine	%
ρ	Density of the Gas	lb_m/ft^3
		kg/m^3
Φ	Flow Coefficient	
ω	Angular Speed	radian/sec

Chapter 14

MAINTENANCE TECHNIQUES

Philosophy of Maintenance

Maintenance, defined as the "upkeep of property," is one of the most important operations in a plant. The manufacture and maintenance of turbomachinery are totally different. The first involves the shaping and assembly of various parts to required tolerances, while the second, maintenance, involves restoration of these tolerances through a series of intelligent compromises. The crux of maintenance technique is in keeping the compromises intelligent.

Maintenance is not a glamorous procedure; however, its importance is second to none. Maintenance procedures are always controversial, since the definition of "upkeep" varies with the individual interpretation of each maintenance supervisor. The latitude of maintenance ranges from strict planning and execution, inspection and overhaul, accompanied by complete reports and accounting of costs, to the operation of machinery until some failure occurs, and then making the necessary repairs.

Modern day turbomachinery is built to last between 30 and 40 years. Thus, the keeping of basic maintenance records and critical data is imperative for a good maintenance program. Economic justification is always the controlling factor for any program, and maintenance practices are not different.

Maintenance costs can be minimized by, and are directly related to, good operation; likewise, better operating results can be obtained when the equipment is under the control of a planned maintenance program. Improper operation of mechanical equipment can be as much or more the cause of its deterioration and failure as is actual, normal mechanical wear. Thus, operation and maintenance go together.

Combining the practice of preventive maintenance and total quality control and total employee involvement results in an innovative system for equipment maintenance that optimizes effectiveness, eliminates breakdowns, and promotes autonomous operator maintenance through day-to-day activities. This concept, known as Total Productive Maintenance (TPM), was conceived by Seiichi Nakajima and is well documented in his book "Introduction of TPM" and is highly recommended reading for all involved in the maintenance area.

A new maintenance system is introduced based on the new mantra for the selection of all equipment "Life Cycle Cost". This new system, especially for major power plants, is based on the combination of Total Condition Monitoring, and the maintenance principles of Total Productive Maintenance, and is called the "Performance-Based Total Productive Maintenance System."

The general maintenance system is fragmented and can be classified into many maintenance concepts. The following are five Ps of maintenance for major power plants, petro-chemical corporations, and other process-type industries leading to the ultimate maintenance system:

1. Panic Maintenance Based on Breakdowns
2. Preventive Maintenance
3. Performance-Based Maintenance
4. Performance-Productive Maintenance
5. Performance-Based Total Productive Maintenance (PTPM).

Performance-Based Total Productive Maintenance consists of the following elements:

1. Performance-Based Total Productive Maintenance aims to maximize equipment efficiency and time between overhaul (overall performance effectiveness).
2. Performance-Based Total Productive Maintenance aims to maximize equipment effectiveness (overall effectiveness).
3. Performance-Based Total Productive Maintenance establishes a thorough system of PM for the equipment's entire life span.
4. Performance-Based Total Productive Maintenance is implemented by various departments (engineering, operations, maintenance).
5. Performance-Based Total Productive Maintenance involves every single employee, from top management to workers on the floor.
6. Performance-Based Total Productive Maintenance is based on the promotions of PM through *motivation management*: autonomous small group activities.

The word "total" in "Performance-Based Total Productive Maintenance" has four meanings that describe the principal features of PTPM:

1. *Total overall performance effectiveness* (referred to in point 1 above) indicates PTPM's pursuit of maximum plant efficiency and minimum downtime.
2. *Total overall performance effectiveness* (referred to in point 2 above) indicates PTPM's pursuit of economic efficiency or profitability.
3. *Total maintenance system* (point 3) includes maintenance prevention (MP) and maintainability improvement (MI) as well as preventive maintenance.
4. *Total participation of all employees* (Points 4, 5, and 6) includes autonomous maintenance by operators through small group activities.

Table 14-1 shows the relationship between PTPM, productive maintenance and preventive maintenance.

Table 14-1. Benefits of Various Maintenance Systems Maintenance

	Performance-Based Total Productive Maintenance	Performance Productive Maintenance	Performance-Based Maintenance	Preventive Maintenance	Panic Maintenance
Economic Efficiency	Yes	Yes	Yes	Yes	No
Economic and Time Efficiency	Yes	Yes	Yes	No	No
Total System Efficiency	Yes	Yes	No	No	No
Autonomous Maintenance by Operators	Yes	No	No	No	No

Performance-Based Total Productive Maintenance eliminates the following seven major losses:

Down time:

1. Loss of time due to unnecessary overhauls based only on time intervals.
2. Equipment failure from breakdowns.
3. Loss of time due to spare part unsuitability or insufficient spares.
4. Idling and minor stoppages due to the abnormal operation of sensors, or other protective devices.
5. Reduced output — due to discrepancies between designed and actual operating conditions.

Defect:

1. Process defects — due to improper process conditions that do not meet machinery design requirements.
2. Reduced yield — from machine startup to stable production due to inability of machine to operate at proper design conditions.

Maximization of Equipment Efficiency and Effectiveness

High machine efficiency and availability can be attained by maintaining the health of the equipment. Total Performance Condition Monitoring can play a major part here as it provides early warnings of potential failures and performance deterioration. Figure 14-1 shows the concept of a Total Performance Condition Monitoring System.

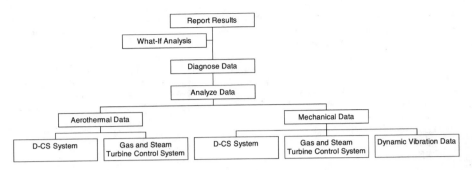

Figure 14-1. Total Performance-Based Condition Monitoring System

Pure preventive maintenance alone cannot eliminate breakdowns. Breakdowns occur due to many factors, such as design and or manufacturing errors, operational errors, and wearing out of various components. Thus, changing out components at fixed intervals does not solve the problems and in some cases adds to the problem. A study at a major nuclear power station indicated that nearly 35% of the failures occurred within a month of a major turnaround. Figure 14-2 shows the life characteristics of a major piece of turbomachinery.

The goal of any good maintenance program is "Zero Breakdown." To achieve this goal, there are five counter measures. These are listed below:

1. Maintaining well-regulated basic conditions (cleaning, lubricating, and bolting).
2. Adhering to proper operating procedures.
3. Total condition monitoring (performance, mechanical and diagnostic based).
4. Improving weaknesses in design.
5. Improving operation and maintenance skills.

The interrelationship between these five items is shown in Figure 14-3.

The division of labor between operations and maintenance is shown in Figure 14-4. It is the primary responsibility of the production department to establish and regulate basic operating conditions, and it is the primary responsibility of the maintenance department to improve defects in design. The other tasks are shared between the two departments.

The successful implementation of Total Productive Maintenance requires:

1. Elimination of the six big losses to improve equipment effectiveness.
2. An autonomous maintenance program with total condition monitoring.
3. A scheduled maintenance program for the maintenance department.
4. Increased skills of operations and maintenance personnel.
5. An initial equipment management program.

Maintenance Techniques • 507

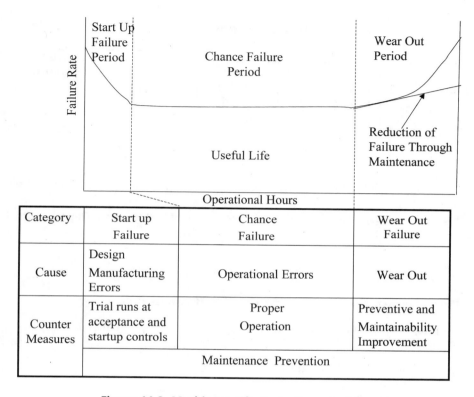

Figure 14-2. Machinery Life Cycle Characteristics

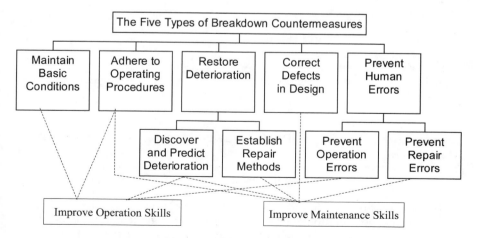

Figure 14-3. Breakdown Countermeasures

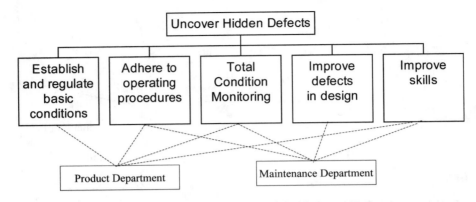

Figure 14-4. Responsibilities of the Operations and Maintenance Departments

Organization Structures

Performance-Based Total Productive Maintenance Program

Typically, successful implementation of PTPM in a large plant takes 3 years. Implementation calls for:

1. Changing people's attitudes
2. Increasing motivation
3. Increasing competency
4. Improving the work environment.

The four major categories in developing a Performance-Based Total Productive Maintenance program are:

1. Preparation for the PTPM program
2. Preliminary implementation
3. PTPM implementation
4. Stabilization of the program.

Implementation of a Performance-Based Total Productive Maintenance

There are several steps involved in implementation of a PTPM program.

1. Announcement of decision to implement PTPM
   A formal presentation must be made by top management introducing the concepts, goals, and benefits of PTPM. Management commitment must be made clear to all levels of the organization.

2. Educational campaign

   The training and promotion of PTPM philosophy is a must. This is useful to reduce the resistance to change. The education should cover how PTPM will be beneficial to both the corporation and the individuals.

3. Creation of organization to promote PTPM

   The PTPM promotional structure is based on an organizational matrix. Obviously, the optimal organizational structure would change from organization to organization. In large corporations, PTPM promotional headquarters must be formed and staffed. Thus, any questions can be addressed here on a corporate level.

4. Establishment of basic PTPM goals

   Establishing mottos and slogans can do this. All goals must be quantifiable and precise specifying:

   - Target (what)
   - Quantity (how much)
   - Time frame (when)

5. Master plan development for PTPM

   A master plan must be created. Total Condition Monitoring equipment should be designed and equipment purchased.

6. Initiation of PTPM

   This represents a "kickoff" stage. At this point, the whole staff must start to get involved.

7. Improvement of equipment effectiveness

   This should start with a detailed design review of the plant machinery. A performance analysis of the plant should point to a specific area known to have problems (i.e. section of plant), must be selected and focused on, project teams should be formed, and assigned to each train. An analysis should be conducted that addresses the following:

   7.1. *Define the problem.* Examine the problem (loss) carefully; compare its symptoms, conditions, affected parts, and equipment with those of similar cases.

   7.2. *Do a physical analysis of the problem.* A physical analysis clarifies ambiguous details and consequences. All losses can be explained by simple physical laws. For example, if scratches are frequently produced in a process, friction or contact between two objects should be suspected. (Of the two objects, scratches will appear in the object with the weaker resistance.) Thus, by examining the points of contact, specific problem areas and contributing factors are revealed.

   7.3. *Isolate every condition that might cause the problem.* A physical analysis of breakdown phenomena reveals the principles that control their occurrence and uncovers the conditions that produce them. Explore all possible causes.

7.4. *Evaluate equipment, material, and methods.* Consider each condition identified in relation to the equipment, jigs and tools, material, and operating methods involved and draw up a list of factors that influence the conditions.

7.5. *Plan the investigation.* Carefully plan the scope and direction of investigation for each factor. Decide what to measure and how to measure it and select the datum plane.

7.6. *Investigate malfunctions.* All items planned in Step 5 must be thoroughly investigated. Keep in mind optimal conditions to be achieved and the influence of slight defects. Avoid the traditional factor analysis approach; do not ignore malfunctions that might otherwise be considered harmless.

7.7. *Formulate improvement plans.* Define consultants who could do redesign the given piece of equipment. Discuss with manufacturers your plans.

8. Establishment of autonomous maintenance program for operators
   This is focused against the classic "operations" versus "maintenance" battle. Operators here must be convinced that they should maintain their own equipment. For example, an attitude has to be developed for operators to understand and act on the reports produced by the on-line performance condition monitoring systems.

9. Setup of scheduled maintenance program
   Scheduled maintenance conducted by the maintenance department must be smoothly coordinated with autonomous maintenance done by the plant operators. This can be done by frequent meetings and plant audits. In most plants, an undeclared conflict exists between the operations and maintenance groups. This arises from the false perception that these two groups have conflicting goals. The **PTPM** philosophy will go a long way in bringing these groups together.

10. Training for improvement of operation and maintenance skills
    This is a key part of **PTPM**. Ongoing training in advanced maintenance techniques, tools and methods must be done. This could cover areas such as:

    - Bearings and seals
    - Alignment
    - Balancing
    - Vibration
    - Troubleshooting
    - Failure analysis
    - Welding procures
    - Inspection procedures
    - NDT

11. Equipment management program
    Startup problems, solutions, design changes should be clearly documented and available for a good equipment management plan. All items

that can reduce life cycle costs (LCC) should be considered. These include:

- Economic evaluation at the equipment-investment stage.
- Consideration of MP or maintenance-free design and economic LCC.
- Effective use of accumulated MP data.
- Commissioning control activities.
- Thorough efforts to maximize reliability and maintainability.

12. Final implementation of PTPM
   This stage involves the refinement of PTPM and the formulation of new goals that meet specific corporate needs.

Maintenance Department Requirements

To ensure the success of the PTPM program, the maintenance department must be well equipped and trained. The following eight basic categories are prerequisite to the proper functioning of the maintenance department under the PTPM:

1. Training of personnel
2. Tools and equipment
3. Inspections
4. Condition and life assessment
5. Spare parts inventory
6. Redesign for higher machinery reliability
7. Maintenance scheduling
8. Maintenance communication.

Training of Personnel. Training must be the central theme. The days of the mechanic armed with a ball-peen hammer, screwdriver and a crescent wrench are gone. More and more complicated maintenance tools must be placed in the hands of the mechanic, and he must be trained to utilize them.

People must be trained, motivated and directed so that they gain experience and develop, not into mechanics, but into highly capable technicians. While good training is expensive, it yields great returns. Machinery has grown more complex, requiring more knowledge in many areas. The old, traditional craft lines must yield before complicated equipment maintenance needs. A joint effort by craftsmen is necessary to accomplish this.

1. Type of personnel
   1.1. Maintenance engineer
      In most plants, the maintenance engineer is a mechanical engineer with training in the turbomachinery area. His needs are to convert what he has learned in the classroom into actual hands-on solutions. He must be well versed in a number of areas such as performance analysis, rotor dynamics, metallurgy, lubrication systems and general shop practices. His training must be well

planned so that he can pick up these various areas in steps. His training must be a combination of a hands-on approach coupled with the proper theoretical background. He should be well versed in the various ASME Power Test Codes. Table 14-2 is a listing of some of the applicable codes to combine cycle power plants. Attendance at various symposiums where users of machinery get together to discuss problems should be encouraged. It is not uncommon to find a solution to a problem at these types of round table discussions.

1.2. Foremen and lead machinist

These men are the key to a good maintenance program. They should be sent frequently to training schools to enhance their knowledge. Some plants have one foreman who is an "in-house serviceman"; he supervises no personnel, but acts as an in-house consultant on maintenance jobs.

1.3. Machinist/millwright

The machinist should be encouraged to operate most of the machinery in the plant maintenance shop. By rotating him among various jobs, his learning and development is accelerated. He should then become as familiar with a large compressor as a small pump. Encouragement should be given to the machinist to learn balancing operations and to participate in the solution of problems.

Spreading around the hardest jobs develops more competent people and is the basis of any PTPM program. Restricting a man to one type of work will probably make him an expert in that area, but his curiosity and initiative, prime motivators, will eventually fade.

Table 14-2. Performance Test Codes

1. ASME, Performance Test Code on Overall Plant Performance, ASME PTC 46 1996, American Society of Mechanical Engineers 1996.
2. ASME, Performance Test Code on Test Uncertainty: Instruments and Apparatus PTC 19.1, 1988.
3. ASME, Performance Test Code on Gas Turbines, ASME PTC 22 1997, American Society of Mechanical Engineers 1997.
4. ASME, Performance Test Code on Gas Turbine Heat Recovery Steam Generators, ASME PTC 4.4 1981, American Society of Mechanical Engineers Reaffirmed 1992.
5. ASME, Performance Test Code on Steam Turbines, ASME PTC 6 1996, American Society of Mechanical Engineers 1996.
6. ASME, Performance Test Code on Steam Condensing Apparatus, ASME PTC 12.2 1983, American Society of Mechanical Engineers 1983.
7. ASME, Performance Test Code on Atmospheric Water Cooling Equipment PTC 23, 1997.
8. ASME Gas Turbine Fuels B 133.7M Published: 1985 (Reaffirmed year: 1992).
9. ISO, Natural Gas — Calculation of Calorific Value, Density and Relative Density International Organization for Standardization ISO 6976-1983(E)

2. Types of training
   2.1. Update training
   This training is mandatory for all maintenance personnel, so that they may keep abreast of this high-technology industry. Personnel must be sent to manufacturers-conducted schools. These schools, in turn, should be encouraged to cover some basic machinery principles, as well as their own machinery. In-house seminars should be provided with in-house personnel and consultants at the plant. Engineers should be sent to various schools so that they may be exposed to the latest technology.
   An in-house website, cataloging experiences and special maintenance techniques, should be updated and available for the entire corporation especially maintenance and operation personnel. These websites should be full of illustrations, short, and to the point.
   A small library should be adjacent to the shop floor, with field drawings, written histories of equipment, catalogs, API specifications, and other literature pertinent to the machine maintenance field. Drawings and manuals should be transferred to the electronic digital media as soon as possible. Access to the Internet on the maintenance and production area computers is a must as many manufacturers post helpful operational and maintenance hints on their websites. API specifications, which govern mechanical machinery, are listed in Table 14-3.
   Manufacturers' instructions books are often inadequate and need to be supplemented. The rewriting of maintenance manuals on such subjects as mechanical seals, vertical pumps, hot-tapping machines, and gas and steam turbines are not uncommon. The turbine overhaul manuals transferred on CDs could consist of (1) step-by-step overhaul procedures, developed largely from the manufactures training school, (2) hundreds of photographs, illustrating the step-by-step procedures on various types of gas and steam turbines, (3) an arrow diagram showing the sequences of the procedures, and (4) typical case histories.
   Detailed drawings on CDs are developed to aid in maintenance, such as a contact seal assembly, because the "typical" dimensionless drawing supplied by the OEM is not adequate to correctly assemble the compressor seals. Many other assembly drawings should be developed to facilitate the overall maintenance program. Videotaped programs are being developed on seals, bearings, and rotor dynamics, which will be a tremendous asset to most company maintenance programs.
   2.2. Practical training
   The engineers in the maintenance group should be encouraged to gather pertinent vibration and aerothermal data and analyze the machinery. ASME performance specifications, which govern all types of power plants and other critical equipment, are listed in Table 14-2. They should be encouraged to work closely at the various maintenance schedules and turnarounds so that they are familiar with the

514 • COGENERATION AND COMBINED CYCLE POWER PLANTS

Table 14-3. Mechanical Specifications

ASME Basic Gas Turbines B 133.2; Published: 1977 (Reaffirmed year: 1997)
ASME Gas Turbine Control And Protection Systems B133.4; Published: 1978 (Reaffirmed year: 1997)
ASME Gas Turbine Installation Sound Emissions B133.8; Published: 1977 (Reaffirmed: 1989)
ASME Measurement Of Exhaust Emissions From Stationary Gas Turbine Engines B133.9; Published: 1994
ASME Procurement Standard For Gas Turbine Electrical Equipment B133.5; Published: 1978 (Reaffirmed year: 1997)
ASME Procurement Standard For Gas Turbine Auxiliary Equipment B133.3; Published: 1981 (Reaffirmed year: 1994)
ANSI/API Std 610, Centrifugal Pumps for Petroleum, Heavy Duty Chemical and Gas Industry Services, 8th Edition, August 1995 (-1995)
API Std 611, General Purpose Steam Turbines for Petroleum, Chemical, and Gas Industry Services, 4th Edition, June 1997
API Std 613, Special Purpose Gear Units for Petroleum, Chemical and Gas Industry Services, 4th Edition, June 1995
API Std 614, Lubrication, Shaft-Sealing, and Control-Oil Systems and Auxiliaries for Petroleum, Chemical and Gas Industry Services, 4th Edition, April 1999
API Std 616, Gas Turbines for the Petroleum, Chemical and Gas Industry Services, 4th Edition, August 1998
API Std 617, Centrifugal Compressors for Petroleum, Chemical and Gas Industry Services, 6th Edition, February 1995
API Std 618, Reciprocating Compressors for Petroleum, Chemical and Gas Industry Services, 4th Edition, June 1995
API Std 619, Rotary-Type Positive Displacement Compressors for Petroleum, Chemical, and Gas Industry Services, 3rd Edition, June 1997
API Publication 534, Heat Recovery Steam Generators, 1st Edition, January 1995
API RP 556, Fired Heaters & Steam Generators, 1st Edition, May 1997
ANSI/API Std 670, Vibration, Axial-Position, and Bearing-Temperature Monitoring Systems, 3rd Edition, November 1993
API Std 671, Special Purpose Couplings for Petroleum Chemical and Gas Industry Services, 3rd Edition, October 1998
API Std 672, Packaged, Integrally Geared Centrifugal Air Compressors for Petroleum, Chemical, and Gas Industry Services, 3rd Edition, September 1996
API Std 677, General-Purpose Gear Units for Petroleum, Chemical and Gas Industry Services, 2nd Edition, July 1997, Reaffirmed March 2000
API Std 681, Liquid Ring Vacuum Pumps and Compressors, 1st Edition, February 1996
ISO 10436:1993, Petroleum and Natural Gas Industries — General Purpose Steam Turbine for Refinery Service, 1st Edition.

machinery. They should be sent to special training sessions where hands-on experience can be gained.

After the completion of basic machinist training, the machinist should continue his training with on-the-job experiences. His skills should be tested and he should be encouraged to take on different tasks.

To develop the skills of in-house personnel, as much repair work as possible should utilize plant personnel. Encouraging the participation of the machinist in the solution of difficult problems often results in the machinist seeking information on his own. References to API and ASME specifications should not be uncommon on the shop floor. Today's machinist and mechanic must be computer literate. Internet training must be provided with some basic training on word processing and spreadsheet programs.

2.3. Basic machinist training

Most of the basic training can be developed and conducted by in-plant personnel. This training can be highly detailed and tailored precisely to meet individual plant requirements. Training must be carefully planned and administered to fit the requirements of different machinery in the plant.

Many plants have a full-time training program, and personnel for conducting training at this basic level. Good maintenance practices should be inculcated into the young machinist from the beginning. He should be taught that all clearances should be carefully checked, and noted both before and after reassembly. He should learn the proper care in the handling of instrumentation, and the care in placing and removing seals and bearings. A base course on the major turbomachinery principles is a must, so there is basic understanding of what these machines do and how they function. The young machinist should also be exposed to basic machinery-related courses such as:

- Reverse indicator alignment
- Gas and steam turbine overhaul
- Compressor overhaul
- Mechanical seal maintenance
- Bearing maintenance
- Lubrication system maintenance
- Single-plane balancing.

Tools and Shop Equipment. A mechanic must be supplied with the proper tools to facilitate his jobs. Many special tools are required for different machines, so as to ensure proper disassembly and reassembly. Torque wrenches should be an integral part of his tools, as well as of his vocabulary.

The concepts of "finger tight" and "hand tight" can no longer be applied to high-speed, high-pressure machinery. A recent major explosion at an oxygen plant, which resulted in a death, was traced back to gas leakage due to improper torquing. A good dial indicator and special jigs for taking reverse indicator dial readings is a must. The jigs must be specially made for the various compressor and turbine trains. Special gear and wheel pullers are usually necessary.

Equipment for heating wheels in the field for assembly and disassembly are needed; specially designed gas rings are often used for this purpose.

A maintenance shop should have the traditional horizontal and vertical lathes, mills, drill presses, slotters, bores, grinders, and a good balancing machine.

A balancing machine can pay for itself in a very short time in providing a fast turnaround and accurate dynamic balance. Techniques to check balance of gear-type couplings for the large high-speed compressors and turbine drives as a unit should be developed. This leads to the solving of many vibration-related problems. High-speed couplings should be routinely check-balanced.

By dynamically balancing most parts, seal life and bearing life is greatly improved, even on smaller equipment. Dynamic balancing is needed on pump impellers, as the practice of static balance is woefully inadequate. Vertical pumps must be dynamically balanced; the long, slender shafts are highly susceptible to any unbalanced-induced vibration.

This assembly and disassembly of rotors must be in a clean area. Horses or equivalents should be available to hold the rotor. The rotor should rest on the bearing journals, which must be protected by soft packing, or the equivalent, to avoid any marring of the journals. To accomplish uniform shrink fits, the area should have provisions for heating and/or cooling. A special rotor-testing fixture should be provided; this is very useful in checking for wheel wobbles, wheel roundness, and shaft trueness. Rotors in long-term storage should be stored in a vertical position in temperature-controlled warehouses.

Spare Parts Inventory

The problem of spare parts is an inherent phase of the maintenance business. The high costs of replacement parts, delivery and, in some instances, poor quality, are problems faced daily by everyone in the maintenance field. The cost of spare parts for a major power plant or refinery runs into many millions of dollars.

The inventory of these plants can run into over 20,000 items, including over 100 complete rotor systems. The field of spare parts is changing rapidly and is much more complex than in the past. A group of plants have gotten together in a given region and formed part banks.

Many pieces of equipment are made up of unitized components from several different vendors. The traditional attitude has been to look to the packaging vendor as the source of supply. Many vendors refuse to handle requests for replacement parts on equipment not directly manufactured by them. More and more specialty companies are entering the equipment parts business; some are supplying parts directly to OEM companies for resale as their "own" brand. Others supply parts directly to the end-user. The end-user must develop multiple sources of supply for as many parts as possible.

Gaskets, turbine carbon packing, and mechanical seal parts can be purchased from local sources. Shafts, sleeves, cast parts can be purchased from local sources. Shafts, sleeves, cast parts such as impellers, are becoming increasingly available from specialty vendors. All this competition is causing the OEMs to alter their spare parts system to improve service and reduce prices, which is definitely a bright spot in the picture. The quality control of both OEM and some specialty houses leaves much to be desired. In turn, this causes many plants to have an in-house quality control person checking all incoming parts, a concept that is highly recommended.

Inspections

As with any power equipment, a combined cycle power plant requires a planned program of planned inspections with repair or replacement of damaged components. Most plants follow the maintenance and inspection schedules suggested by the OEM.

Condition and Life Assessment

Condition and life assessment is significant for all types of plants, and especially combined cycle power plants. The most important aspect of a plant is high Availability and Reliability, in some cases, even more significant than higher efficiency.

Availability. The Availability of a power plant is the percent of time the plant is available to generate power in any given period at its acceptance load. The Acceptance Load or the Net Established Capacity would be the net electric power generating capacity of the Power Plant at design or reference conditions established as result of the Performance Tests conducted for acceptance of the plant. The actual power produced by the plant would be corrected to the design or reference conditions and is the actual net available capacity of the Power Plant. Thus it is necessary to calculate the effective forced outage hours which are based on the maximum load the plant can produce in a given time interval when the plant is unable to produce the power required of it. The effective forced outage hours is based on the following relationship:

$$EFH = HO \times \frac{(MW_d - MW_a)}{MW_d} \qquad (14-1)$$

where

MW_d = Desired Output corrected to the design or reference conditions. This must be equal to or less than the plant load measured and corrected to the design or reference conditions at the acceptance test.

MW_a = Actual maximum acceptance test produced and corrected to the design or reference conditions.

HO = Hours of operation at reduced load.

The Availability of a plant can now be calculated by the following relationship, which takes into account the stoppage due to both forced and planed outages, as well as the forced effective outage hours:

$$A = \frac{(PT - PM - FO - EFH)}{PT} \qquad (14-2)$$

where

PT = Time Period (8760 hrs/year)
PM = Planned Maintenance Hours
FO = Forced Outage Hours
EFH = Equivalent forced outage hours

Reliability. The reliability of the plant is the percentage of time between planed overhauls and is defined as:

$$R = \frac{(PT - FO - EFH)}{PT} \qquad (14-3)$$

Availability and reliability have a very major impact on the plant economy. Reliability is essential in that when the power is needed it must be there. When the power is not available, it must be generated or purchased and can be very costly in the operation of a plant. Planned outages are scheduled for non-peak periods. Peak periods is when the majority of the income is generated, as usually there are various tiers of pricing depending on the demand. Many power purchase agreements have clauses, which contain capacity payments, thus making plant availability critical in the economics of the plant.

The Gas turbine is the key for the combined cycle power plants. The new technology, with higher pressure ratio and higher firing temperature, has led to the building of large gas turbines producing nearly 300 MW, and reaching gas turbine efficiencies in the mid 40s. The availability factor for units with mature technology, below 100 MW, are between 94% and 97%, while the bigger units above 100 MW have availability factors of 85% to 89%. The bigger units produce twice the output, but the availability factor has decreased from 95% to 85%; a decrease of seven to ten points for all manufacturers. Two comments should be taken into account. Well managed, bigger plants with four or more units can reach up to 2% higher values over time, especially by shortening the time for revisions. The differences in all other parameters are reasonable small and should be considered in detail for individual projects.

The increase in unit size and complexity, together with the higher turbine inlet temperature and higher pressure ratio, has lead to an increase in overall plant efficiency. The increase in efficiency of 5% to 7% has, in many cases, led to an availability decrease of the same amount or even more as seen in Figure 14-5. A 1% reduction in plant availability could cost US$500,000 in income on a 100-MW plant, thus in many cases offsetting gains in efficiency.

Reliability of a plant depends on many parameters, such as the type of fuel, the preventive maintenance programs, the operating mode, the control systems, and the firing temperatures.

Redesign for Higher Machinery Reliability

Low reliability of units gives rise to high maintenance costs. Low reliability is usually a greater economic factor than the high maintenance costs. In many large power plants, refineries, and petrochemical complexes, about one-third of the failures are due to machinery failure; it is therefore necessary to redesign parts of a machine to improve reliability.

The maintenance practice of one large refinery is to replace gas turbine control systems with state-of-the-art electronics and "plug-in" concepts for ease of

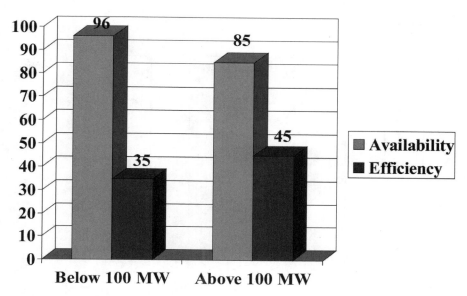

Figure 14-5. Comparison of Availability and Efficiency for Large Frame Type Gas Turbines

maintenance. These installations have been highly successful in that maintenance has been minimal, and can usually be accomplished on-stream.

In addition, turbine performance, speed control and flexibility are greatly improved. The original design has been supplemented to include a self-contained alarm system, a semi-automatic sequential start system, and a complete trip and protection system, as well as the electronic controls. The cost of this system is substantially less than the cost of a similar device offered by the OEM on new machines.

To be able to define the problem areas in a plant, one must first define the contributions various areas contribute to the overall decrease in plant reliability. Figure 14-6 shows the effect of the various major components that contribute to plant problems. The new gas turbines, due to their higher temperature and higher pressure ratios, are the major problem areas in combined cycle power plants.

The gas turbines' major limitations on the life are the combustor cans, first-stage turbine nozzles and first-stage turbine blades as seen in Figure 14-7. The effect of dry low NO_x Combustors have been very negative on the availability of a combined cycle power plant, especially those with dual fuel capability. Flashback problems are very major problems, as they tend to create burning in the pre-mix section of the combustor, and cause failure of the pre-mix tubes. These pre-mix tubes are also very susceptible to resonance vibrations.

The HRSG is usually not a major problem in the downtime of a combined cycle power plant. The problems in the HRSG are due to the high temperatures in the superheater and low temperatures in the exit stack, especially if the fuel has a

520 • COGENERATION AND COMBINED CYCLE POWER PLANTS

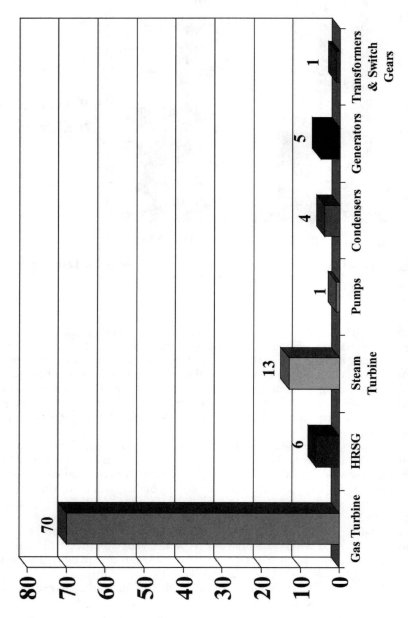

Figure 14-6. Contributions of Various Major Components to Plant Down Time

Maintenance Techniques • 521

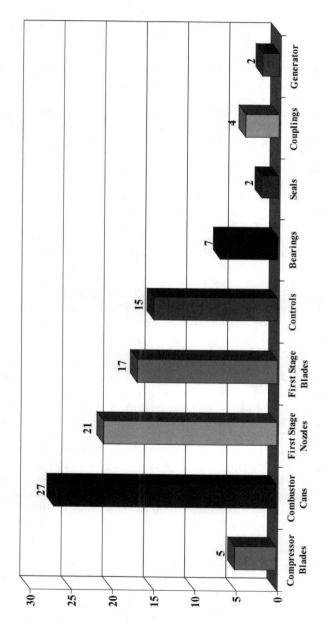

Figure 14-7. Contributions of Various Major Components to Gas Turbine Down Time

sulfur content. Figure 14-8 shows some of the major contributors to the down time in HRSG.

The steam turbine's major problems usually arises in the low-pressure turbine blades and diaphragms, which are susceptible to erosion problems as seen in Figure 14-9. Casing problems, such as deformation of the low pressure stage casing, due often to water flashing from the condenser creating large temperature gradients, cause major downtime as the casings have to be remachined at the site. The failures in the steam turbines, especially in the area of blades and diaphragms, are due to many factors as shown in Table 14-4.

The pumps usually do not contribute significantly in lowering the plant availability due to the fact that pumps in most cases are redundant. The problems with pumps, usually the boiler feed water pumps, are usually a series of impellers on a long shaft are with the rotor assembly. Figure 14-10 shows the major components that create problems in a pump, such as problems with the rotor assembly and the rotor impeller. The rotor assembly can suffer from problems due to the dynamic balance of the rotor, especially those that operate above the first critical and the cavitation problems in the rotor impeller.

Bearing failures are one of the major causes of failures in turbomachinery. The changing of various types of radial bearings from cylindrical and/or pressure dam babbitted sleeve bearings to tilting pad journal bearing is becoming common in the industry. In most cases, this gives better stability, eliminates oil whirl, and under misalignment condition, is more forgiving.

Thrust bearing changes, from the simple, tapered land thrust bearings to tilting pad thrust bearings with leveling links (Kingsbury type), is another area of common change. These types of bearings absorb sudden load surges and liquid slugs.

A major plant replaces the entire large journal and thrust bearings in their main machinery to tilting pad bearings in their plant as a matter of practice.

Material changes of the babbit are sometimes undertaken. Changing from the more common steel-backed babbitted bearings to the copper alloys, with this babbitted pads, conducts surface heat away at a faster rate, thus increasing the load carrying capacity. In some instances, a 50% to 100% load carrying capacity improvement can be achieved. Some equipment manufacturers are offering bearing-upgrading kits for their machine in service.

Design of turbine blades to obtain higher efficiency and damping has been done. In some cases, this has improved efficiency by 8% to 10%, and stopped failures in these blades. Steam injection has been utilized in gas turbines to improve efficiency and to increase the power output. Redesign of various bleed-off ports has reduced tip stalls and their accompanying blade failures.

In some steam turbines, the shrink fit provided is inadequate for the steam operating conditions, and causes the wheels to move on the shaft. To increase the shrink fit would create a phenomenon called hysteric whirl; thus, the solution was to redesign the shaft and provide locking rigs for each wheel.

Today's machinery, which is pushing the state-of-the-art in design, needs more than "simple fixes." This is one major reason why so much redesign takes place in the field. Maintenance engineers are no longer just required to repair, they are required in many cases to make revisions. Continual improvements and updating of the machinery is required to obtain the long runs and high efficiencies desirable in today's turbomachinery.

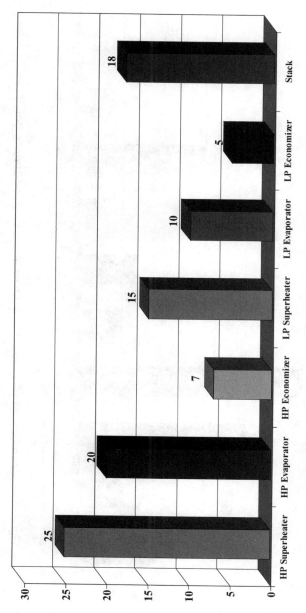

Figure 14-8. Contributions of Various Major Components to HRSG Down Time

524 • COGENERATION AND COMBINED CYCLE POWER PLANTS

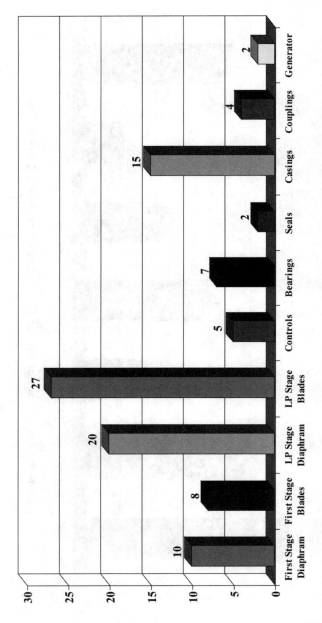

Figure 14-9. Contributions of Various Major Components to Steam Turbine Down Time

Maintenance Techniques • 525

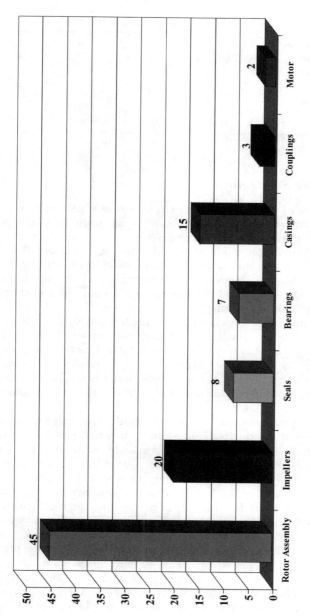

Figure 14-10. Contributions of Various Major Components to Pump Down Time

Table 14-4. Contributors of Failure Modes

Component	Corrosion	Creep	Erosion	Fatigue	Rubbing	High Load	FOD	Water Impingement
1st-stage Nozzles			X	X			X	
1st-stage Blades	X	X		X	X	X	X	
LP Diaphragms	X		X	X				
LP Blades			X	X		X	X	X
Casing Deformation		X						X
Turbine Split Line	X	X						X

Maintenance Scheduling

The scheduling of maintenance inspections and overhauls is an essential part of the total maintenance philosophy. As we move from "breakdown" or "panic" maintenance towards a Performance-Based Total Productive Maintenance System, total condition monitoring and diagnostics becomes an integral part of both operation and maintenance. Total condition monitoring and diagnostic examines both the mechanical and performance of the machinery and then carries out diagnostics. Condition monitoring systems, which are only mechanical systems without performance inputs, give less than half of the picture and can be very unreliable. Unscheduled maintenance is very costly and should be avoided. To properly schedule overhauls, both mechanical and performance data must be gathered and evaluated. As indicated earlier, we want to consider repairs during a planned "turnaround", not "random" repairs which are frequently done on an "emergency" basis and where, due to time restraints, techniques are sometimes used which are questionable and should only be used in emergencies.

To plan for a "turnaround," one must be guided by the operating history of the given plant and, if it is the first "turnaround," by conditions found in other plants utilizing the same or closely similar process and machinery. This is how the time between subsequent "turnarounds" has been extended to 3 years or more in many instances. By utilizing the operating history and inspection at previous "turnarounds" at this or similar installations, one can get a fair idea of what parts are most likely to be found deteriorated and, therefore, must be replaced and/or repaired, and what other work should be done to the unit while it is down. It should be pointed out that, with modern turbomachinery, items such as bearings, seals, filters and certain instrumentation, which are precision-made, are seldom, if ever, repaired except in an emergency; such items are replaced with new parts.

This means that parts must be ordered in advance for the "turnaround" and other work must be planned so that the whole operation may proceed smoothly and without holdups that could have been foreseen. This usually means close collaboration with the manufacturer or consultant and the OEM (or specialty service shop) so that handling facilities, service men, parts, cleaning facilities, inspection facilities, chrome plating and/or metalizing facilities, balancing facilities, and in some cases, even heat treatment facilities are available and will be open for production at the proper time required. This is the planning, which must be done in detail before the shutdown with sufficient lead-time available in order to have replacement parts available at the job site.

The old maxim "if it ain't broke don't fix it" is very applicable in today's machinery. A study conducted at a major nuclear power facility found that 30% of the failures occurred after a major turnaround. This is why Total Condition Monitoring is necessary in any Performance-Based Total Productive Maintenance System, and which leads to overhauls being planned on proper data evaluation of the machinery rather than on a fixed interval.

Maintenance Communications

It is not uncommon to hear the complaint that the maintenance department has "never been informed as to what is happening in the plant." If this is a common complaint, the maintenance manager needs to examine the communications in his department. The following are seven practical suggestions for improving communications:

1. Operation and service manuals
2. Continuous updating of drawing and print files
3. Updating of training materials
4. Pocket guides
5. Written memos, inter office e-mails
6. Seminars
7. Website postings.

Each of the items listed, if properly employed, can transmit knowledge to the person who must keep the plant's machinery running. How well the information is transmitted depends entirely on the communication skills applied to the preparation of the materials.

Operation and Service Manuals. To be of real value to the mechanic and operation, a service manual must be indexed to permit quick location of needed information. The manual must be written in simple, straightforward language, have illustrations, sketches, or exploded views adjacent to pertinent text, and have minimum references to another page or section. Major sections or chapters should be tabbed for quick location.

Most often, a mechanic or serviceman refers to a manual because of a problem. Problems seem to happen during a production run. It is essential, therefore, that he be able to find the needed information quickly. The mechanic should not be delayed by wordy, irrelevant text. The objective of any manual is to be an effective, immediate source of service information.

The assignment of a non-technical person to write a manual is shortsighted and costlier in the long run. A well written manual is continuously in use. Good manuals need not be complicated. In fact, the simpler the better. Manuals should be readable and understandable, whether they are compiled in-house or outside.

Drawing and Print File. A good print file is a vital tool for any maintenance organization. Reference files in a large or multi-plant company can be particularly burdensome for several reasons:

- Prints are bulky and difficult to store properly
- Control of use is necessary
- Files must be kept up-to-date
- Handling and distribution of new or revised prints is usually expensive.

A practical solution is to digitize the drawings and place them on CDs available to the maintenance and operation department. A good digital file reduces search time and helps the departments do a better job of keeping the machinery operating at their peak efficiency with minimal downtime.

Training Materials. Like any other written or audio-visual maintenance tool, training materials of all kinds are basically communication devices, and to be effective, should be presented in a simple straightforward, attractive, and professional manner.

Once the need for specific maintenance training has been determined, a program must be developed. If the training need applies to a proprietary machine or one that is unique to a very few industries, it might be necessary to contact companies who specialize in custom digital programs on CDs, slide/tape, movie, videotape, or written training programs. The cost may shock the uninitiated, but after shopping around, the company may find that it can recover far more than the initial cost in tangible benefits over a relatively short period.

Pocket guide. When a new maintenance form or procedure is introduced, a quick reference pocket guide can promote understanding and accuracy. The key to effectiveness is a deliberate design to provide maximum illustrations or examples in simple language. If it cannot be prepared in-house, outside help should be sought. Professionalism is essential to good communications.

Written memos. One of the most effective devices for improving maintenance communications is a newsletter or internal memo. The memo's success depends heavily on communicating formal tips and techniques in the mechanics language and using photos, sketches, and drawings generously to get the message across.

Everyone in the maintenance department should be encouraged to contribute ideas on a better way to do a task or a solution to a nagging problem related to the maintenance or operation of production equipment. Each contributor should be given credit by name and location for his or her effort. Very few workers can resist a bit of pride in seeing their names attached to an article that is seen by virtually everyone in the company.

Seminars and workshops. College or industry-sponsored seminars, continuing education courses, and workshops are means of upgrading or sharpening skills of maintenance people. Such an approach serves a two-fold purpose. First, it communicates the company's good faith in the person's ability to benefit from the experience, and by acceptance, the worker shows willingness to improve his or her usefulness to the company. The seminars are very useful in disseminating knowledge. They also provide forum for gripes and meaningful solutions. Discussion groups in these seminars and workshops are very important, as participants share experiences and solutions to problems. The knowledge gained from these seminars is very useful.

APPENDIX A

EQUIVALENT UNITS

Length

$12\,\dfrac{\text{in.}}{\text{ft}}$ $6080.2\,\dfrac{\text{ft}}{\text{naut.mi}}$ $5280\,\dfrac{\text{ft}}{\text{mi}}$ $0.3937\,\dfrac{\text{in.}}{\text{cm}}$ $30.48\,\dfrac{\text{cm}}{\text{ft}}$ $10^4\,\dfrac{\mu\text{m}}{\text{cm}}$

$3\,\dfrac{\text{ft}}{\text{yd}}$ $1.152\,\dfrac{\text{mi}}{\text{naut.mi}}$ $10^{10}\,\dfrac{\text{A}}{\text{m}}$ $2.54\,\dfrac{\text{cm}}{\text{in.}}$ $3.28\,\dfrac{\text{ft}}{\text{m}}$ $1.609\,\dfrac{\text{km}}{\text{mi}}$

Area

$144\,\dfrac{\text{in.}^2}{\text{ft}^2}$ $43,560\,\dfrac{\text{ft}^2}{\text{acre}}$ $640\,\dfrac{\text{acres}}{\text{mi}^2}$ $10.76\,\dfrac{\text{ft}^2}{\text{m}^2}$ $929\,\dfrac{\text{cm}^2}{\text{ft}^2}$ $6.452\,\dfrac{\text{cm}^2}{\text{in.}^2}$

Volume

$1728\,\dfrac{\text{in.}^3}{\text{cu ft}}$ $7.481\,\dfrac{\text{gal}}{\text{cu ft}}$ $43,560\,\dfrac{\text{cu ft}}{\text{acre-ft}}$ $3.7854\,\dfrac{\text{L}}{\text{gal}}$ $28.317\,\dfrac{\text{L}}{\text{cu ft}}$ $35.31\,\dfrac{\text{cu ft}}{\text{m}^3}$

$231\,\dfrac{\text{in.}^3}{\text{gal}}$ $8\,\dfrac{\text{pt}}{\text{gal}}$ $10^3\,\dfrac{\text{L}}{\text{m}^3}$ $61.025\,\dfrac{\text{in.}^3}{\text{L}}$ $10^3\,\dfrac{\text{cm}^3}{\text{L}}$ $28,317\,\dfrac{\text{cm}^3}{\text{cu ft}}$

Density

$1728\,\dfrac{\text{lb/cu ft}}{\text{lb/in.}^3}$ $32.174\,\dfrac{\text{lb/cu ft}}{\text{slug/cu ft}}$ $0.51538\,\dfrac{\text{g/cm}^3}{\text{slug/cu ft}}$ $16.018\,\dfrac{\text{kg/m}^3}{\text{lb/cu ft}}$ $1000\,\dfrac{\text{kg/m}^3}{\text{g/cm}^3}$

Angular

$2\pi = 6.2832\,\dfrac{\text{rad}}{\text{rev}}$ $57.3\,\dfrac{\text{deg}}{\text{rad}}$ $\dfrac{1}{2\pi}\,\dfrac{\text{rpm}}{\text{rad/min}}$ $9.549\,\dfrac{\text{rpm}}{\text{rad/sec}}$

Time

$$60\,\frac{\text{s}}{\text{min}} \qquad 3600\,\frac{\text{s}}{\text{hr}} \qquad 60\,\frac{\text{min}}{\text{hr}} \qquad 24\,\frac{\text{hr}}{\text{day}}$$

Speed

$$88\,\frac{\text{fpm}}{\text{mph}} \qquad 0.6818\,\frac{\text{mph}}{\text{fps}} \qquad 0.5144\,\frac{\text{m/s}}{\text{knot}} \qquad 0.3048\,\frac{\text{m/s}}{\text{fps}} \qquad 0.44704\,\frac{\text{m/s}}{\text{mph}}$$

$$1.467\,\frac{\text{fps}}{\text{mph}} \qquad 1.152\,\frac{\text{mph}}{\text{knot}} \qquad 1.689\,\frac{\text{fps}}{\text{knot}} \qquad 152.4\,\frac{\text{cm/min}}{\text{ips}}$$

Force, Mass

$$16\,\frac{\text{oz}}{\text{lb}_m} \qquad 32.174\,\frac{\text{lb}_m}{\text{slug}} \qquad 444,820\,\frac{\text{dynes}}{\text{lb}_f} \qquad 2.205\,\frac{\text{lb}_m}{\text{kg}} \qquad 9080665\,\frac{\text{N}}{\text{kg}_f}$$

$$1000\,\frac{\text{lb}_f}{\text{kip}} \qquad 32.174\,\frac{\text{poundals}}{\text{lb}_f} \qquad 980.665\,\frac{\text{dynes}}{\text{g}_f} \qquad 14.594\,\frac{\text{kg}}{\text{slug}} \qquad 4.4482\,\frac{\text{N}}{\text{lb}_f}$$

$$2000\,\frac{\text{lb}_m}{\text{ton}} \qquad 7000\,\frac{\text{grains}}{\text{lb}_m} \qquad 453.6\,\frac{\text{g}}{\text{lb}_m} \qquad 10^5\,\frac{\text{dynes}}{\text{N}} \qquad 1\,\frac{\text{kilopound}}{\text{kg}}$$

$$14.594\,\frac{\text{kg}}{\text{slug}} \qquad 28.35\,\frac{\text{g}}{\text{oz}} \qquad 453.6\,\frac{\text{gmole}}{\text{pmole}} \qquad 907.18\,\frac{\text{kg}}{\text{ton}} \qquad 1000\,\frac{\text{kg}}{\text{metric ton}}$$

Pressure

$$14.696\,\frac{\text{psi}}{\text{atm}} \qquad 101,325\,\frac{\text{N/m}^2}{\text{atm}} \qquad 13.6\,\frac{\text{kg}}{\text{mm Hg}\,(0°\text{C})} \qquad 51.715\,\frac{\text{mm Hg}\,(0°\text{C})}{\text{psi}} \qquad 47.88\,\frac{\text{N/m}^2}{\text{psf}}$$

$$29.921\,\frac{\text{in. Hg}\,(0°\text{C})}{\text{atm}} \qquad 10^5\,\frac{\text{N/m}^2}{\text{bar}} \qquad 13.57\,\frac{\text{in. H}_2\text{O}\,(60°\text{F})}{\text{in. Hg}\,(60°\text{F})} \qquad 703.07\,\frac{\text{kg/m}^2}{\text{psi}} \qquad 6894.8\,\frac{\text{N/m}^2}{\text{psi}}$$

$$33.934\,\frac{\text{ft H}_2\text{O}\,(60°\text{F})}{\text{atm}} \qquad 14.504\,\frac{\text{psi}}{\text{bar}} \qquad 0.0361\,\frac{\text{psi}}{\text{in. H}_2\text{O}\,(60°\text{F})} \qquad 0.0731\,\frac{\text{kg/cm}^2}{\text{psi}} \qquad 760\,\frac{\text{torr}}{\text{atm}}$$

$$1.01325\,\frac{\text{bar}}{\text{atm}} \qquad 10^6\,\frac{\text{dynes/cm}^2}{\text{bar}} \qquad 0.4898\,\frac{\text{psi}}{\text{in. Hg}\,(60°\text{F})} \qquad \frac{9.869}{10^7}\,\frac{\text{atm}}{\text{dyne/cm}^2} \qquad 133.3\,\frac{\text{N/m}^2}{\text{torr}}$$

$$33.934 \frac{\text{ft H}_2\text{O (60°C)}}{\text{atm}} \quad 760 \frac{\text{mm Hg (0°C)}}{\text{atm}} \quad 406.79 \frac{\text{in. H}_2\text{O (39.2°F)}}{\text{atm}}$$

$$0.1 \frac{\text{dyne/cm}^2}{\text{N/m}^2} \quad 1.0332 \frac{\text{kg/cm}^2}{\text{atm}}$$

Energy and Power

$$778.16 \frac{\text{ft-lb}}{\text{Btu}} \quad 2544.4 \frac{\text{Btu}}{\text{hp-hr}} \quad 5050 \frac{\text{hp-hr}}{\text{ft-lb}} \quad 1 \frac{\text{J}}{\text{W-s}} \frac{\text{J}}{\text{N-m}} \quad 0.01 \frac{\text{bar-dm}^3}{\text{J}}$$

$$550 \frac{\text{ft-lb}}{\text{hp-s}} \quad 42.4 \frac{\text{Btu}}{\text{hp-min}} \quad 1.8 \frac{\text{Btu/lb}}{\text{cal/gm}} \quad 1 \frac{\text{kW-s}}{\text{kJ}} \quad \frac{16.021}{10^{12}} \frac{\text{J}}{\text{MeV}}$$

$$33{,}000 \frac{\text{ft-lb}}{\text{hp-min}} \quad 3412.2 \frac{\text{Btu}}{\text{kW-hr}} \quad 1800 \frac{\text{Btu/pmole}}{\text{kcal/gmole}} \quad 1 \frac{\text{V-amp}}{\text{W-s}} \quad \frac{1.6021}{10^{12}} \frac{\text{erg}}{\text{eV}}$$

$$737.562 \frac{\text{ft-lb}}{\text{kW-s}} \quad 56.87 \frac{\text{Btu}}{\text{kW-min}} \quad 2.7194 \frac{\text{Btu}}{\text{atm-cu ft}} \quad 10^7 \frac{\text{ergs}}{\text{J}} \quad \frac{11.817}{10^{12}} \frac{\text{ft-lb}}{\text{MeV}}$$

$$1.3558 \frac{\text{J}}{\text{ft-lb}} \quad 251.98 \frac{\text{cal}}{\text{Btu}} \quad 4.1868 \frac{\text{kJ}}{\text{kcal}} \quad 3600 \frac{\text{kJ}}{\text{kW-hr}} \quad 0.746 \frac{\text{kW}}{\text{hp}}$$

$$1.055 \frac{\text{kJ}}{\text{Btu}} \quad 101.92 \frac{\text{kg-m}}{\text{kJ}} \quad 0.4300 \frac{\text{Btu/pmole}}{\text{J/gmole}} \quad 860 \frac{\text{cal}}{\text{W-hr}} \quad 1.8 \frac{\text{Btu}}{\text{chu}}$$

$$37.29 \frac{\text{kJ/m}^3}{\text{Btu/ft}^3} \quad 0.948 \frac{\text{Btu}}{\text{kW-sec}} \quad 2.33 \frac{\text{kJ/kg}}{\text{Btu/lb}_m}$$

Entropy, Specific Heat, Gas Constant

$$1 \frac{\text{Btu/pmole-R}}{\text{cal/gmole-K}} \quad 1 \frac{\text{Btu/lb-R}}{\text{gal/cm-K}} \quad 1 \frac{\text{Btu/lb-R}}{\text{kcal/kg-K}} \quad 0.2389 \frac{\text{Btu/pmole-R}}{\text{J/gmole}}$$

$$4.187 \frac{\text{kJ/kg-K}}{\text{Btu/lb-R}}$$

Universal Gas Constant

$$1545.32 \frac{\text{ft-lb}}{\text{pmole-R}} \quad 8.3143 \frac{\text{kJ}}{\text{kmole-K}} \quad 0.7302 \frac{\text{atm-ft}^3}{\text{pmole-R}} \quad 82.057 \frac{\text{atm-cm}^3}{\text{gmole-K}}$$

$$1.9859\frac{\text{Btu}}{\text{pmole-R}} \quad 1.9859\frac{\text{cal}}{\text{gmole-K}} \quad 10.731\frac{\text{psi-ft}^3}{\text{pmole-R}} \quad 83.143\frac{\text{bar-cm}^3}{\text{gmole-K}}$$

$$8.3143\frac{\text{J}}{\text{gmole-K}} \quad 8.3149 \times 10^7 \frac{\text{erg}}{\text{gmole-K}} \quad 0.08206\frac{\text{atm-m}^3}{\text{kgmole-K}}$$

$$0.083143\frac{\text{bar-l}}{\text{gmole-K}}$$

Newton's Proportionality Constant k (as a conversion unit)

$$32.174 \text{ fps}^2\left(\frac{\text{lb}}{\text{slug}}\right) \quad 386.1 \text{ ips}^2\left(\frac{\text{lb}}{\text{p sin}}\right) \quad 9.80665\frac{\text{m}}{\text{s}^2}\left(\frac{\text{N}}{\text{kg}}\right) \quad 980.655\frac{\text{cm}}{\text{s}^2}\left(\frac{\text{dynes}}{\text{g}}\right)$$

Miscellaneous Constants

Speed of light Avogadro Constant Planck Constant

$$c = 2.9979 \times 10^8 \frac{\text{m}}{\text{s}} \quad N_A = 6.02252 \times 10^{23}\frac{\text{molecules}}{\text{gmole}} \quad h = 6.6256 \times 10^{-34} \text{ J-s}$$

Boltzmann Constant Gravitational Constant Normal mole volume

$$k = 1.38054 \times 10^{-23}\frac{\text{J}}{\text{K}} \quad G = 6.670 \times 10^{-11}\frac{\text{N-m}^2}{\text{kg}^2} \quad 2.24136 \times 10^{-2}\frac{\text{m}^3}{\text{gmole}}$$

APPENDIX B

Specific Heat of Air at Low Pressures

T, °R	C_p, Btu/lb$_m$ °R	C_v, Btu/lb$_m$ °R	γ (C_p/C_v)
400	0.2393	0.1707	1.402
450	0.2394	0.1708	1.401
500	0.2396	0.1710	1.401
550	0.2399	0.1713	1.400
600	0.2403	0.1718	1.399
650	0.2409	0.1723	1.398
700	0.2416	0.1730	1.396
750	0.2424	0.1739	1.394
800	0.2434	0.1748	1.392
900	0.2458	0.1772	1.387
1,000	0.2486	0.1800	1.381
1,100	0.2516	0.1830	1.374
1,200	0.2547	0.1862	1.368
1,300	0.2579	0.1894	1.362
1,400	0.2611	0.1926	1.356
1,500	0.2642	0.1956	1.350
1,600	0.2671	0.1985	1.345
1,700	0.2698	0.2013	1.340
1,800	0.2725	0.2039	1.336
1,900	0.2750	0.2064	1.332
2,000	0.2773	0.2088	1.328
2,100	0.2794	0.2109	1.325
2,200	0.2813	0.2128	1.322
2,300	0.2831	0.2146	1.319
2,400	0.2848	0.2162	1.317

Specific Heats of Products of Combustion (400% Theoretical Air; Fuel $(CH_2)_n$; Molecular Weight = 28.9553)

T, °R	C_p, Btu/lb$_m$ °R	C_v, Btu/lb$_m$ °R	γ (C_p/C_v)
800	0.2483	0.1797	1.382
850	0.2496	0.1810	1.379
900	0.2510	0.1825	1.376
950	0.2526	0.1840	1.373
1,000	0.2542	0.1856	1.369
1,100	0.2575	0.1890	1.363
1,200	0.2609	0.1924	1.357
1,300	0.2644	0.1958	1.350
1,400	0.2679	0.1993	1.344
1,500	0.2712	0.2026	1.339
1,600	0.2743	0.2057	1.333
1,700	0.2774	0.2088	1.328
1,800	0.2802	0.2116	1.324
1,900	0.2830	0.2144	1.320
2,000	0.2855	0.2166	1.316
2,100	0.2878	0.2192	1.313
2,200	0.2900	0.2214	1.310
2,300	0.2920	0.2234	1.307
2,400	0.2938	0.2253	1.304
2,500	0.2956	0.2270	1.302
2,600	0.2973	0.2287	1.300
2,700	0.2988	0.2302	1.298
2,800	0.3002	0.2316	1.296
2,900	0.3016	0.2330	1.294
3,000	0.3029	0.2343	1.293
3,200	0.3052	0.2366	1.290
3,400	0.3073	0.2387	1.287
3,600	0.3092	0.2407	1.285
3,800	0.3109	0.2423	1.283
4,000	0.3126	0.2440	1.281

BIBLIOGRAPHY

CHAPTER 1 — AN OVERVIEW OF POWER GENERATION

[1] Alderfer, R., Eldridge, M., Starrs, T., 2000, "Making Connections: Case Studies of Interconnection Barriers and their Impact on Distributed Power Projects," NREL/SR-200-28053.
[2] Boyce, M. P., July 1995, Chapter 1, "An Overview of Gas Turbines," *Gas Turbine Engineering Handbook*, 7th Edition, Gulf Publishing Company.
[3] Clean Air Act, 1990, United States Environmental Protection Agency, Washington, D.C.
[4] "Cogeneration System Package for Micro-Turbines," 2000 Sales Literature-Ingersoll-Rand Corporation, Portsmouth, New Hampshire.
[5] "Distributed Generation: Understanding The Economics," May/June 2000 Distributed Power.
[6] "Kyoto Protocol of 1997," 1997, *United Nations Framework Convention on Climate Change*, N.Y., N.Y., United Nations.
[7] Leo., A. J., Ghezel-Ayagh, H., Sanderson, R., "Ultra High Efficiency Hybrid Direct Fuel Cell/Turbine Power Plant," ASME Paper No. 2000-GT-0552, ASME.
[8] "Simple Cycle Micro Turbine Power Generation System," 2000, Sales Literature-Capstone Micro Turbine, Chatsworth, CA.
[9] "Solid Oxide Fuel Cell, Passing The Learning Curve," May/June 2000 Distributed Power.

CHAPTER 2 — CYCLES

[1] Boyce, M. P., November/December 2000, *Advanced Cycles for Combined Cycle Power Plants*, Russia Gas Turbo-Technology Publication.
[2] Boyce, M. P., September/October 2000, *Turbo-Machinery for the Next Millennium*, Russia Gas Turbo-Technology Publication.
[3] Boyce, M. P., Meher-Homji, C. B, Lakshminarasimha, A. N., "Gas Turbine and Combined Cycle Technologies for Power and Efficiency enhancement in Power Plants," ASME Paper No. 94-GT-435, ASME.
[4] Boyce, M. P., July 1995, Chapter 2, "Theoretical and Actual Cycle

Analysis," *Gas Turbine Engineering Handbook*, 7th Edition, Gulf Publishing Company.

[5] Chodkiewicz, R., Porochnicki, J., Potapczyk, A., 1998, "Electric Power And Nitric Acid Coproduction — A New Concept In Reducing The Energy Costs." *Powergen Europe'98*, Milan, Italy, Vol. iii, pp. 611-625.

[6] Chodkiewicz, R., "A Recuperated Gas Turbine Incorporating External Heat Sources in the Combined Gas-Steam Cycle," ASME Paper No. 2000-GT-0593, ASME.

[7] Holden, P., Moen, D., DeCorso, M., "Alabama Electric Cooperative Compressed Air Energy Storage (CAES) Plant Improvements," ASME Paper No. 2000-GT-0595, ASME.

[8] Kehlhofer, R. H., et. al., 1999, *Combined Cycle Gas & Steam Turbine Power Plants*, 2nd Edition, PennWell, Tulsa, Oklahoma.

[9] Lane, A. W., Hoffman, P. A., 1998, "The U.S. Dep. of Energy Advanced Turbine System. Program," ISROMAC-7, Hawaii, D. O. E.

[10] Miller, H. F., 1989, *Blade Erosion — FCCU Power Recovery Expanders*, D-R Turbo Products Division, Olean, N.Y.

[11] Nakhamkin, M., "Increasing Gas Turbine or Combined Cycle Power Production With Compressed Air to Meet Peak Power Demands," ASME Paper No. 2000-GT-0596, ASME.

[12] Ram, N., "The Single-Shaft Combined Cycle Myth," ASME Paper No. 2000-GT-0594, ASME.

[13] Wieler, C. L., 1998, *WR-21 Intercooled Recuperated Gas Turbine*, http://www.Gas-Turbines.Com.Randd/Icr-Wrds.Htm.

[14] *Ullman Encyclopaedia of Industrial Chemistry*, 1991, Vol. A17.

CHAPTER 3 — PERFORMANCE AND MECHANICAL EQUIPMENT STANDARDS

[1] ANSI/API, August 1995, *Centrifugal Pumps for Petroleum, Heavy Duty Chemical and Gas Industry Services*, 8th Edition, API Std 610, API.

[2] ANSI/API, November 1993, *Vibration, Axial-Position, and Bearing-Temperature Monitoring Systems*, 3rd Edition, API Std 670, API.

[3] API, January 1995, *Heat Recovery Steam Generators*, 1st Edition, Publication 534, API.

[4] API, May 1997, *Fired Heaters & Steam Generators*, 1st Edition, RP 556, API.

[5] API, June 1997, *General Purpose Steam Turbines for Petroleum, Chemical, and Gas Industry Services*, 4th Edition, API Std 611, API.

[6] API, June 1995, *Special Purpose Gear Units for Petroleum, Chemical and Gas Industry Services*, 4th Edition, API Std 613, API.

[7] API, April 1999, *Lubrication, Shaft-Sealing, and Control-Oil Systems and Auxiliaries for Petroleum, Chemical and Gas Industry Services*, 4th Edition, API Std 614, API.

[8] API, August 1998, *Gas Turbines for the Petroleum, Chemical and Gas Industry Services*, 4th Edition, API Std 616, API.

[9] API, February 1995, *Centrifugal Compressors for Petroleum, Chemical and Gas Industry Services*, 6th Edition, API Std 617, API.

[10] API, June 1995, *Reciprocating Compressors for Petroleum, Chemical and Gas Industry Services*, 4th Edition, API Std 618, API.

[11] API, June 1997, *Rotary-Type Positive Displacement Compressors for Petroleum, Chemical, and Gas Industry Services*, 3rd Edition, API Std 619, API.

[12] API, October 1998, *Special Purpose Couplings for Petroleum Chemical and Gas Industry Services*, 3rd Edition, API Std 671, API.

[13] API, September 1996, *Packaged, Integrally Geared Centrifugal Air Compressors for Petroleum, Chemical, and Gas Industry Services*, 3rd Edition, API Std 672, API.

[14] API, July 1997 (Reaffirmed March 2000), *General-Purpose Gear Units for Petroleum, Chemical and Gas Industry Services*, 2nd Edition, API Std 677, API.

[15] API, February 1996, *Liquid Ring Vacuum Pumps and Compressors*, 1st Edition, API Std 681, API.

[16] ASME, 1977 (Reaffirmed 1997), *Basic Gas Turbines*, B133.2, ASME.

[17] ASME, 1978 (Reaffirmed 1997), *Gas Turbine Control And Protection Systems*, B133.4, ASME.

[18] ASME, 1985 (Reaffirmed 1992), *Gas Turbine Fuels*, B133.7M, ASME.

[19] ASME, 1977 (Reaffirmed 1989), *Gas Turbine Installation Sound Emissions*, B133.8, ASME.

[20] ASME, 1994, *Measurement Of Exhaust Emissions From Stationary Gas Turbine Engines*, B133.9, ASME.

[21] ASME, 1981 (Reaffirmed 1994), *Procurement Standard For Gas Turbine Auxiliary Equipment*, B133.3, ASME.

[22] ASME, 1978 (Reaffirmed 1997), *Procurement Standard For Gas Turbine Electrical Equipment*, B133.5, ASME.

[23] ASME, 1997, *Performance Test Code on Atmospheric Water Cooling Equipment*, ASME PTC 23, ASME.

[24] ASME, 1981 (Reaffirmed 1992), *Performance Test Code on Gas Turbine Heat Recovery Steam Generators*, ASME PTC 4.4, ASME.

[25] ASME, 1997, *Performance Test Code on Gas Turbines*, ASME PTC 22, ASME.

[26] ASME, 1996, *Performance Test Code on Overall Plant Performance*, ASME PTC 46, ASME.

[27] ASME, 1983, *Performance Test Code on Steam Condensing Apparatus*, ASME PTC 12.2, ASME.

[28] ASME, 1996, *Performance Test Code on Steam Turbines*, ASME PTC 6, ASME.

[29] ASME, 1988, *Performance Test Code on Test Uncertainty: Instruments and Apparatus*, ASME PTC 19.1, ASME.

[30] ISO 10436:1993 *Petroleum and Natural Gas Industries — General Purpose Steam Turbine for Refinery Service*, 1st Edition, ISO.
[31] ISO, 1983, *Natural Gas — Calculation of Calorific Value, Density and Relative Density*, ISO 6976-(E), International Organization for Standardization.
[32] Table of Physical Constants of Paraffin Hydrocarbons and other components of Natural Gas — Gas Producers Association Standard 2145-94.

CHAPTER 4 — AN OVERVIEW OF GAS TURBINES

[1] Abidat, C., Baines, N. C., Firth, M. R., 1992, "Design of a highly loaded mixed flow turbine," *Proc. Inst. Mechanical Engineers, Journal Power 8 Energy*, **206**:95-107.
[2] Anderson, R. J., Ritter, W. K., Dildine, D. M., 1947, "An Investigation of the Effect of Blade Curvature on Centrifugal Impeller Performance," NACA TN-1313.
[3] Arcoumanis, C., Martinez-Botas, R. F., Nouri, J. M., Su, C. C., 1997. "Performance and exit flow characteristics of mixed flow turbines," *International Journal of Rotating Machinery*, **3**(4):277-293.
[4] Baines, N. A., Hajilouy-Benisi, A., Yeo, J. H., 1994, "The pulse flow performance and modeling of radial inflow turbines." IMechE, Paper No. a405/017.
[5] Balje, O. E., Binsley, R. L., 1960, "Axial Turbine Performance Evaluation," *Journal of Engineering for Power, ASME Transactions*, **90A**:217-232, ASME.
[6] Balje, O. E., 1964, "A Study of Reynolds Number Effects in Turbomachinery," *Journal of Engineering for Power, ASME Transactions*, **86**(A):227, ASME.
[7] Balje, O. E., 1968, "Axial Cascade Technology and Application to Flow Path Designs," *Journal of Engineering for Power, ASME Transactions*, **90A**:309-340, ASME.
[8] Balje, O.E., 1952, "A Contribution to the Problem of Designing Radial Turbomachines," *Transactions of the ASME*, **74**:451, ASME.
[9] Ballal, D. R., Lefebvre, A. H., "A Proposed Method for Calculating Film Cooled Wall Temperatures in Gas Turbine Combustor Chambers," ASME Paper No. 72-WA/HT-24, ASME.
[10] Bammert, K., Rautenberg, M., "On the Energy Transfer in Centrifugal Compressors," ASME Paper No. 74-GT-121, ASME.
[11] Barker, T., Jan/Feb 1995, "Siemens' New Generation," Turbomachinery International.
[12] Behning, F. P., Schum, H. J., Szanca, E. M., 1971, "Cold-Air Investigation of a Turbine with Transpiration-Cooled Stator Blades, IV-Stage Performance with Wire-Mesh Shell Blading," NASA, TM X-2176, NASA.
[13] Benign, F. O. P., Rust, H. O. W., Jr., Moffitt, T. P., 1971, "Cold-Air Investigation of a Turbine with Transpiration-Cooled Stator Blades, III —

Performance of Stator with Wire-Mesh Shell Blading," NASA, TM X-2166, NASA.

[14] Benisek, E., 1998, "Experimental and analytical investigation for the flow field of a turbocharger turbine," IMechE, Paper No. 0554/027/98.

[15] Benson, R. S., 1970, "A Review of Methods for Assessing Loss Coefficients in Radial Gas Turbines," *International Journal of Mechanical Sciences*, **12**:905-932.

[16] Bernstien, H.L., 1998, "Materials Issues for users of Gas Turbines," *Proceedings of the 27th Texas A&M Turbomachinery Symposium*.

[17] Boyce, M .P., Oct. 1972, "New Developments in Compressor Aerodynamics," *Proceedings of the 1st Turbomachinery Symposium, Texas A&M*.

[18] Boyce, M. P., Oct. 1988, "Rerating of Centrifugal Compressors — Part I." Diesel and Gas Turbine Worldwide. 46-50.

[19] Boyce, M. P., Jan.-Feb. 1989, "Rerating of Centrifugal Compressors — Part II." *Diesel and Gas Turbine Worldwide*. pp. 8-20.

[20] Boyce, M. P., October/November 1999, "Cutting Edge Turbine Technology," Middle East Electricity.

[21] Boyce, M. P., Bale, V. S., "A New Method for the Calculations of Blade Loadings in Radial-Flow Compressors," ASME Paper No. 71-GT-60, ASME.

[22] Boyce, M. P., Bale, Y. S., Sept. 1972, "Diffusion Loss in a Mixed-Flow Compressor," *Intersociety Energy Conversion Engineering Conference, San Diego*, Paper No. 729061.

[23] Boyce, M. P., Desai, A. R., Aug. 1973, "Clearance Loss in a Centrifugal Impeller," *Proc. of the 8th Intersociety Energy Conversion Engineering Conference*, Paper No. 7391 26, p. 638.

[24] Boyce, M. P., Nishida, A., May 1977, "Investigation of Flow in Centrifugal Impeller with Tandem Inducer," ASME Paper, Tokyo, Japan, ASME.

[25] Boyce, M. P., Sept. 1993, "Principles of Operation and Performance Estimation of Centrifugal Compressors," *Proceedings of the 22nd Turbomachinery Symposium, 14-16 161-78, Dallas, TX*.

[26] Boyce, M. P., "A Practical Three-Dimensional Flow Visualization Approach to the Complex Flow Characteristics in a Centrifugal Impeller," ASME Paper No. 66-GT-83, ASME.

[27] Boyce, M. P., "Secondary flows in Axial-Flow Compressors with Treated Blades," AGARD-CCP-214 pp. 5–1 to 5–13.

[28] Boyce, M. P., "Transonic Axial-Flow Compressor," ASME Paper No. 67-GT-47, ASME.

[29] Boyce, M. P., June 1978, "How to Achieve On-Line Availability of Centrifugal Compressors," *Chemical Weekly*, pp. 115-127.

[30] Boyce, M. P., Schiller, R. N., Desai, A. R., "Study of Casing Treatment Effects in Axial-flow Compressors," ASME Paper No. 74-GT-89, ASME.

[31] Brown, L.E., 1972, "Axial Flow Compressor and Turbine Loss Coefficients: A Comparison of Several Parameters," *Journal of Engineering for Power, ASME Transactions*, **94A**:193-201, ASME.

[32] Clarke, J. S., Lardge, H. E., "The Performance and Reliability of Aero-Gas Turbine Combustion Chambers," ASME Paper No. 58-GTO-13, ASME.

[33] Dalla, B., Ralph, A., Nickolas, S. G., Weakley, C. K., Lundberg, K., Caron, T. J., Chamberlain, J., Greeb, K., "Field Test of a 1.5 MW Industrial Gas Turbine with a Low Emissions Catalytic combustion System," ASME Paper No. 99-GT-295, ASME.

[34] Dallenback, F., Jan. 1961, "The Aerodynamic Design and Performance of Centrifugal and Mixed-Flow Compressors," *SAE International Congress*.

[35] Dawes, W., 1995, "A Simulation of the Unsteady Interaction of a Centrifugal Impeller with its Vaned Diffuser: Flows Analysis," *ASME Journal of Turbomachinery*, **117**:213-222, ASME.

[36] Deniz, S., Greitzer, E. Cumpsty, N., "Effects of Inlet Flow Field Conditions on the Performance of Centrifugal Compressor Diffusers Part 2: Straight-Channel Diffuser," ASME Paper No. 98-GT-474, ASME.

[37] Domercq, O., Thomas, R., "Unsteady Flow Investigation in a Transonic Centrifugal Compressor Stage," AIAA Paper No. 97-2877, AIAA.

[38] Dutta, P., Cowell, L. H., Yee, D. K., Dalla Betta, R. A., "Design and Evaluation of a Single-Can Full Scale Catalytic combustion System for Ultra-Low Emissions Industrial Gas Turbines," ASME 97-GT-292, ASME.

[39] Editor, August 1994, "Steam cooled 60 Hz W501G generates 230 MW," Modern Power Systems.

[40] Faires, V. M., Simmang, C. M., 1978, "Reactive Systems," *Thermodynamics*, 6th Edition, pp. 345-347, The Macmillan Co., New York.

[41] Farmer, R., May/ June 1995, "Design 60% net efficiency in Frame 7/9H steam cooled CCGT," Gas Turbine World.

[42] Filipenco, V., Deniz, S., Johnston, J., Greitzer, E., Cumpsty, N., 1998, "Effects of Inlet Flow Field Conditions on the Performance of Centrifugal Compressor Diffusers Part 1: Discrete Passage Diffuser," ASME Paper No. 98-GT-473, ASME.

[43] Gehring, S., Riess, W., March 1999, "Through flow Analysis for cooled Turbines," *London 3rd Conference on Turbomachinery — Fluid Dynamics and Thermodynamics*.

[44] Giamati, C. C., Finger, H. B., 1965, "Design Velocity Distribution in Meridional Plane," NASA SP 36, Chapter VIII, p. 255, NASA.

[45] Glassman, A. J., Moffitt, T. P., 1972, "New Technology in Turbine Aerodynamics," *Proceedings of the 1st Turbomachinery Symposium*, p. 105 Texas A&M University.

[46] Graham, R. W., Guentert, E. C., 1965, "Compressor Stall and Blade Vibration," NASA SP 36, Chapter XI, p. 311, NASA.

[47] Grahman, J., Jones, R. E., Mayek, C. J., Niedzwicki, R. W., "Aircraft Propulsion," Chapter 4, NASA SP-259, NASA.

[48] Greenwood, S. A., September 2000, "Low Emission Combustion Technology for Stationary Gas Turbine Engines," *Proceedings of the 29th Turbomachinery Symposium*.

[49] Hatch, J. E., Giamati, C. C., Jackson, R. J., 1954, "Application of Radial Equilibrium Condition to Axial-flow Turbomachine Design Including Consideration of Change of Entropy with Radius Downstream of Blade Row," NACA RM E54A20.

[50] Hawthorne, W. R., Olsen, W .T., Editors, 1960, *Design and Performance of Gas Turbine Plants*, Vol. II, pp. 563-590. Princeton University Press.

[51] Herrig, L. J., Emery, J. C., Erwin, J. R., 1955, "Systematic Two Dimensional Cascade Tests of NACA 65 Series Compressor Blades at Low Speed," NACA R.M. E 55HII.

[52] Hilt, M. B., Johnson, R. H., 1972, "Nitric Oxide Abatement in Heavy Duty Gas Turbine Combustors by Means of Aerodynamics and Water Injection," ASME Paper, No. 72-GT-22, ASME.

[53] Horlock, J. H., 1973, *Axial Flow Compressors*, Robert E. Krieger Publishing Company.

[54] Horlock, J. H., 1966, *Axial Flow Turbines*, London, Butterworth and Company Ltd.

[55] Horner, M. W., August 1996, "GE Aeroderivative Gas Turbines — Design and Operating Features" *39th GE Turbine State-of-the-Art Technology Seminar.*

[56] Johnston, R., Dean, R., 1966, "Losses in Vaneless Diffusers o Centrifugal Compressors and Pumps," *ASME Journal of Basic Engineering*, **88**:49-60, ASME.

[57] Karamanis, N., Martinez-Botas, R.F., Su, C.C., 2000, "Mixed Flow Turbines: Inlet and Exit flow under steady and pulsating conditions," ASME Paper No. 2000-GT-470, ASME.

[58] Klassen, H. A., Jan. 1975, "Effect of Inducer Inlet and Diffuser Throat Areas on Performance of a Low-Pressure Ratio Sweptback Centrifugal Compressor," NASA TM X-3148, Lewis Research Center, NASA.

[59] Knoernschild, E. M., 1961, "The Radial Turbine for Low Specific Speeds and Low Velocity Factors," *Journal of Engineering for Power, Transactions of the ASME*, **83**(A):1-8, ASME.

[60] Koller, U., Monig, R., Kosters, B., Schreiber, H-A, "Development of Advanced Compressor Airfoils for Heavy-Duty Gas Turbines Part I: Design and Optimization", ASME Paper No. 99-GT-95, ASME.

[61] Lakshminarayana, B., 1996, *Fluid Dynamics and Heat Transfer of Turbomachinery*, John Wiley & Sons Inc., New York.

[62] Lavoie, R., McMordie, B. G., April 1994, "Measuring Surface Finish of Compressor Airfoils protected by Environmentally resistant Coatings," *30th Annual Aerospace/Airline Plating and Metal Finishing Forum.*

[63] Lieblein, S., Schwenk, F. C., Broderick, R. L., 1953, "Diffusion Factor for Estimating Losses and Limiting Blade Loading in Axial-Flow Compressor Blade Elements," NACA RM No. 53001.

[64] Maurice, L. Q. W., Blust, J.W., 1999, "Emission from Combustion of Hydrocarbons in a Well Stirred Reactor," AIAA.

[65] McMordie, B. G., March 2000, "Impact of Smooth Coatings on the

Efficiency of Modern Turbomachinery," *Cincinnati, Ohio, 2000 Aerospace/ Airline Plating & Metal Finishing Forum.*

[66] Mellor, G., 1957, "The Aerodynamic Performance of Axial Compressor Cascades with Application to Machine Design," Sc. D. Thesis, M.I.T. Gas Turbine Lab, M.I.T. Rep. No. 38.

[67] Moffitt, T. P., Prust, H. W., Jr., Szanca, E. M., Schum, H. J., 1971, "Summary of Cold-Air Tests of a Single-Stage Turbine with Various Stator Cooling Techniques," NASA, TM X-52969, NASA.

[68] O'Brien, W. J., 1975, "Temperature Measurement for Gas Turbine Engines," SAE Paper No. 750207, SAE.

[69] Owczarek, J. A., 1968, *Fundamentals of Gas Dynamics*, pp. 165-197, International Textbook Company, Pennsylvania.

[70] Paul, T. C., Schonewald, R. W., Marolda, P. J., August 1996, "Power System for the 21st Century — H Gas Turbine Combined Cycles," *39th GE Turbine State-of-the-Art Technology Seminar.*

[71] Petrovic, M., Riess, W., 1995, "Through-Flow Calculation in Axial Turbines at Part Load and Low Load. Is," *Erlangen, Conference on Turbomachinery — Fluid Dynamics and Thermodynamics.*

[72] Phillips, M., 1997, "Role of Flow Alignment and Inlet Blockage on Vaned Diffuser Performance," Report No. 229, Gas Turbine Laboratory, Massachusetts Institute of Technology.

[73] Prust, H. W., Jr., Schum, H. J., Szanca, E. M., 1970, "Cold-Air Investigation of a Turbine with Transpiration-Cooled Stator Blades, I — Performance of Stator with Discrete Hole Blading," NASA, TX X-2094, NASA.

[74] Prust, H. W., Jr., Behning, F. P., Bider, B., 1970, "Cold-Air Investigation of a Turbine with Stator Blade Trailing Edge Coolant Ejection, II — Detailed Stator Performance," NASA, TM X-1963, NASA.

[75] Prust, H. W., Jr., Schum, H. J., Behning, F. P., 1968, "Cold-Air Investigation of a Turbine for High-Temperature Engine Application, II — Detailed Analytical and Experimental Investigation of Stator Performance," NASA, TN D-4418, NASA.

[76] Rodgers, C., Shapiro, L., "Design Considerations for High-Pressure-Ratio Centrifugal Compressors," ASME Paper No. 73-GT-31, ASME.

[77] Rodgers, C., Oct. 1966, "Efficiency and Performance Characteristics of Radial Turbines," SAE Paper 660754, SAE.

[78] Rodgers, C., Jan. 1961, "Influence of Impeller and Diffuser Characteristics and Matching on Radial Compressor Performance," SAE Preprint 268B, SAE.

[79] Rodgers, C., "Effect of Blade Numbers on the Efficiency of a Centrifugal Impeller," ASME Paper No. 2000-GT-0455, ASME.

[80] Rodgers, C., "The Performance of Centrifugal Compressor Channel Diffusers," ASME Paper No. 82-GT-10, ASME.

[81] Schilke, P. W., August 1996, "Advanced Gas Turbine Materials and Coatings," *39th GE Turbine State-of the-Art Technology Seminar.*

[82] Schlatter, J. C., Dalla Betta, R. A., Nickolas, S. G., Cutrone, M. B., Beebe, K. W., Tsuchiya, T., "Single-Digit Emissions in a full Scale Catalytic Combustor," ASME Paper No. 97-GT-57.

[83] Schlichting, H., 1962, *Boundary Layer Theory*, 4th Edition, pp. 547–550, McGraw-Hill Book Co.

[84] Senoo, V., Nakase, V., "An Analysis of Flow Through a Mixed Flow Impeller," ASME Paper No. 71-GT-2, ASME.

[85] Shahpar, S., "A Comparative Study of Optimization Methods for Aerodynamic Design of Turbomachinery Blades," ASME Paper No. 2000-GT-523, ASME.

[86] Shepherd, D.G., 1956, *Principles of Turbomachinery*, The Macmillan Company, New York.

[87] Shouman, A. R., Anderson, J. R., 1964, "The Use of Compressor-Inlet Prewhirl for the Control of Small Gas Turbines," *Journal of Engineering for Power, Transactions* of the ASME, **86**(A):136-140, ASME.

[88] Stewart, W. L., 1954, "Investigation of Compressible Flow Mixing Losses Obtained Downstream of a Blade Row," NACA RM E54I20.

[89] Szanca, E. M., Schum, H. J., Behnong, F. P., 1970, "Cold-Air Investigation of a Turbine with Transpiration-Cooled Stator Blades, II — Stage Performance with Discrete Hole Stator Blades," NASA, TM X-2133, NASA.

[90] Szanca, E. M., Schum, H. J., Prust, H. W., Jr., 1970, "Cold-Air Investigation of a Turbine with Transpiration-Cooled Stator Blades, I — Performance of Stator with Discrete Hole Blading," NASA, TM X-2094, NASA.

[91] Talceishi, K., Matsuura, M., Aoki, S. Sato, T., 1989, "An Experimental Study of heat transfer and Film Cooling on Low Aspect Ratio Turbine Nozzles," ASME Paper No. 89-GT-187, ASME.

[92] Thompson, W. E., 1972, "Aerodynamics of Turbines," *Proceedings of the 1st Turbo-machinery Symposium*, p. 90, Texas A&M University.

[93] Traupel, W., 1988, *Thermische Turbomaschinen*, Vol. 1., Springer-Verlag, Berlin.

[94] Valenti, M., September 1998, "A Turbine for Tomorrows Navy," ASME Mechanical Engineering.

[95] Vavra, M. H., March, 1968, "Radial Turbines," Pt. 4., AGARD-VKI Lecture Series on Flow in Turbines (Series No. 6).

[96] Vincent, E.T., 1950, *Theory and Design of Gas Turbines and Jet Engines*, New York, McGraw-Hill.

[97] Wallace, F. J., Pasha, S. G. A., 1972. *Design, construction and testing of a mixed-flow Turbine*.

[98] Warnes, B. M., Hampson, L. M., "Extending the Service Life of Gas Turbine Hardware," ASME Paper No. 2000-GT-559, ASME.

[99] Whitney, W. J., 1969, "Analytical Investigation of the Effect of Cooling Air on Two- Stage Turbine Performance," NASA, TM X-1728, NASA.

[100] Whitney, W. J., 1968, "Comparative Study of Mixed and Isolated Flow

Methods for Cooled Turbine Performance Analysis," NASA, TM X-1572, NASA.

[101] Whitney, W. J., Szanca, E. M., Behning, F. P., 1969, "Cold-Air Investigation of a Turbine with Stator Blade Trailing Edge Coolant-Ejection, I — Overall Stator Performance," NASA, TM X-1901, NASA.

[102] Whitney, W. J., Szanca, E. M., Bider, B., Monroe, D. E., 1968, "Cold-Air Investigation of a Turbine for High-Temperature Engine Application III — Overall Stage Performance," NASA, TN D-4389, NASA.

[103] Whitney, W. J., Szanca, E. M., Moffitt, T. P., Monroe, D. E., 1967, "Cold-Air Investigation of a Turbine for High-Temperature Engine Application," I — Turbine Design and Overall Stator Performance, NASA, TN D-3751, NASA.

[104] Winterbone, D. E., Nikpour, B., Alexander, G. L., 1990. "Measurement of the performance of a radial inflow turbine in conditional steady and unsteady flow." IMechE, Paper No. 0405/015.

[105] Wood, M. I., March 1999, "Developments in Blade Coatings: Extending the life of blades? Reducing Lifetime costs?," CCGT Generation, IIR Ltd.

[106] Wu, C. H., 1952, "A General Theory of Three-Dimensional Flow in Subsonic and Supersonic Turbomachines of Axial, Radial, and Mixed-Flow Type," NACA TN-2604.

[107] Yee, D. K., Lundberg, K., Weakley, C. K., "Field Demonstration of a 1.5 MW Industrial Gas Turbine with a Low Emissions Catalytic Combustion System," ASME Paper No. 2000-GT-88, ASME.

CHAPTER 5 — AN OVERVIEW OF STEAM TURBINES

[1] Cotton, K. C., 1993, *Evaluating and Improving Steam Turbine Performance*, Cotton Fact, Inc., Rexford, NY.

[2] Craig, H. R. M., Hobson, G., 1973, "The Development of Long Last-Stage Turbine Blades," *GEC Journal of Science and Technology*, **40**(2):65-71.

[3] Craig, H. R. M., Kalderon, D., 1973, "Research and Development for Large Steam Turbines," *Proc. American Power Conference*.

[4] Leyzerovich, A., 1997, *Large Power Steam Turbines, Volume 1: Design and Operation, Volume 2: Operations*, PennWell Books, Tulsa OK.

[5] McCloskey, T. H., et. al., 1999, "Turbine Steam Path Damage: Theory & Practice, Volume 1:Turbine Fundamentals," EPRI.

[6] McCloskey, T. H., et. al., 1999, "Turbine Steam Path Damage: Theory & Practice, Volume 2:Damage Mechanisms," EPRI.

[7] Petrovic, M., Riess, W., "Off-Design Flow Analysis and Performance Prediction of Axial Turbines," ASME Paper No. 97-GT-55, ASME.

[8] Petrovic, M., Riess, W., 1997, "Off-Design Flow Analysis of LP Steam Turbines," *Amsterdam, 2nd Conference on Turbomachinery — Fluid Dynamics and Thermodynamics*.

[9] Sanders, W. P., December 1998, *Turbine Steam Path Engineering for*

Operations and Maintenance Staff, Turbo-Technic Services Incorporated, Toronto Ontario, Canada.
[10] Trumpler, W. E., Owens H. M., "Turbine Blade Vibration and Strength," Transactions of the ASME, **77**:337-341, ASME.

CHAPTER 6 — AN OVERVIEW OF PUMPS

[1] Boyce, M. P., 1977, Chapter 10, "Transport and Storage of Fluids-Pumping of Liquids and Gases," *Perry's Chemical Engineers' Handbook*, 7th Edition, McGraw-Hill.
[2] Brown, R. D., 1975, *Vibration Phenomena in Boiler Feed Pumps Originating from Fluid Forces, Vibrations and Noise in Pump Fan and Compressor Installations*, CP9, Mech. Eng. Publ., Ltd., New York.
[3] Corley, J. E., 1978, "Subsynchronous Vibration in a Large Water Flood Pump," *Proceedings of the Seventh Turbomachinery Symposium, College Station, Texas, Texas A&M University.*
[4] Fraser, W. H., "Recirculation in Centrifugal Pumps," ASME Winter Meeting 81-WA- 465, ASME.
[5] Hergt, P., Krieger, J., 1970 "Radial Forces in Centrifugal Pumps with Guide Vanes," London, I. Mech. E., Convention on Advanced Class Boiler Feed Pumps.
[6] Massey, I. C., 1985, "Subsynchronous Vibration Problems in High Speed Multistage Centrifugal Pumps," *Proceedings of the Fourteenth Turbomachinery Symposium.*

CHAPTER 7 — HEAT RECOVERY STEAM GENERATORS

[1] Aalborg Industries Inc., 2000, "High Performance Heat Recovery Steam Generators," Erie, PA.
[2] Boyce, M. P., Meher-Homji, C. B., Focke, A. B., Nov. 1984, "An Overview of Cogeneration Technology Design Operations and Maintenance," *Proc. of the 13th TurboMachinery Symposium, Houston, TX, 13-15, 3-24, Texas A & M University.*
[3] Brady, M. F., 1999, "Differences Between once Through Steam Generators and Drum-Type HRSG's and Their Suitability for Barge Mounted Combined Cycles," Asia, POWER-Gen.
[4] Duffy, T. E., 2000, "Heat Recovery for Steam Injected Gas Turbine Application," Cambridge, Ontario, Innovative Steam Technologies.
[5] Duffy, T. E., 2000, "Once Through Heat Recovery Steam Generators Evaluation Criteria for Combined Cycles," Cambridge, Ontario, Innovative Steam Technologies.
[6] Ganapathy, V., August 1987, "HRSGs for Gas Turbine Application," Hydrocarbon Processing.

[7] George, N. S., et al., "Dynamic Behavior of a Vertical Natural Circulation Two Pressure Stage HRSG Behind a Heavy Duty Gas Turbines," ASME Paper No. 2000-GT-0592, ASME.

[8] Jeffs, E., January/February 1998, "ABB Brings GT 24 and Once-Through Boiler to New England Merchant Plant," Turbomachinery International.

[9] Johns, W. D., 1995, "Enhanced Combined Cycle Technology," *Eleventh Symposium on Industrial Applications of Gas Turbines*.

CHAPTER 8 — CONDENSERS AND COOLING TOWERS

[1] ASME, 1983, *Performance Test Code on Steam Condensing Apparatus*, ASME PTC 12.2, ASME.

[2] Aull, R. J., Wallis, J. S., 2000, Brentwood Industries, Sales Documentation.

[3] Burger, R., Chapter 6, "Thermal Evaluation Cooling Tower," *Cooling Tower Technology Textbook*, 3rd Edition.

[4] Burger, R., July, 2000, "Cooling Tower Fill: The Neglected Moneymaker," Hydrocarbon Processing, Cooling Tower Institute Material Standard STD-136.

[5] Meek, G., 1967, "Cellular Cooling Tower Fill," CTI Paper TP-32A.

[6] Phelps, P., 1979, "Cooling Tower — Waste Heat Superstar," CTI Paper TP 76-06.

CHAPTER 9 — GENERATORS, MOTORS AND SWITCH GEARS

[1] ASME, 1978 (Reaffirmed 1997), *Procurement Standard For Gas Turbine Electrical Equipment*, B133.5, ASME.

[2] Daugherty, R. H., 1997, "Chapter 29 Electric Motors and Auxiliaries," *Perry's Chemical Engineers' Handbook*, 7th Edition, McGraw-Hill.

[3] Hargett, Y. S., "Large Steam turbine Driven Generators," Large Steam Turbine Generator Department-Schenectady N.Y.

[4] McNeely, M., May/June 2000, "New Switchgear Targeted at DG Applications," Distributed Power.

[5] Nippes, P. I., 2000, "Synchronous Machinery," *The Electric Power Engineering Handbook*, CRC Press LLC.

[6] Wright, J., "A Practical Solution to Transient Torsional Vibration in Synchronous Motor Drive Systems," Pub. 75-DE-15, ASME.

CHAPTER 10 — FUELS, FUEL PIPING AND FUEL STORAGE

[1] Bahr, D. W., Smith, J. R., Kenworthy, N. J., "Development of Low Smoke Emission Combustors for Large Aircraft Turbine Engines," AIAA Paper No. 69-493.

[2] Boyce, M. P., 1997, Chapter 10, "Transport and Storage of Fluids — Process —

Plant Piping," *Perry's Chemical Engineers' Handbook*, 7th Edition, McGraw-Hill.

[3] Boyce, M. P., Trevillion, W., Hoehing, W. W., March 1978 (Reprint), "A New Gas Turbine Fuel," Diesel & Gas Turbine Progress.

CHAPTER 11 — BEARINGS, SEALS AND LUBRICATION SYSTEMS

[1] Abramovitz, S., December, 1977, "Fluid Film Bearings, Fundamentals and Design Criteria and Pitfalls," *Proceedings of the 6th Turbomachinery Symposium*, pp. 189–204, Texas A & M University.

[2] API, April 1999, *Lubrication, Shaft-Sealing, and Control-Oil Systems and Auxiliaries for Petroleum, Chemical and Gas Industry Services*, 4th Edition, API Std 614, API.

[3] Boyce, M. P., Morgan, E., White, G., 1978, "Simulation of Rotor Dynamics of High- Speed Rotating Machinery," Madras, India, pp. 6–32, *Proceedings of the First International Conference in Centrifugal Compressor Technology*.

[4] Clapp, A. M., 1972, "Fundamentals of Lubricating Relating to Operating and Maintenance of Turbomachinery," *Proceedings of the 1st Turbomachinery Symposium, Texas A&M University*.

[5] Egli, 1935, "The Leakage of Steam through Labyrinth Seals," *Transactions of the ASME*, pp. 115-122.

[6] Fuller, D. D., 1956, *Theory & Practice of Lubrication for Engineers*, Wiley Inter-science.

[7] Herbage, B. S., October 1972, "High Speed Journal and Thrust Bearing Design," *Proceedings of the 1st Turbomachiery Symposium*, pp. 56-61. Texas A&M University.

[8] Herbage, B., December, 1977, "High Efficiency Fluid Film Thrust Bearings for Turbomachinery," *6th Proceedings of the Turbomachinery Symposium*, pp. 33-38, Texas A&M University.

[9] King, T. L., Capitao, J. W., October 1975, "Impact on Recent Tilting Pad Thrust Bearing Tests on Steam Turbine Design and Performance," *Proceedings of the 4th Turbomachinery Symposium*, pp. 1-8, Texas A&M University.

[10] Leopard, A. J., December 1977, "Principles of Fluid Film Bearing Design and Application," *Proceedings of the 6th Turbomachinery Symposium*, pp. 207-230, Texas AM University.

[11] Reynolds, O., 1886, *Theory of Lubrication, Part I*, Trans. Royal Society, London.

[12] "Rolling Bearing Damage," 1995, FAG Publication No. WL 82 102/2 Esi.

[13] "Rolling Bearings," 1996, Fundamentals, Types, Design, FAG Publication No. WL 43 1190 EA.

[14] Shapiro, W., Colsher, R., December, 1977, "Dynamic Characteristics of Fluid Film Bearings," *Proceedings of the 6th Turbomachinery Symposium*, pp. 39-53, Texas A&M University.

[15] Tessarzik, J. M., Badgley, R. H., Anderson, W. J., February 1972, "Flexible Rotor Balancing by the Exact-Point Speed Influence Coefficient Method," *Transactions of the ASME, Institute of Engineering for Industry*, **94** B(1):148, ASME.

CHAPTER 12 — CONTROL SYSTEMS, AND CONDITION MONITORING

[1] ASME, 1978 (Reaffirmed 1997), *Gas Turbine Control And Protection Systems*, B133.4, ASME.

[2] Boyce, M. P., Cox, W. M., August 1997, "Condition Monitoring Management-Strategy," *Presented at The Intelligent Software Systems in Inspection and Life Management of Power and Process Plants in Paris, France.*

[3] Boyce, M. P., Herrera, G., Sept. 1993, "Health Evaluation of Turbine Engines Undergoing Automated FAA Type Cyclic Testing," *Presented at the SAE International Ameritech '93. Costa Mesa, CA, 27-30. SAE Paper No. 932633, SAE.*

[4] Boyce, M. P., Venema, J., June 1997, "Condition Monitoring and Control Center," *Presented at the Power Gen Europe in Madrid, Spain, Power Gen.*

[5] Boyce, M. P., July/August 1999, "Condition Monitoring of Combined Cycle Power Plants," pp. 35-36, Asian Electricity.

[6] Boyce, M. P., December 1994, "Control and Monitoring an Integrated Approach," pp. 17-20, Middle East Electricity.

[7] Boyce, M. P., Gabriles, G. A., Meher-Homji, C. B., 3-5 Nov. 1993, "Enhancing System Availability and Performance in Combined Cycle Power Plants by the Use of Condition Monitoring," *Presented at the European Conference and Exhibition Cogeneration of Heat and Power, Athens, Greece.*

[8] Boyce, M. P., Gabriles, G. A., Meher-Homji, C.B., Lakshminarasimha, A.N. Meher-Homji, F. J., 14-16 Sept. 1993, "Case Studies in Turbomachinery Operation and Maintenance using Condition Monitoring," *Proc. of the 22nd Turbomachinery Symposium. Dallas, TX,* pp. 101-12, Texas A & M University.

[9] Boyce, M. P., March, 1999, "How to Identify and Correct Efficiency Losses through Modeling Plant Thermodynamics," *Proceedings of the CCGT Generation Power Conference, London, U.K.*

[10] Boyce, M. P., March/April 1996, "Improving Performance with Condition Monitoring," *Power Plant Technology Economics and Maintenance*, pp. 52-55.

[11] Meher-Homji, C. B., Boyce, M. P. Lakshminarasimha, A. N., Whitten, J. A. Meher-Homji, F. J., Sept. 21-23, 1993, "Condition Monitoring and Diagnostic Approaches for Advanced Gas Turbines," pp. 347-55, *Proc. ASME Cogen Turbo Power 1993. 7th Congress and Exposition on Gas Turbines in Cogeneration and Utility.* Sponsored by ASME in participation of BEAMA. IGTI-Vol. 8 Bournemouth, United Kingdom, ASME.

CHAPTER 13 — PERFORMANCE TESTING OF A COMBINED CYCLE POWER PLANT

[1] ASME, 1981 (Reaffirmed 1992), *Performance Test Code on Gas Turbine Heat Recovery Steam Generators*, ASME PTC 4.4, ASME.
[2] ASME, 1983, *Performance Test Code on Steam Condensing Apparatus*, ASME PTC 12.2 1, ASME.
[3] ASME, 1985 (Reaffirmed 1992), *Gas Turbine Fuels*, B 133.7M., ASME.
[4] ASME, 1988, *Performance Test Code on Test Uncertainty: Instruments and Apparatus*, ASME PTC 19.1, ASME.
[5] ASME, 1996, *Performance Test Code on Overall Plant Performance*, ASME PTC 46, ASME.
[6] ASME, 1996, *Performance Test Code on Steam Turbines*, ASME PTC 6, ASME.
[7] ASME, 1997, *Performance Test Code on Gas Turbines*, ASME PTC 22, ASME.
[8] ASME, 1997, *Performance Test Code on Atmospheric Water Cooling Equipment*, PTC 23, ASME.
[9] Boyce, M. P., August 1999, "Performance Characteristics of a Steam Turbine in a Combined Cycle Power Plant," *Proceedings of the 6th EPRI Steam Turbine Generator /Workshop*, EPRI.
[10] Boyce, M. P., July, 1999, "Performance Monitoring of Large Combined Cycle Power Plants," *Proceedings of the ASME 1999 International Joint Power Generation Conference, San Francisco CA*. Vol. 2 pp. 183-190, ASME.
[11] ISO, 1983, *Natural Gas — Calculation of Calorific Value, Density and Relative Density*, International Organization for Standardization, ISO 6976-1983(E).
[12] Table of Physical Constants of Paraffin Hydrocarbons and other components of Natural Gas — Gas Producers Association Standard 2145-94.

CHAPTER 14 — MAINTENANCE TECHNIQUES

[1] Boyce, M. P., July 1999, "Managing Power Plant Life Cycle Costs," pp. 21-23, International Power Generation.
[2] Herbage, B. S., 1977, "High Efficiency Film Thrust Bearings for Turbomachinery," pp. 33-38, *Proceedings of the 6th Turbomachinery Symposium, Texas A&M University*.
[3] Nakajima, Seiichi, *Total Productive Maintenance*, Productivity Press, Inc.
[4] Nelson, E., 1973, "Maintenance Techniques for Turbomachinery," *Proceedings of the 2nd Turbomachinery Symposium, Texas A&M University*.
[5] Sohre, J., "Reliability Evaluation for Trouble-Shooting of High-Speed Turbomachinery," *ASME Petroleum Mechanical Engineering Conference, Denver, CO.*, ASME.

[6] Sohre, J., Sept. 1968, "Operating Problems with High-Speed Turbomachinery — Causes and Correction," *23rd Annual Petroleum Mechanical Engineering Conference*.

[7] VanDrunen, G., Liburdi, J., 1977, "Rejuvenation of Used Turbine Blades by Host Isostatic Processing," pp. 55-60, *Proceedings of the 6th Turbomachinery Symposium, Texas A&M University*.

INDEX

A

Absolute velocity, 159, 163, 165, 166, 197, 201, 237, 240, 487
Absorption coolers, 16
Absorption Cooling Systems, 6, 48
Absorption refrigeration, 48
Accelerometers, 128, 130, 219
Acid gas corrosion, 93, 306
Acoustic velocity, 159, 231–232, 481
Actual, 11, 45, 66, 68, 71, 107, 110, 115, 128, 173, 175, 177, 186, 209, 222, 230, 258, 268, 325, 347, 377–378, 382, 456, 464, 476, 483, 487, 495, 503, 505, 511, 517
Adiabatic, 52, 58, 68, 159, 160, 183, 190–191, 231, 314, 454, 480–481, 483
Adiabatic process, 52, 58, 158, 159, 231, 480–481
Adiabatic processes, 58
Advanced combined cycle power plants, 7
Advanced Gas Turbine, Combined Cycle Power Plant, 34
Advanced gas turbine cycles, 7, 63
Aero-derivative, 99, 101, 104, 110, 124, 137, 150, 434, 469
Aero-dynamic Cross Coupling Whirl, 124
Aerothermal Analysis, 452, 454
Affinity laws, 257, 270
AGMA, 129
Air Inlet Filter, 467
Air Pollution, 178
Air-Cooled Condensers, 307, 309

Air-cooled first-stage nozzle and blade, 151
Air-cooled generators, 328–329
Aircraft-Derivative Gas Turbine, 138, 150
Alabama Electric Cooperative, 64
Alarm/System Logs, 454
Alternating Current Squirrel-Cage Induction, 324
American Petroleum Institute (API), 118
American Water Works Association, 387
Ammonia, 23, 48
Amplification factor, 125
Analysis Programs, 454
Annular combustors, 147, 178, 189, 354
Annular, 147, 151, 155, 157, 172–173, 175–178, 189, 249, 252, 268, 324, 354, 374, 388, 415
ANSI/API 610, 121
ANSI/API 670, 122
API Publication 534, 122
API RP 556, 122
API Standard 620, 389, 393
API Standards 650, 393
API Std 613, 120
API Std 614, 121
API Std 616, 118
API Std 618, 120
API Std 619, 120
API Std 670, 122, 131
API Std 671, 122
API Std 672, 123
API Std 677, 121

API Std 681, 123
Approach Temperature, 38, 91, 282
ASME, 113–120, 123, 128, 131, 296, 389, 391, 464–466, 470, 472, 474, 512–514, 515
ASME B 133.2, 118
ASME B133.4, 118
ASME B133.5, 119
ASME B 133.7M, 118
ASME B133.8, 119
ASME B133.9, 119
ASME Performance Test Codes, 113, 466, 486
ASME PTC 4.4 1981, 115, 464
ASME PTC 4.4, 115, 464, 470
ASME PTC 6 1996, 116, 464
ASME PTC 12.2 1983, 116, 464
ASME PTC 12.2, 116, 464, 474
ASME PTC 22 1997, 114, 464
ASME PTC 46 1996, 113, 464
Asymmetrical stage, 166
Atmospheric Tanks, 387–388
Atmospheric, 29, 31, 33, 117, 123, 180, 257, 275–276, 380–382, 387–389, 393, 417, 424, 426–427, 450, 474
Auto-ignition, 187–189
Automatic transfer switch, 347–348
Automatic transfer switching equipment, 347
Automatic Voltage Regulating System, 338
Automatic voltage regulation, 330, 336
Automotive regenerators, 55
Auxiliary systems, 120, 145–146, 151
Availability, 4, 21, 23, 28, 33, 35–36, 101, 113, 118, 138, 141, 145–146, 147, 286, 289, 295, 347, 355, 357, 433, 452, 457, 461, 463, 505, 517–519
Axial flow compressor, 141, 147, 150, 155–157, 160, 165–167, 197
Axial flow pumps, 255, 260, 264
Axial-Flow Turbines, 197

B

Backpressure, 28–29, 31, 33, 37–38, 40, 91
Backward-curved, 171

Backward-swept vanes, 171
Ball, 37, 124, 395, 397, 401, 511
Barrel roller, 397, 401
Base, 4, 6, 11, 13, 16, 19–20, 24, 27–28, 30, 35, 37–38, 40, 44, 48, 91, 101, 106, 111–112, 114, 128, 131, 138, 141, 147, 150, 159, 166, 169, 199, 209, 217–219, 226–227, 230–231, 244, 259–260, 263, 276, 282, 284, 287, 289–290, 293, 294, 299, 307, 313, 324, 347, 350, 353–354, 358, 362, 367, 376, 378, 383, 393, 407, 420, 422, 425, 433, 440, 444–446, 449, 451–454, 456, 458, 460, 464, 466, 470, 474, 477, 481, 483, 486, 487, 490, 495, 497, 499, 503–505, 507, 515–517
Bearing Lubrication Oil, 337
Bearing rings, 397, 399, 402
Bearings, 45, 121, 124–127, 129–130, 132, 141, 147, 266, 331, 337, 395, 397, 399, 401–402, 404–407, 410, 413–414, 417, 424–425, 428, 448, 456, 510, 513, 515
Biased Differential, 342–343
BioMass Systems, 13, 21
Black start, 323
Blade coatings, 146, 217, 253
Blade Damping, 251
Blade life, 145, 217, 250
Blade roots, 245
Blast furnace gas, 353
Blending, 353, 367, 369
Bottoming cycle, 7, 23, 29–31, 89, 275, 362
Brayton Cycle, 7, 31, 43–44, 47, 275, 480
Brayton-Rankine Cycle, 82, 86
Buchholz, 342, 344
Buffered gas, 417
Bunker C oil, 254
Bypass diverter damper, 297

C

Cages, 399, 401
Camber of the blades, 165
Can-annular combustors, 147, 178, 189
Can-annular, 147, 151, 177–178, 189

Capacity, 1, 3, 7–8, 12, 21, 24, 35, 56, 79, 115, 131, 141, 145, 167, 177, 208, 256–257, 260, 263–264, 268, 272, 313, 379, 382–383, 397, 399, 401–402, 404–405, 414, 420, 422, 426, 429, 457, 517–518, 524
Capacity payments, 145, 518
Capital cost, 33, 47, 138, 340, 369
Carbon deposits, 175, 360, 374
Carbon Monoxide, 19, 175, 181, 180
Carnot cycle, 58, 73
Casing of the HRSG, 283
Catalytic cleanup, 180
Catalytic combustion, 42, 193–194
Catalytic converters, 42, 180
Catalytic reactor, 193
Catalytica, 19, 192
Catastrophic oxidation, 362
Cavern recharging, 66
Cavern, 64, 66
Cavitation, 255–256, 260, 272, 524
Centrifugal compressor, 48, 153, 155–157, 167, 170, 196–197, 268, 327
Centrifugal Flow Compressors, 167
Centrifugal pumps, 121, 255, 264–265, 270, 423
Centrifuges, 128, 427, 437
Chemical Water Treatment, 321
Chlorofluorocarbon (CFC), 48
Choke point, 156, 447
Circular casing, 268
Circulation pumps, 280, 287
Circulation ratio, 287
Circumferential grooved, 402
Cleanliness, 116–117, 252, 308–309, 355, 357, 401, 451
Cleanliness factor, 116–117, 308–309, 451
Coal, 3–4, 13, 23, 33, 119, 253–254, 298–299, 317, 353, 362, 369, 427–428
Coalescers, 299
Coatings, 137, 146, 197, 214, 217–219, 253, 293, 393, 449
Coefficient of performance, 52
Cogeneration, 6, 16–17, 23–25, 27–29, 32–33, 37–38, 99, 113–114, 118, 151, 240, 244, 275, 294, 353, 433, 472

Coke oven gas, 353
Cold casing, 283
Collector, 167, 425
Combined cycle, 1, 3–4, 6–7, 23–24, 30–33, 35, 40, 43, 83, 86, 89, 91–93, 99, 101, 104, 107, 110, 112–113, 115–116, 118, 121–123, 137–138, 149, 150, 180, 194, 199, 208, 212–213, 227, 264–265, 270, 275–276, 285–287, 290, 294, 296–298, 304, 309, 323–324, 328, 351, 353–354, 382, 433–434, 438–440, 445–447, 449–450, 457, 463–466, 469, 472, 476–477, 492, 495, 497, 500, 517–519
Combined cycle plant, 4, 30–33, 35, 43, 86, 92, 99, 275, 290, 328, 446, 477, 492
Combined cycle power plant, 3–4, 6–7, 23, 30–31, 33, 35, 40, 92–93, 99, 101, 104, 107, 110, 112–113, 115–116, 118, 121–123, 137, 147, 194, 199, 213, 227, 264–265, 270, 275–276, 285–286, 294, 296, 298, 304, 309, 332–324, 353, 382, 433–434, 438–440, 446, 457, 464–466, 472, 476, 492, 495, 497, 500, 517–519
Combined generation, 23
Combined heat power, 28, 155
Combustion Analysis, 453
Combustion efficiency, 173, 181
Combustion instability, 187, 189
Combustor, 27, 31–33, 40, 42–43, 59, 61–64, 66, 73, 75, 79, 86, 124–125, 127–128, 146, 147, 151, 153, 157, 161, 171–178, 181–183, 185–194, 214–215, 219, 273, 275, 323, 354, 357, 374, 434, 445–446, 449, 453, 456, 467, 469, 486, 490, 519
Combustor Design, 175, 178, 193, 490
Combustor Module, 469
Combustor performance, 173, 177
Compound-Flow, 240
Compressor, 7, 16, 18–19, 28, 31, 43–45, 47–48, 52, 57–59, 61–64, 66, 68–71, 73, 75, 79, 82, 86, 118, 120–121, 123–124, 128, 137, 141, 146–147, 150–151, 153, 155–157,

160–163, 165–168, 170–175, 177–178, 186, 188, 192, 196–197, 199, 207–208, 218–219, 268, 273, 275, 299, 323–324, 327, 338, 352, 362, 370, 376–377, 404, 410, 416, 424, 437–438, 444–449, 451, 454, 456, 458, 460, 463, 467–469, 475, 481–483, 486–487, 490, 499, 501–502, 512–513, 515–516
Compressed Air Energy Storage, 63
Compressed air injection, 61
Compressor washing, 376
Condensate-Polishing Systems, 313
Condenser, 24, 28–29, 31, 48, 86, 91, 93, 100–101, 110–111, 116, 121, 221, 224, 227, 240, 245, 254, 264–265, 270, 272, 276, 307–309, 313–314, 321, 438, 440, 445–446, 450–452, 465, 472, 474–475, 492, 497, 499–500
Condensers, 29, 93, 100–101, 110–111, 116, 272, 307–309, 321, 445, 450
Condenser dearation, 91
Condensing steam turbine, 25, 28, 31, 100, 244–245, 275, 450
Condition-monitoring system, 436, 445, 452, 455–456, 459–461,
Constant Speed Motors, 325
Contact angle, 397, 401
Continuity equation, 230
Continuous oil flow, 429–430
Control systems, 130, 138, 145–146, 425, 433–434, 452, 455, 457, 519
Control-vortex prewhirl, 170
Convection cooling, 207–209
Convergent-divergent nozzle, 232, 234
Cooling, 6, 12, 23, 25, 27, 29, 37, 45, 47–48, 52, 56–59, 63, 71, 73, 86, 91, 93, 100, 110–111, 116–117, 120–121, 129, 131, 137–138, 145–147, 151, 172, 176–177, 180–181, 189–190, 197, 199, 207–209, 212–214, 216, 254, 263, 265, 270, 272, 290, 295, 297, 307–309, 312, 314, 316–317, 319–321, 327–329, 331–333, 337–338, 340, 342, 344–345, 354, 372, 374, 402, 429, 437, 445, 449–

452, 460, 467, 470, 474, 483, 487, 497, 500, 516
Cooling Towers, 110, 117, 254, 307, 314, 319–320, 445
Cooling Water Pumps, 270, 272
Copper-backing, 414
Corona, 334
Corrected, 358, 454, 464, 490–491, 517
Corrected fuel flow, 491
Corrected power, 491
Corrected speed, 491
Corrected temperature, 491
Corrosion Analysis, 453
Corrosion fatigue resistance, 250
Corrosion, 73, 79, 82, 86, 90, 92–93, 111, 146–147, 177, 213–214, 216–217, 218–219, 250–255, 260, 272, 279, 281, 283, 293, 299, 304–306, 312–313, 321, 357–358, 360, 362, 369–370, 374, 377, 384, 392, 407, 425–426, 445, 449, 452–453
Corrosion-inhibiting, 362
Corrosivity, 355, 357, 453
Cost of a combined cycle plant, 35
Coulomb Whirl, 124, 407
Coupling lubrication, 429
Couplings, 99, 120, 122, 127–128, 130, 405, 429–431, 516
Critical speeds, 124, 129, 405
Cross-Compound Turbine, 244
Crude, 23, 100, 137, 353–355, 359, 362, 364, 366–367, 369
Curtis-Type Impulse Turbine, 237
Cycle analysis, 66, 86, 455, 463
Cylindrical rollers, 397

D

D-CS, 338, 433–434, 437–438, 452, 455, 457
Dampers, 28, 296, 299
Dearation, 90–91
Dearator, 91, 492
Degree of reaction, 165–166, 194–195, 203, 234
Deposition, 218–219, 353, 357–358, 369–370, 373
Deposition and fouling tendencies, 355

Desalination plants, 25
Design, 6, 16, 18–19, 21, 23, 25, 28, 31, 33, 38, 40, 52, 62, 64, 66, 68, 71, 82–83, 86, 89, 91, 93, 99, 112, 114, 118, 121–124, 126–127, 129–132, 137–138, 141, 145–147, 150–152, 155, 161, 163, 167, 170, 173, 175–178, 180–181, 183, 187–188, 193, 197, 201, 207, 209, 212, 216–217, 227, 230, 233–234, 237, 240, 244–245, 247, 249, 251–253, 255–260, 263–265, 268, 270, 272, 275–276, 279–287, 289–290, 294–295, 298–299, 308–309, 316–317, 319–320, 323–326, 329–331, 335, 347, 353–355, 360, 366, 373–374, 384–385, 387–389, 393, 395, 399, 401, 402, 404–407, 413–415, 417, 427, 429–431, 440, 450–454, 456–461, 463–465, 474, 476, 490–492, 494–495, 499, 505–507, 509–511, 515, 517, 519
Diagnosis, 453
Diagnostic Analysis, 455
Diesel and Gasoline Engines, 13, 15
Diesel Cycle, 7
Diesel engines, 13, 15, 127, 187, 323, 352
Diffuser, 79, 86, 153, 161, 167, 170–171, 176, 264, 268
Diffuser casing, 268
Diffusion type blading, 167
Direct fuel cell, 19
Direct water fogging, 45
Directional solidification, 216
Directionally solidified blades, 197, 216
Distillate fuel, 119, 353, 370
Distillate oil fuels, 66
Distributed Generation, 8, 11–13, 15, 21, 345
Diverters, 296, 299
DLE, 181, 183, 185, 187, 189
Double-Flow Turbines, 244
Downtime of, 519
Drift eliminators, 317
Droop, 440
Drum-type HRSG, 111, 281, 290
Dry low NOx, 40, 146, 181, 183, 445–446, 453, 469, 519

Dry Low NOx Combustors, 40, 146 445–446, 469, 519
Dual Fuel Nozzles, 62
Duct burners, 27, 40, 284, 304
Duplex stainless steels, 253
Durability, 304
Dynamic pressure transducers, 32, 219, 445, 453

E

Economizer, 31, 38, 90–91, 111, 272, 276–277, 279, 282–283, 287, 290, 294–296, 304, 314, 450, 470, 472, 492–495
Eductor, 417
Effective forced outage hours, 35, 517
Efficiency, 6, 7, 11, 16, 19–21, 25, 27–28, 33, 35, 44–45, 48, 52, 54, 58–59, 62, 66, 68, 71, 73, 75, 77, 82, 84, 86, 91–93, 101, 104, 107, 110–111, 113–115, 126–127, 133, 137–138, 141, 151–153, 155, 157, 167, 171, 173–175, 177, 180–181, 187, 197, 200–201, 207, 212–214, 222–224, 226–227, 230, 234, 237, 240, 244–245, 249, 252–253, 255, 257, 268, 269, 287, 296, 298–301, 313, 324, 326, 328–330, 355, 362, 370, 382, 385, 394, 404, 414–415, 417, 433, 446–450, 452, 455–461, 463–464, 467, 469, 474, 476, 481–483, 486–487, 490, 495, 497, 504–505, 517, 519
Electric tracing, 380, 382–387
Electrical motor, 113, 323
Electrostatic separators, 369
Elevated Tanks, 387, 388
Elliptical, 402
Emergency oil pump, 422
Emissions, 13, 16, 23, 32, 42, 79, 82, 119, 146, 175, 180, 181, 183, 188–190, 194, 453
Enclosures, 101, 112, 120, 338
Energy equation, 158, 230, 477, 480
Environmental considerations, 146
Environmental Effects, 40

Environmental Protection Agency, 180
EPRI, 445
Erosion Resistance, 250, 253
Euler turbine equation, 164, 165, 480
Europe, 6, 11, 23, 25, 104, 138, 147, 151, 178, 251, 253, 440
Evaporative coolers, 45, 47
Evaporative Cooling, 37, 47–48, 52, 63, 64, 295
Evaporative Regenerative Cycle, 82
Evaporator, 37–38, 40, 48, 82, 90–91, 93, 270, 272, 276–277, 279, 282–284, 286–287, 289–290, 295, 304, 450, 470, 472, 487, 492, 494–495
Evaporators, 38, 284, 287, 289
Excitation System, 327, 330, 336–338
Expander Module, 469
External, 17, 23, 61, 124, 153, 173, 223, 253, 276, 279, 280, 283, 289, 296, 308, 313, 338, 343, 389, 397, 420, 427, 430, 445, 451
Extraction Flow Turbine, 244
Extraction type steam turbines, 275

F

Failures, 132, 219, 250, 382, 385, 431, 440, 505, 507, 519, 524, 526–527
Federal Energy Regulatory Commission, 24
Feedback, 290, 433–434
Feedforward, 433–434
Feedwater, 31, 37, 90–91, 116, 221, 224, 276, 280, 290, 304, 312–314, 321, 438, 450, 492
Feedwater heater, 221, 224, 476, 492
Feedwater tank, 90–91
Fills, 21, 317, 319–320
Film cooling, 176–177, 207–209, 212
Filter house, 297–299
Filter Selection, 426
Filtration, 126, 131, 298–299, 357, 370, 376, 424, 430, 447, 454
Fin density, 283, 287, 290
Finned Tubing, 283

Fir tree, 245
Firing Temperature, 7, 12, 28, 32, 36, 44–45, 63, 137–138, 141, 145–146, 193, 197, 199–200, 209, 214, 217, 289, 354, 362, 438, 440, 446–447, 449, 460, 463, 469–470, 476, 481, 482–483, 486–487, 491, 495, 500, 518–519
Fixed Roof Tanks, 388
Flash-back, 188
Flexible diaphragm coupling, 127
Flexible shaft, 124
Floating Roof Tanks, 389
Fog, 45, 48, 52
For low Btu gases, 354
Forced Circulation, 40, 283–284, 286–287
Forced circulation system, 40, 284
Forced-vortex prewhirl, 169
Foremen and lead machinist, 512
Forward-curved, 170
Forward-swept, 170
Fouling resistance, 116, 309
Frame Type, 15, 99, 101, 104, 110, 124, 173, 178, 208, 309, 402
Free-vortex prewhirl, 169
Frequency response, 440
Fuel, 4, 6, 8, 13, 15, 17–21, 23–25, 27–28, 33, 35, 37, 40, 54, 58, 62, 66, 68–69, 71, 75, 79, 86, 93, 99–101, 104, 110–111, 115, 118–120, 125, 127–128, 137–138, 145–146, 153, 155, 171–177, 180–194, 218, 220, 226, 244, 253–254, 265, 270, 273, 281, 283–287, 289, 294–295, 350, 353–355, 357–360, 362, 364, 366–367, 369–375, 377–379, 393, 434, 436–439, 446–449, 451, 454–457, 459–460, 464, 469–470, 475, 481, 483, 486, 491, 493, 499, 500
Fuel Cell Technology, 13, 17–19
Fuel Economics, 377
Fuel Pumps, 270, 273
Fuel Tanks, 380–386
Fuel treatment, 353, 355, 358, 362, 364, 374, 377, 437
Full Admission Turbines, 228, 234
Fundamental natural frequency, 124

G

Gas Producers Association, 117
Gas Turbine Exhaust, 27–28, 36, 40, 52, 91, 115, 276, 284, 295, 475, 492
Gas Turbines, 4, 6, 7, 10, 13, 15, 21, 23, 25, 27–28, 32, 36, 40, 48, 52, 62, 66, 71, 73, 101, 104, 107, 110, 112, 114–115, 118–120, 123–126, 137–138, 146–147, 150–152, 156–157, 167, 171, 175, 178, 180, 187, 189, 191, 194, 199, 207–208, 212, 214, 216, 219, 223, 227, 265, 280, 283–284, 294–295, 323, 376, 378, 395, 402, 415, 425, 440, 446–455, 460, 463–466, 470, 482, 486, 518–519, 526
Gasifier, 19, 146, 151, 219, 438, 469, 483, 486–487, 490–491
Gasifier turbine, 19, 151, 219, 438, 469, 486, 487, 490
Gas-Turbine Performance Calculation, 482
Gear Pumps, 265
Gear-type coupling, 127, 429–430
Gear-type pumps, 255, 272
Generator bearings, 331
Germany, 24
Graphic User Interface (GUI), 452–453
Grease-packed, 429–430

H

Half-frequency whirl, 407
Head, 1, 3, 52, 117, 131, 137, 152–153, 156, 164–165, 167–168, 171, 187, 195, 255–260, 263–266, 268–269, 272–273, 276, 296, 300, 304, 306, 309, 320, 353, 364, 366, 370, 389, 391–392, 417, 419, 422, 423, 424, 437, 454, 456, 509
Heat added, 43, 82, 493
Heat balance, 37, 307, 315–316, 450, 481, 486–487,
Heat exchangers, 29, 54–55, 90, 175, 253, 277, 453
Heat rate, 7, 27, 33, 104, 116, 138, 196, 226, 244, 298, 465, 467, 469, 474–477, 490, 500
Heat recovery steam generator, 28, 31, 59, 89, 104, 107, 111, 115, 122, 275, 323, 445–446, 449, 463–464, 492
Heat recovery unit, 19, 38, 40, 284
Heat Tracing, 378–380, 382, 384–385
Heated Compressed Air, 59, 61
Heat-Recovery Steam Generator, 463–464, 492
Heating value, 20–21, 173, 285, 355, 357, 359, 451, 456, 483, 499
Heavy Fuels, 104, 110, 254, 283, 366–367, 370, 378
Helical gear pump, 265
High-Efficiency Filters, 126, 298–299
High-pressure compressor, 150, 161
High-pressure turbine, 68, 71, 150, 224, 276
High-pressure turbine stage, 276
High-voltage insulation, 334
Historical Data Management, 455
Hot casing, 283
Hot corrosion, 146–147, 214, 216–217
HP and IP Nozzles, 247
HP Circulating Pumps, 270
HP Feed Water Pumps, 270–271
HP steam-turbine power, 495
HRSG Effectiveness, 296, 451, 470, 495, 499
HRSG Horizontal, 276–277, 287, 289
HRSG Once through Steam Generators, 276, 279–280
HRSG Reliability, 304
HRSG Vertical, 276–277, 279, 283, 287, 289
HRSG, 25, 28, 31, 35, 38, 43, 59, 61–62, 86, 89–94, 96, 100, 104, 107, 110–112, 115, 118, 122, 270, 272, 275–277, 279–281, 283–287, 289–290, 293–297, 299, 303–304, 352, 434, 438–440, 444–445, 448–449, 451–452, 465–466, 470, 472, 475, 492–495, 499, 500, 519, 524
Huff and Puff, 299
Humidified, 59, 61
Hybrid, 7, 18–19, 21, 25, 52, 364
Hybrid Power Plant, 7
Hybrid system, 19, 21, 52

Hydrodynamic Whirl, 124
Hydroelectric, 3, 21
Hydrogen, 17, 19, 175, 182, 313, 328–329, 330, 358, 373
Hydrogen-cooled generators, 328–329
Hysteretic Whirl, 124

I

Ice, 3–4, 6, 11–13, 17, 19, 23–24, 33, 35, 38, 52, 58, 71, 91, 104, 110, 113, 115, 118–123, 129, 131–132, 137–138, 146, 151, 156, 168, 171, 177, 194, 200, 204, 207, 218–219, 224, 230, 234, 237, 244–245, 249, 253, 255, 261, 263–266, 282, 299, 296, 304, 308, 320, 324, 337–339, 341, 344–348, 350–351, 354, 358, 369–370, 373, 375–377, 382, 384, 395, 397, 407, 413, 415, 417, 420, 423, 425–426, 428–430, 433, 445, 448–449, 451, 455, 458–460, 503, 505, 511–512, 515–516, 518–520, 524, 527–528
IGVs, 169
Impeller, 153, 167–168, 170–171, 194, 197, 255, 257, 260, 263–266, 268, 272, 312, 372, 404, 516, 524
Impeller eye, 167
Impingement cooling, 207, 209
Impulse and Reaction Combination, 228, 230
Impulse Turbines, 194, 228, 230
Impulse type, 197, 230, 237, 450
IN 738 blades, 216
Incidence angle, 165, 168
Independent power producers, 24, 147
Inducer, 167–170
Industrial cogeneration, 25
Industrial Heavy-Duty Gas Turbines, 138, 147
Inertial Filters, 298–299
Injection of Steam in, 62–63
Inlet Cooling, 45, 47, 52, 297
Inlet fogging, 48
Inlet guide vanes, 31, 89, 160, 166–167, 186, 275, 354, 438, 449, 481

Insulation, 40, 218, 279–280, 283, 286, 297, 303, 330, 334, 340–342, 344, 380, 382–385, 392–393
Interconnection, 11, 13, 24
Intercooled Regenerative Reheat Cycle, 73
Intercooled simple cycle, 71
Intercooler, 58, 64, 71, 73, 120
Intercooling, 48, 56–59, 71, 73, 86
Intercooling regenerative cycle, 71
Intermediate-pressure turbine stage, 31, 226, 276
IP-LP Circulating Pump, 270, 272
Isaac Babbitt, 407
Isentropic expansion, 222
Isentropic processes, 43, 231, 480
ISO 10436:1993, 122
ISO 6976-1983 (E), 117
Isobaric process, 43
Isolated Phase Bus-Duct, 330, 335, 338
Isothermal, 59, 73
Isothermal compression, 59, 73

J

Jet gas turbines, 150
Journal bearings, 337, 395, 402, 404–405

K

Kinetic, 43, 57, 165, 170, 195, 200–201, 221, 228, 230, 237, 240, 417, 480
Kinetic energy, 57, 165, 170, 195, 200–201, 221, 228, 230, 237, 240, 417, 480
Knockout drums, 128

L

Labyrinth lands, 417
Labyrinth Seals, 124, 415–417
Larson-Miller parameter, 217, 449, 456
Larson-Miller, 217, 449, 456
Latent heat of vaporization, 47, 59
Leading-edge lockup, 406
Lemon bore, 402

Life Cycle Analysis, 455, 463
Life cycle cost, 35, 111, 137, 294, 446, 459–461, 463–464, 503
Life Cycle Costs, 459–461, 463–464, 511
Liquid fuels, 115, 118, 128, 285–286, 353, 355, 357, 362, 437, 448
Lithium-bromide, 48
Ljongstrom turbine, 228
Load rating, 401
Losses, 8, 12, 24, 56, 66, 85–86, 117, 175, 206, 213–214, 226, 234–235, 237, 245, 253, 257, 268, 329, 332–333, 334, 336, 340, 342, 362, 388, 404, 414, 446, 449, 451–452, 467–468, 470, 474–475, 480, 499–500, 505, 507, 509
Losses in a Combined Cycle Power Plants, 446
Low humidity, 47
Low NOx combustors, 32, 40, 125, 146, 189, 219, 445–446, 469, 519
Low-Btu gas turbines, 21
Low-cycle fatigue, 216–217, 290
Lower heating value, 20–21, 173, 285, 451, 483, 499, 500, 501
Low-pressure compressor, 150, 161
Low-pressure turbine, 31, 33, 150, 226, 276, 450–451, 500, 524
Low-pressure turbine stage, 31, 226, 276
LP steam-turbine power, 497
Lubricant Selection, 425
Lubrication Management Program, 430–431
Lubrication Oil, 337, 420, 436–438
Lubrication Pumps, 270, 272
Lubrication systems, 121–122, 126, 130–131, 134, 146, 156, 226, 272, 395, 436, 444, 511
Lukens Inc., 391
Luminosity, 362

M

Mach number, 159–160, 166–167, 169–170, 231–232, 234, 481, 501
Magnesium, 110, 366, 369–370, 374–375, 384, 448

Main fuel, 6, 185, 193–194
Main fuel injector, 193–194
Maintenance, 4, 12, 15, 21, 32–33, 35–36, 48, 104, 110–113, 118, 122, 138, 145–146, 151, 178, 266, 290, 299, 320–321, 324, 345, 351, 355, 369, 377–378, 383, 426, 430–431, 433–434, 444–446, 452, 456–460, 463, 469, 503–507, 509–513, 515–517, 519, 526–528
Maintenance Communications, 527–528
Maintenance costs, 12, 15, 32–33, 138, 377–378, 383, 444, 459–460, 463, 503, 519
Maintenance engineer, 511
Maintenance engineers, 526
Maintenance Scheduling, 434, 511, 526
Martensitic stainless steel, 251
Materials, 19–21, 111, 122, 128, 137, 146, 191, 207, 214, 250–252, 255, 260–261, 263, 266, 283, 293, 306, 312, 336, 388, 392–393, 399, 407, 414, 417, 425, 445, 449, 456, 460, 470, 527–528
Maximum unbalance, 125
Maximum work, 45, 68, 71, 137
Mechanical Analysis, 453–454
Mechanical damping, 249
Mechanical Refrigeration, 47–48, 52
Meridional velocity, 163
MI Cables, 384
Micro-Turbines, 13, 15–16, 138, 537
Mid compressor Flashing of Water, 59
Mineral oils, 425
Misalignment, 122, 124, 129, 401, 404–405, 413, 524
Mixed Fills, 319
Mixed-Flow Turbine, 197, 545
Momentum Equation, 164, 480
Monitoring Software, 453
Motors, 113, 120, 127, 147, 272, 309, 323–327, 331, 352, 422, 455
Multi-Pressure Steam Generators, 281
Multi-shaft combined cycle power, 99, 101, 112
Multi-Stage Impulse, 237

N

NACA, 166
NASA, 127, 146, 166
Natural, 4, 6, 13, 15, 19, 25, 35, 40, 66, 93, 100, 104, 110, 117, 119–120, 122, 124, 129, 137, 175, 245, 253, 281, 283, 286–287, 289–290, 297, 340, 351, 354–355, 358–359, 362, 378, 405, 449,
Natural circulation, 40, 283, 286–287, 289
Natural gas, 4, 6, 13, 15, 19, 25, 35, 66, 93, 100, 104, 110, 117, 119–120, 122, 137, 175, 253, 281, 283, 353–355, 358–359, 362, 378, 449
Natural gas reciprocating engine, 13, 15
NEC (NFPA 70), 348
Needle rollers, 397
Negative prewhirl, 168
Net positive suction head, 256
No. 2 distillate, 66, 110, 353
Noise, 23, 112, 119, 146–147, 255, 280, 296–297, 304, 395,
Non-contacting probes, 128
Non-contacting Seals, 414–415
Non-salient pole, 336
Norway, 21
Notch Sensitivity, 250
NOx Emissions, 32, 42, 79, 82, 146, 181, 189–190, 194, 280, 453
Nuclear power plants, 33

O

Oil Contamination, 425–426
Oil coolers, 131, 423–424, 426–428
Oil-filled, 341, 429
Oil Sampling and Testing, 425
Oil Whirl, 124, 404, 407, 410, 524
On-line monitoring, 146, 219, 433
On-line turbine wash, 110
Open Tanks, 388, 393
Once through heat recovery steam generator, 107, 111
Once through steam generator (OTSG), 279, 290

Operation and maintenance cost, 33, 290
Optimization Analysis, 455
Optimum pressure, 45, 57, 68, 71, 96, 141
Optimum pressure ratio, 45, 68, 71, 141
OSTG, 107
Overall thermal efficiency, 16, 77, 197, 481, 490
Overcurrent, 343–344
Oxides of Nitrogen, 180

P

Palladium, 193
Parsons Turbine, 228, 230
Partial Admission Turbines, 228, 234
Peaking, 12, 33, 63, 101, 112, 138, 345, 347–349, 449
Performance-Based Total Productive Maintenance, 504–505, 508, 526–527
Performance analysis, 436, 463–464, 509, 511
Performance Curves, 257, 269, 436, 452, 464, 474
Performance Maps, 453–454, 456, 458
Photovoltaic, 13, 20–21
Physical Constants of Paraffin Hydrocarbons, 117
Pilot fuel, 185
Pinch point, 37–38, 91, 282, 289, 470
Plain Journal, 402
Plant Location, 101, 104
Plant Losses, 499
Plant Power Optimization, 457
Pocket guide, 527–528
Polytropic compression, 57
Positive displacement compressors, 120, 156
Positive displacement-type pump, 252
Positive prewhirl, 168
Potassium, 353, 358, 362
Potential energy, 43, 228, 480, 501
Pour point, 357–359, 393
Power, 1, 3–4, 6–8, 11–13, 15–21, 23–25, 27–28, 30–31, 33, 35–37, 40, 48, 56, 59, 61–64, 66, 68–69, 71, 73, 79,

86, 89, 91–93, 99, 101, 104, 107, 110–123, 127–129, 132, 137–138, 141, 145–147, 150–151, 155, 173–174, 177, 181, 187, 189–190, 193–194, 196, 199, 203, 207, 209, 213–214, 219, 221, 224, 226–227, 230, 240, 245, 252–255, 257, 260, 263–266, 268, 270, 272, 275–276, 279, 285–287, 289, 294–296, 298, 304, 309, 317, 323–326, 329–331, 339–342, 344–354, 362, 371–372, 375, 382, 384–386, 395, 404, 407, 414, 416, 422, 433–434, 438–440, 444–448, 450–454, 456–457, 459, 463–467, 469, 472, 474–477, 481, 483, 486, 490, 492, 495, 497, 499, 500–501, 504, 507, 512–513, 516–519, 526–527
Power-Factor, 326
Practical training, 513
Pre-burner, 192–194
Pre-heater, 31, 61, 90, 276–277
Pre-mixing chamber, 185
Prefilters, 299
Preload, 405–406
Pressure dam, 402, 524
Pressure ratio, 28, 33, 44–45, 68, 71, 73, 77, 82, 86, 137–138, 141, 147, 151, 153, 156–157, 165–167, 171, 173, 197, 208, 230, 371–372, 438, 446–447, 449, 454, 463, 481–482, 490, 518–519
Pressure Tanks, 389
Pressurized fluidized beds, 33
Preventive maintenance programs, 145, 519
Preventive, 145, 427, 445, 503–504, 506, 519,
Process gas, 7, 118–119, 353, 355, 417, 422
Process Pumps, 263
Producer gas, 353
Proximity Probes, 133
Psychometric chart, 48, 52, 315
PTC 19.1, 1988, 114
PTC 23, 1997, 117
Public Utility Regulatory Policies Act, 24

Pump, 28–29, 40, 48, 75, 82–83, 90, 100, 117, 121, 123, 131, 147, 221–224, 253, 255–257, 259–266, 268–273, 277, 280, 284, 287, 289–290, 309, 320, 324, 352–353, 360, 364, 366–367, 379, 388, 395, 402, 420–424, 428, 437–438, 445, 449–450, 454–455, 492–494, 512–513, 516, 524
Purging, 127
PURPA efficiency, 25, 27
PURPA, 24–25, 27
Pyrometer technology, 219
Pyrometers, 219, 438, 445, 453

R

Raceway, 397, 401
Radial vanes, 170
Radial-flow turbines, 16, 227
Radial-Inflow Turbine, 153, 194, 196
Rain Screens, 298
Rankine Cycle, 7, 31, 43, 82, 86, 221, 223, 231, 254, 275, 446, 495
Rateau Type Impulse Turbine, 240
Reaction Turbine, 194–195, 197, 204, 206–207, 230, 234, 450
Reaction type, 194, 197, 230, 240
Reciprocating, 13, 15, 48, 73, 120
Reciprocating engines, 13, 15, 73
Recovery ratio, 40, 285
Recuperative gas turbine, 7
Recuperative heat exchanger, 56, 152
Recuperative, 7, 54, 56, 152
Recuperator, 18–19, 64, 152, 294
Reference velocity, 173, 176
Refinery gas, 353
Refrigerated inlet, 47–48, 52, 468
Regeneration, 28, 52, 71, 224, 313, 321
Regeneration Effect, 52
Regenerative, 28, 36, 54, 58, 64, 69, 71, 73, 82, 86, 118, 141, 152, 171, 173, 180, 223–224, 264, 294, 364
Regenerative cycle, 28, 54, 69, 71, 82, 86, 118, 141, 180
Regenerative gas turbine, 58, 171
Regenerative heat exchanger, 54, 152
Regenerative mode, 36, 294

Regenerative Pumps, 264
Regenerative reheat cycle, 73, 86, 223
Regenerator, 16, 27–28, 36, 54, 55, 56, 58–59, 69–71, 75, 82, 86, 141, 151–152, 171–172
Regenerator and reheater, 59
Regenerator effectiveness, 54, 56
Reheat cycle, 33, 58, 71, 73, 86, 223–224
Reheat Effects, 57
Reheat of the steam, 227
Relative velocity, 163, 165, 167–168, 200, 203, 234
Reliability, 11–12, 23, 33, 36, 38, 40, 118, 123, 133, 137–138, 145–146, 167, 177, 188, 276, 284, 286, 304, 321, 324, 328–330, 340, 345, 379, 382, 384–385, 395, 415–416, 429, 455, 457, 459–460, 511, 517–519
Reliability of Combustors, 177
Reservoir, 127, 131, 217–218, 387, 395, 420, 423, 424, 426–429
Residual Fuel, 19, 100, 253, 353, 362, 449
Residual unbalance, 125, 130
Restricted Earth Fault, 342–343
River Hydro-Turbines, 13, 21
Roller bearings, 395, 397, 401
Roof Tanks, 388–389
Rotor unbalance, 124, 129
Rotor, 21, 99, 112, 123–125, 127, 129, 132, 160–161, 163, 165–167, 170, 194–195, 197, 200–201, 203, 204, 206–207, 213, 226–228, 234, 237, 240, 244–245, 247, 249, 252, 265, 323–327, 331–332, 335–338, 352, 395, 405, 410, 413–414, 416–417, 455, 480, 511, 513, 516, 524
Rotor velocity, 163

S

Saturation pressure, 29
Saturation temperature, 37–38, 91, 281, 282
Screw Pumps, 265
Scroll, 167, 170, 197
Seal Leakage, 415
Seals, 90, 124, 130, 245, 263, 299, 402, 405, 414–417, 426–427, 448, 510, 513, 515, 526
Self-Cleaning, 299
Self-equalizing tilting-pad thrust bearing, 413
Seminars and workshops, 528–529
Serrated, 245, 287
Service Manuals, 527
Serviceability, 146
Severity charts, 410
Shrouds, 245, 249–250
Side combustors, 147, 151, 172, 178
Silencers, 101, 126–127, 146, 299, 304
Silo type combustors, 147
Simple cycle, 7, 15, 28, 36, 56, 58–59, 66, 68, 71, 73, 77, 83, 86, 112, 141, 151, 214, 294, 490
Simple cycle gas turbine, 7, 15, 28, 56, 141
Single crystal blades, 128, 197
Single line diagram, 351
Single stage impulse, 237
Single-crystal blades, 216
Single-shaft combined cycle power plant, 99, 101, 104, 107, 110, 112
Site Configuration, 101
Skin-Effect Current Tracing, 384
Slip-ring assembly, 331
Slip-rings, 337
Smoke, 146, 175, 178, 358, 373–374
Sodium, 128, 314, 321, 353, 358, 362, 367
Solar Energy-Photovoltaic Cells, 13, 20
Solid oxide fuel cell, 18–19
Spare Parts Inventory, 511
Specific gravity, 255, 269, 353, 360, 367
Specific heat, 29, 37–38, 44, 58, 231, 257, 294, 308, 340, 467, 480, 486, 497,
Specific speed, 259–260
Splash Fills, 319–320
Split-shaft cycle, 68–69, 71, 86
Split-shaft gas turbine, 486
Split-shaft simple cycle, 68
Spur gears, 265
Squealer blades, 124
Squirrel-cage, 319, 324–326
Stack exhaust temperature, 37

Stack temperatures, 93, 357
Standard, 129
Start-up, 101, 113, 125, 283, 287, 289, 313, 323, 366, 367, 370, 420, 434
Stator, 147, 160–161, 165–166, 207, 213–214, 226, 324–325, 327, 329, 331–336, 414
Stator Magnetic Core, 331–332
Stator Windings, 325, 329, 331–334
Steam cooling, 147, 189–190, 199, 207–209, 213–214
Steam Injection, 59, 61–63, 73, 77, 79, 82, 84, 86, 115, 120, 181, 212, 444, 526
Steam Injection Cycle, 73, 82, 84
Steam Regenerative-Reheat Cycle, 223
Steam turbine plant, 4, 7, 33, 286, 440
Steam Turbine, 4, 7, 19, 25, 28–31, 33, 35, 43, 82–83, 86, 89, 90–94, 96, 99–100, 104, 107, 110–113, 116, 118, 122, 127, 133, 138, 147, 212, 221–224, 226–228, 230–231, 234, 240, 244–245, 252–253, 268, 272, 275–277, 286–287, 290, 295, 307, 323, 327–328, 337, 352, 395, 421–422, 424–425, 426, 434, 436–439, 440, 444–446, 449–453, 455, 463–465, 472, 474–476, 487, 494–495, 497, 499–500, 513, 515
Steam-Tracing, 380, 382–383, 385–386
Steel-backing, 414
Stiff shaft, 124
Stoichiometric, 173, 175–176, 181, 189
Stone wall, 157
Storage of Liquids, 387
Strouhal number, 296
Suction specific speed, 260
Sulfur, 37, 93, 180, 283, 306, 321, 357–358, 360, 524
Sump Pumps, 121, 263
Superheater, 31, 40, 90, 93, 94, 111, 227, 276–277, 279, 286, 289–290, 296, 304, 450, 470, 492, 494–495, 519
Supplementary Firing, 31, 40, 281, 284, 286, 444

Support, 12, 23, 69, 111, 129, 226, 247, 252, 254, 276–277, 287, 289, 304, 320, 331–332, 334–335, 337, 340, 342, 388, 392–393, 395, 405, 413–414
Surge point, 156
Surge-to-choke margin, 167
Switchgear, 345
Symmetrical stage, 166
Synchronous, 113, 129, 323, 325–327, 335, 352
Synchronous motors, 323, 325–327
Synchronous rotor, 335
Synchronous whirl, 129
Synthetic oil, 424–425

T

Tandem-Compound Turbine, 240, 244
Tank Volume, 389, 392
Tanks, 131, 366, 370, 377, 387–389, 392–393, 424
Tap Changer, 341, 343–344
Tapered roller, 397
Tapered-land thrust bearing, 413, 524
The Brayton-Rankine Cycle, 82, 86
The HP Circulating Pump, 272
Theoretical head, 170–171
Thermal barrier coating, 190, 218–219
Thermal efficiency, 16, 58–59, 71, 73, 77, 82, 84, 86, 92, 114, 137–138, 141, 187, 197, 222, 224, 226–227, 446, 481, 490
Thermal Energy Storage Systems, 47, 52
Thermal energy storage, 47, 52
Three-lobe bearing, 404
Thrust Bearings, 45, 127, 141, 395, 401, 410, 413–414, 456, 524
Tie-wires, 249
Tilting pad bearings, 124, 404–405, 524
Titanium, 177, 252, 450
Tools and Shop Equipment, 515
Tooth pitting index, 129
Topping cycles, 30
Total life-cycle cost, 35
Total Productive Maintenance, 503–505, 507–508, 526–527

Tracing Systems, 380, 382, 384–385
Trailing edge slots, 213
Training Materials, 527, 528
Training of Personnel, 511
Transformer, 100, 107, 119, 327, 330, 335–345, 352, 383
Transpiration Cooling, 207, 209
Transposed, 334, 464, 491–492
Transpose power output, 492
Transition transfer control strategy, 346
Treatment of Condensate, 312
Trending and Prognosis, 453
T-shape, 245
Tubular, 172, 177–178, 313, 320
Turbine Blade Cooling, 207, 209
Turbine efficiency, 28, 62, 213, 244, 252–253, 414, 448, 467, 474, 476, 482, 487, 490
Turbine firing temperature, 44, 141, 146, 214, 438, 447, 449, 463, 476, 481, 483, 486–487, 491, 500
Turbine inlet temperature, 28, 33, 36, 45, 48, 58, 68, 77, 82, 86, 138, 141, 147, 151, 155, 173–174, 190–191, 196, 207, 213, 438, 469, 486, 519
Turbine Pumps, 264, 268
Turbine Wash, 110, 374
Turbochargers, 196
Turbomachinery, 163, 194, 414–415, 425, 427–430, 445, 503, 507, 511, 515, 524, 526
Turbosplash Fill, 320
Turnaround, 146, 507, 513, 516, 526, 527
Turning gears, 127
Type of Fuel, 33, 37, 99, 101, 104, 110, 145, 449, 455–456, 460, 470, 519

U

U.S. Department of Energy, 42, 147, 199
U.S. Department of Energy's (DOE), 42
U.S. Rural Electrification Agency, 8
UL 142, 387–388
UL 58, 387–388
Unburnt Hydrocarbons, 180, 181
United States, 1, 23–24, 138
Update training, 513
Utility cogeneration, 25
Utilization factor, 195, 196, 204

V

Vanadium, 110, 128, 252, 353, 358, 362, 366, 369–370, 375, 448
Vapor pressure, 255–257
Velocity transducers, 128
Vertical Fills, 319
Vibration Analysis, 452–453, 455
Vibration limit, 125
Vibration Measurements, 127, 131
Viscosity, 255, 264, 268, 334, 353, 357, 358, 360, 362, 367, 369, 370–371, 377–379, 401, 404, 406, 425, 430
Volute casings, 268

W

Waste heat boilers, 28
Wasteheat recovery unit, 40
Water Chemistry, 296
Water contamination, 426–427, 437
Water cooled condenser, 100, 111, 245, 309, 472
Water-cooled generators, 328
Water Injection, 59, 62, 82, 86, 181, 475
Water washing systems, 369
Wear, 252, 401, 423, 426, 429–430, 503, 507
Wet combustor, 40, 146, 183
Wet control, 180
What-If Analysis, 452–453
Whirl, 124, 129, 166, 168, 404, 407, 524, 526
Whirling mechanisms, 124
Wind Energy, 13, 21
Wind turbines, 21
Winding Temperature, 343–344
Wobbe Number, 176
Work of the compressor, 483
Work, 23, 28–29, 43, 45, 47, 52, 56–59, 66, 68, 71, 73, 75, 77, 82–84, 86, 91, 93, 127, 131, 137, 141, 146, 155, 165–166, 168, 174, 187, 195, 200, 201, 207,

221–222, 226, 229, 237, 240, 251, 253, 257, 266, 297–298, 303–304, 319, 339, 342, 348–349, 354, 376, 385, 387, 404, 410, 413, 420, 438, 448, 451–453, 469–470, 480–483, 486–487, 497, 501, 504, 508, 512–513, 515, 526, 528–529
Work of turbine, 43
World Bank, 8
Written memos, 527–528

ABOUT THE AUTHOR

Dr. Meherwan P. Boyce, P.E., Fellow ASME & IDGTE, has over 35 years of experience in the field of TurboMachinery in both industry and academia. His industrial experience covers 20 years as Chairman and CEO of Boyce Engineering International, and 5 years as a designer of compressors and turbines for gas turbines for various gas turbine manufacturers. His academic experience covers a 15-year period, which includes the position of Professor of Mechanical Engineering at Texas A&M University and Founder of the TurboMachinery Laboratories and The TurboMachinery Symposium, which is now in its 30th year. He is the author of several books such as the **Gas Turbine Engineering Handbook** (Butterworth & Heinemann), **Cogeneration & Combined Cycle Power Plants** (ASME Press), and **Centrifugal Compressors, A Basic Guide** (PennWell Books). He is a contributor to several Handbooks; his latest contribution is to the **Perry's Chemical Engineering Handbook Seventh Edition** (McGraw Hill) in the areas of **Transport and Storage of Fluids, and Gas Turbines**. Dr. Boyce has taught over 100 short courses around the world attended by over 3000 students representing over 400 companies. He is a **Consultant** to the **Aerospace, Petrochemical** and **Utility Industries** globally, and is a much-requested speaker at Universities and Conferences throughout the world.

Dr. Boyce was the pioneer of On-Line Condition Based Performance Monitoring. He has developed models for various types of Power Plants and Petrochemical Complexes. His programs are being used around the world in Power Plants, Offshore Platforms, and Petrochemical Complexes. He is a consultant for Major Airlines in the area of Engine Selection, Noise and Emissions.

Dr. Boyce has authored more than 100 technical papers and reports on Gas Turbines, Compressors Pumps, Fluid mechanics, and TurboMachinery. He is a **Fellow** of the ASME (USA) and the Institution of Diesel and Gas Turbine Engineers (UK), and member of SAE, NSPE, and several other professional and honorary societies such as Sigma Xi, Pi Tau Sigma, Phi Kappa Phi, and Tau Beta Phi. He is the recipient of the ASME award for Excellence in Aerodynamics and the Ralph Teetor Award of SAE for enhancement in Research and Teaching. He is also a **Registered Professional Engineer** in the **State of Texas**.

Dr. Boyce received his B.S. and M.S. degrees in mechanical engineering from the South Dakota School of Mines and Technology and the State University of New York, respectively, and Ph.D. degree (Aerospace and Mechanical engineering) from the University of Oklahoma.